AF615071

ADVANCES IN
PHARMACOLOGY AND THERAPEUTICS

Volume 3

IONS - CYCLIC NUCLEOTIDES - CHOLINERGY

ADVANCES IN PHARMACOLOGY AND THERAPEUTICS

Proceedings of the 7th International Congress of Pharmacology, Paris 1978

General Editors: J. R. BOISSIER, P. LECHAT and J. FICHELLE, Paris

---

Volume 1 RECEPTORS *Edited by* J. Jacob

Volume 2 NEURO-TRANSMITTERS *Edited by* P. Simon

Volume 3 IONS-CYCLIC NUCLEOTIDES-CHOLINERGY *Edited by* J. C. Stoclet

Volume 4 PROSTAGLANDINS-IMMUNOPHARMACOLOGY *Edited by* B. B. Vargaftig

Volume 5 NEUROPSYCHOPHARMACOLOGY *Edited by* C. Dumont

Volume 6 CLINICAL PHARMACOLOGY *Edited by* P. Duchêne-Marullaz

Volume 7 BIOCHEMICAL CLINICAL PHARMACOLOGY *Edited by* J. P. Tillement

Volume 8 DRUG-ACTION MODIFICATION—COMPARATIVE PHARMACOLOGY *Edited by* G. Olive

Volume 9 TOXICOLOGY *Edited by* Y. Cohen

Volume 10 CHEMOTHERAPY *Edited by* M. Adolphe

(Each volume is available separately)

Satellite symposia of the 7th International Congress of Pharmacology published by Pergamon Press

CEHOVIC & ROBISON: Cyclic Nucleotides and Therapeutic Perspectives

HABERLAND & HAMBERG: Current Concepts in Kinin Research

IMBS: Peripheral Dopaminergic Receptors

LANGER, STRAKE & DUBOCOVICH: Presynaptic Receptors

NAHAS & PATON: Marhiuana: Biological Effects

PASSOUANT: Pharmacology of the States of Altertness

REINBERG & HALBERG: Chronopharmacology

Send to your nearest Pergamon office for further details

ADVANCES IN
PHARMACOLOGY AND THERAPEUTICS

Proceedings of the 7th International Congress
of Pharmacology, Paris 1978

Volume 3

IONS - CYCLIC NUCLEOTIDES - CHOLINERGY

Editor

J.C. STOCLET
Strasbourg

PERGAMON PRESS

OXFORD · NEW YORK · TORONTO · SYDNEY · PARIS · FRANKFURT

U.K. Pergamon Press Ltd., Headington Hill Hall, Oxford OX3 0BW, England

U.S.A. Pergamon Press Inc., Maxwell House, Fairview Park, Elmsford, New York 10523, U.S.A.

CANADA Pergamon of Canada, Suite 104, 150 Consumers Road, Willowdale, Ontario M2 J1P9, Canada

AUSTRALIA Pergamon Press (Aust.) Pty. Ltd., P.O. Box 544, Potts Point, N.S.W. 2011, Australia

FRANCE Pergamon Press SARL, 24 rue des Ecoles, 75240 Paris, Cedex 05, France

FEDERAL REPUBLIC OF GERMANY Pergamon Press GmbH, 6242 Kronberg-Taunus, Pferdstrasse 1, Federal Republic of Germany

---

First edition 1979

British Library Cataloguing in Publication Data

International Congress of Pharmacology, *7th Paris, 1978*
Advances in pharmacology and therapeutics.
Vol. 3 : Ions, cyclic nucleotides, cholinergy.
1. Pharmacology 2. Therapeutics
I. Title II. Boissier, J R III. Lechat, P
IV. Fichelle, J V. Stoclet, J C VI. Ions, cyclic nucleotides, cholinergy
615 RM101 78-41028

ISBN 0-08-023193-4

In order to make this volume available as economically and as rapidly as possible the authors' typescripts have been reproduced in their original forms. This method unfortunately has its typographical limitations but it is hoped that they in no way distract the reader.

Printed in Great Britain by A. Wheaton & Co. Ltd., Exeter

Contents

Introduction ix

Ions

Pharmacology of calcium homeostasis

The control of calcium metabolism in health and disease 3
R.G.G. RUSSELL

Vitamin D and its metabolites: physiology, biochemistry and pharmacology 15
H.F. DELUCA

Clinical pharmacology and application of the active forms of vitamin D 25
J.W. COBURN and A.S. BRICKMAN

Biosynthesis, metabolism and mode of action of parathyroid hormone 37
J.T. POTTS Jr.

Calcitonin: from physiology to therapeutic uses 45
G. MILHAUD

Diphosphonates 61
H. FLEISCH

Summary 73
P.L. MUNSON

Invited lectures

Ca ion and muscle contraction 81
S. EBASHI

The sodium channels in excitable membranes 99
R.D. KEYNES

Cyclic nucleotides

Role of cyclic nucleotides in the regulation of muscular tissue functions

Cyclic nucleotides and muscle function: introduction 107
J.G. HARDMAN

The possible role of cyclic GMP in the actions of hormones and drugs on smooth muscle tone: effects of exogenous cyclic GMP derivatives 113
G. SCHULTZ, K.D. SCHULTZ, E. BÖHME and V.A.W. KREYE

Effect of nitro-compound smooth muscle relaxants and other materials on cyclic GMP metabolism 123
F. MURAD, C.K. MITTAL, W.P. ARNOLD and J.M. BRAUGMLER

Adenylate cyclase in heart and skeletal muscle 133
G.I. DRUMMOND, J. DUNHAM, P. NAMBI and M. SANO

Cyclic nucleotide accumulation and hormonal modulation of cardiac metabolism and contraction 143
J.H. BROWN, L.L. BRUNTON, J. SCOTT HAYES, J.B. REESE and S.E. MAYER

Possible regulation of the calcium permeability of cardiac cell membranes by cyclic nucleotides 153
H. REUTER

Some characteristics of low molecular weight phosphoprotein constituents of cardiac sarcoplasmic reticulum and sarcolemma 161
H. WILL, H.J. MISSELWITZ, T.S. LEVCHENKO and A. WOLLENBERGER

Phosphorylation and regulation of contractile proteins 171
J.T. STULL, D.K. BLUMENTHAL, P. DE LANEROLLE, C.W. HIGH and D.R. MANNING

Evaluation of cyclic AMP and cyclic GMP

A reassessment of the cirteria used to involve cyclic nucleotides in hormone and drug mechanisms 181
J.C. STOCLET

Radioimmunoassay of cyclic AMP, cyclic CMP 193
M.A. DELAAGE, D. ROUX and H.L. CAILLA

The distribution of cyclic nucleotides and their protein kinases in tissues: an immunocytochemical approach 207
A.L. STEINER, Y. KOIDE, W.A. SPRUILL and J.A. BEAVO

Contribution of theoretical simulation to the study of the cyclic AMP system 221
S. SWILLENS and J.E. DUMONT

Invited lecture

Cyclic nucleotides, phosphorylated proteins and drug actions in the central nervous system 231
P. GREENGARD

Cholinergy

Chemical structure and cholinergic activity

The use of affinity constants for studying interactions between pharmacodynamic groups and the muscarinic receptor 253
R.B. BARLOW

Structure and activity of some new muscarine derivatives and acetylcholine analogues 261
P.G. WASER and W. HOPFF

The conformation and flexibility of cholinergic molecules 271
N.V. KHROMOV-BORISOV

An NMR study of the conformations of atropine and scopolamine cations in aqueous solution 281
J. FEENEY, E.A. PIPER and R. FOSTER

The conformation of anticholinergic substances 293
P. PAULING

Biodegradable neuromuscular blocking agents 303
J.B. STENLAKE

Effects of anticholinergic drugs on the actions of anticholinesterases on cat skeletal muscle 313
R.W. BRIMBLECOMBE, M.C. FRENCH and S.N. WEBB

Dual mechanism of the antidotal action of atropine-like drugs in poisoning by organophosphorus anticholinesterases 319
T.D. INCH and D.M. GREEN

Index 327

Introduction

The scientific contributions at the 7th International Congress of Pharmacology were of considerable merit. Apart from the sessions organised in advance, more than 2,200 papers were presented, either verbally or in the form of posters, and the abundance of the latter in the congress hall is a good indication that this particular medium of communication is becoming increasingly attractive to research workers, and offers scope for discussions which combine an elaborate, thorough approach with a certain informality.

It would have been preferable to have published the entire congress proceedings within the framework of the reports. That was, however, physically impossible, and the organisers had to adopt a realistic solution by publishing only the main lectures, symposia and methodological seminars. The amount of material presented necessitated the printing of ten volumes, each volume containing congress topics regrouped according to their relevant content and subject areas. This system of division may give rise to criticism on account of its artificiality, and we readily admit that certain texts could have been placed in more than one volume. We are asking the reader to excuse this arbitrariness, which is due to the editors' personal points of view.

I draw attention to the fact that most of the symposia finish with a commentary which the chairmen had the option of including, presenting their personal opinions on one or several points. We think that such an addition will facilitate reflection, discussion, indeed even controversy.

The launching of the scientific programme for this congress began in September 1975 on returning from the last meeting in Helsinki. Long and delicate discussions took place in the Scientific Programme Committee and with the International Advisory Board. Should it be a pioneer, 'avant-garde' congress? Or one laid out like a balance-sheet? Should we restrict the congress to the traditional bounds of pharmacology, or extend the range of papers to cover the finest discipline? The choice was difficult, and the result has been a blend of the two, which each participant will have appreciated in terms of his training, his tastes, and his own research.

A certain number of options, however, were taken deliberately: wide scope was given to toxicology, from different points of view, and to clinical pharmacology, a subject much discussed yet so badly practised; the founding of two symposia devoted to chemotherapy of parasitic diseases which are still plagues and scourges in certain parts of the world; a modest but firm overture in the field of immunopharmacology, which up until now was something of a poor relation reserved only for clinical physicians; the extension of methodological seminars, in view of the fact that new techniques are indispensable to the development of a discipline.

We have been aware since the beginning that, out of over 4,000 participants who made the journey to Paris, not one could assimilate such a huge body of knowledge. Our wish is that the reading of these reports will allow all of them to become aware of the fantastic evolution of pharmacology in the course of these latter years. If one considers pharmacology as the study of the interactions between a "substance" and a living organism, then there is no other interpretation. Nevertheless, one must admit that there exists a period for describing and analysing a pharmacological effect, and that it is only afterwards that the working mechanism can be specified; a mechanism which will permit these "substances" to be used for the dismantling and breaking down of physiological mechanisms, a process which justifies Claude BERNARD'S term, "chemical scalpel".

The reader will be able to profit equally from more down-to-earth contributions, more applied to therapeutics, and less "noble", perhaps, for the research worker. He will realise then that his work, his research and his creative genius are first and foremost in the service of Man, and will remember this statement from Louis PASTEUR:

> "Let us not share the opinion of these narrow minds who scorn everything in science which does not have an immediate application, but let us not neglect the practical consequences of discovery."

I would like to renew my thanks to my colleagues in the Scientific Programme Committee and also to the members of the International Advisory Board, whose advice has been invaluable. I owe a particular thought to J J BURNS, now the past-president of IUPHAR, who granted me a support which is never discussed, and a staunch, sincere friendship. The Chairmen have effected an admirable achievement in the organisation of their proceedings, and in making a difficult choice from the most qualified speakers. The latter equally deserve our gratitude for having presented papers of such high quality, and for having submitted their manuscripts in good time.

The publisher, Robert MAXWELL, has, as always, put his kindness and efficiency at our service in order to carry out the publication of these reports. But none of it would have been possible without the work and competence of Miss IVIMY, whom I would like to thank personally.

My thanks again to the editors of the volumes who, in the middle of the holiday period, did not hesitate to work on the manuscripts in order to keep to the completion date.

Finally, a big thank you to all my collaborators, research workers, technicians and secretaries who have put their whole hearts into the service of pharmacology. They have contributed to the realisation of our hopes for this 7th International Congress, the great festival of Pharmacology. Make an appointment for the next one, in 1981, in Tokyo.

Jacques R BOISSIER

Chairman

Scientific Programme Committee

Ions

The Control of Calcium Metabolism in Health and Disease

R.G.G. Russell

Department of Chemical Pathology, University of Sheffield Medical School, Beech Hill Road, Sheffield, S10 2RX, U.K.

SUMMARY

In man the body's calcium economy is determined by the relationship between the intestinal absorption of calcium, renal handling of calcium and the movements of calcium in and out of the skeleton. The regulation of calcium metabolism involves three distinct but inter-related aspects: 1) The control of the concentration of calcium in extracellular fluid and tissues; 2) The control of the body's overall calcium balance, i.e., the relationship between gains and losses; and 3) The control of the shape, structure and composition of bone and the way these respond to external factors such as load bearing. These processes are influenced by many factors. The principal calcium regulating hormones are parathyroid hormone, the renal metabolites of vitamin D (notably 1,25-dihydroxy vitamin D3), and calcitonin, although the role of the latter in man is still controversial. The secretion of these major regulating hormones is determined mainly by calcium and phosphate levels. Many other hormones have some influence on calcium metabolism; these include thyroid hormone, prolactin, growth hormone, insulin, the somatomedins, and the adrenal and gonadal steroids. The major clinical disorders of calcium metabolism either involve disturbances in the secretion or action of these hormones, or are due to disturbances in bone metabolism itself. Several drugs have some effect on calcium metabolism but much remains to be learnt about the pharmacology of the skeleton. Perhaps the greatest therapeutic challenge is presented by osteoporosis, which is a major cause of disability and fractures in the elderly.

INTRODUCTION

In recent years there has been a rapid advance in knowledge about calcium metabolism. In particular, the chemical nature of the major calcium regulating hormones has been determined and their mode of action clarified. This recent work has led to a re-examination of many older concepts and the new knowledge has yet to be fully applied to human disease.

DISTRIBUTION OF CALCIUM AND PHOSPHATE

The total body content of calcium is between 1-1.5 kg and for phosphorus 0.7-1 kg for an average adult weighing 70 kg. The skeleton contains about 98% of this calcium and 85% of the phosphorus. Studies with radioisotopes (^{45}Ca and ^{47}Ca) indicate that in normal human adults the exchangeable pool of calcium represents less than 1% of total body calcium. This exchangeable pool of calcium is very important in homeostasis and about half of it lies outside the skeleton. Continuous movement occurs of calcium ions between body fluids, cells and the surface of bone. Skeletal renewal occurs at a rate approximately 1-4% of the total adult skeleton each year. Trabecular bone has a faster turnover than cortical bone.

Extracellular Fluid Calcium

The total concentration of calcium in plasma is normally between 2 to 2.5 mmol/litre. Of this about 1.2 mmol/litre is present as ionised calcium (Ca^{2+}), the remainder is bound to proteins, particularly albumin, and to small ions such as citrate and phosphate. It is the ionic Ca^{2+} which is regulated. Under most conditions Ca^{2+} bears a constant relationship to total plasma calcium. The concentration of extracellular Ca^{2+} is important for neuromuscular function and many other biological events, e.g., blood coagulation.

Intracellular Calcium

In most, if not all, cells, cytosol concentrations of Ca^{2+} are maintained at much lower concentrations than extracellular and are in the range of 10^{-5} to 10^{-6} mols/litre. Within cells, mitochondria are capable of accumulating large amounts of calcium against electrochemical gradients to the point at which intra-mitochondrial deposits of insoluble calcium phosphate can form. Intracellular calcium concentrations are of critical metabolic importance and changes in intracellular calcium frequently accompany changes in cell function, e.g., in response to hormones. There is a complex inter-relationship between intracellular Ca^{2+}, 3'5'-cyclic AMP and cell activation in response to many external stimuli.

Major Fluxes of Calcium and Net Calcium Balance

Three major organs are involved in the regulation of calcium metabolism. These are the gut, kidney and bone. These three organs can account for the major gains and losses of calcium from the external environment and for the exchange of calcium in and out of the extracellular fluid.

The only significant route of entry of calcium into the body is by intestinal absorption. The true absorption of calcium is greater than the net absorption because some calcium is returned to the intestinal lumen in biliary, pancreatic and intestinal secretions. The body loses calcium by urinary excretion and in sweat. The latter is usually ignored in balance studies because the losses are usually small and difficult to measure.

Inspection of the size of the fluxes shows that the kidney handles the largest amounts of calcium per day in terms of filtered load and tubular reabsorption.

In the adult under normal conditions, the body is neither gaining nor losing calcium, so that inflow and outflow are matched precisely (intake = output). In disease states there may be transient or sustained net gains or losses, to produce calcium balances that are positive (intake exceeds output) or negative (output exceeds intake).

During growth there is a net daily gain to provide the calcium necessary for skeletal growth. During pregnancy or lactation, the fetus or child gains calcium from the mother. Under these various conditions, the extra requirements for calcium are met by increased net intestinal absorption and diminished renal excretion, so that a neutral balance can be maintained.

There are two important aspects of calcium homeostasis. The first is concerned with the body's overall balance which is determined by the relative rates of flux through these organs and the complex way in which they are changed by regulating factors, particularly parathyroid hormone (PTH), calcitonin (CT) and vitamin D (Vit D). The second aspect is concerned with the control of the concentration of calcium and phosphate (Pi) in extracellular fluid (ECF); this control can often be independent of the total body economy. The control of ECF concentrations is achieved by changes in the relative sizes of the various fluxes and by the level of Ca^{2+} and P_i at which the various controlling agents are switched on and off

CALCIUM REGULATING HORMONES

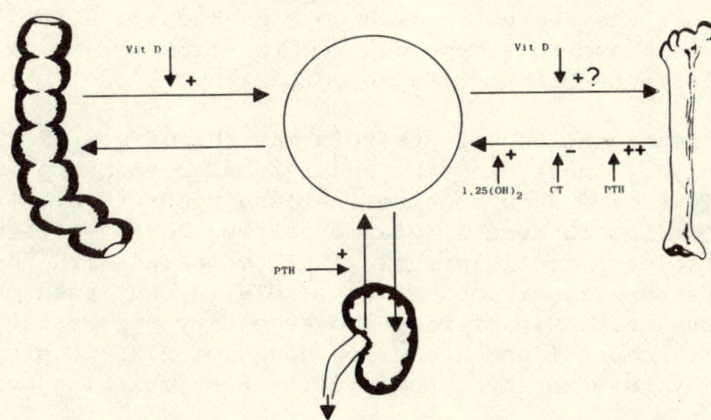

Fig. 1. The principal movements (fluxes) of calcium in an adult human and the major sites of hormonal control

HORMONES ACTING ON CALCIUM METABOLISM

The hormones can be divided into "controlling" hormones and "influencing" hormones. The controllers are the primary calcium regulating hormones, PTH, CT and Vit D metabolites, the secretion of each of which is altered in response to changes in plasma ionised Ca concentration (and phosphate in the case of Vit D). They have their major effects on plasma Ca and phosphate by altering the fluxes of these ions between the ECF pool and gut, kidney and bone. The "influencing" hormones are those other hormones, e.g., thyroid hormones, prolactin, growth hormone, insulin, somatomedins, and adrenal and gonadal steroids, which have effects on calcium metabolism but whose secretion is determined primarily by factors other than changes in plasma calcium and phosphate. These hormones tend to have their major effects on the skeleton itself rather than on the regulation of plasma Ca.

THE MAJOR CONTROLLING HORMONES

Parathyroid Hormone (PTH)

Mammalian PTH consists of a single peptide chain, 84 amino acids long, for which the entire sequence is now known for several mammalian species, including man. It circulates in plasma mainly as fragments of this molecule. Biological activity resides in the first 32 to 34 amino acids reading from the N-terminal end. The major stimulus to its secretion is a fall in plasma Ca^{2+}, whereas a rise in plasma Ca^{2+} suppresses PTH secretion. Other ions, e.g., Mg^{2+}, only play a minor role in regulating secretion.

The major action of PTH is on the kidney, to increase renal tubular reabsorption of calcium and to depress the tubular reabsorption of phosphate. This leads to a rise in plasma Ca and a fall in plasma P_i. The detailed biochemical basis for these actions is not clear, although the effect on phosphate secretion seems to involve activation of an adenylate cyclase to produce cyclic AMP within the renal tubule. This renal receptor mechanism appears to be defective in pseudohypoparathyroidism, since patients with this inherited disorder show no renal response to PTH in terms of increased excretion of phosphate or cyclic AMP.

PTH also diminishes the secretion of H^+ by the kidney, which leads to the excretion of HCO_3 ions and to a hyperchloraemic acidosis, which is often present to some degree in patients with primary hyperparathyroidism.

It is extremely important to note that in man the effect of PTH on plasma calcium is mediated mainly by increasing the renal tubular reabsorption of calcium. The better known effect of PTH on bone to increase resorption is relatively less important in producing changes in plasma calcium because the net fluxes of Ca in and out of the skeleton are relatively small compared with the large fluxes across the kidney. The bone resorbing effect of PTH is best seen only under pathological conditions, e.g., in primary and secondary hyperparathyroidism. Nevertheless, there is some evidence that low doses of PTH, possibly within the physiological range, may have an important effect in stimulating bone formation and bone turnover.

The effect of PTH to increase intestinal absorption of calcium is now thought to be an indirect action brought about by the ability of PTH to increase the renal synthesis of 1,25-dihydroxycholecalciferol.

Calcitonin (CT)

CT is a peptide hormone, which contains 32 amino acids with a cysteine-dependent disulphide bridge between positions 1 and 7. Although there are striking differences between the amino acid composition of CTs from different species, CT from one species commonly exerts biological effects in others. CT is secreted by specialised cells, designated C-cells, which are part of the APUD cell series derived embryologically from the neural crest. In man, C-cells reside mainly in the thyroid.

CT is secreted in response to a raised plasma Ca^{2+} but other factors also stimulate its secretion, e.g., gastrointestinal peptide hormones, alcohol and β-adrenergic agents. Like PTH it is heterogenous in the peripheral circulation.

The most important action of CT in mammals is to inhibit bone resorption and thereby lower plasma calcium and phosphate. Large doses of CT also increase renal excretion of calcium, and gut secretion of calcium, sodium and phosphate and alter the soft tissue distribution of these ions, but these are probably not significant effects under physiological conditions. There is some evidence that in other species CT may have other roles, for example, in regulating the availability of Ca from bone during egg laying in birds and in controlling Ca and sodium homeostasis in migratory fish.

The exact physiological role of CT in man is unclear. Some, but not all, radioimmunoassays can detect circulating CT in normal man and can show enhanced secretion in response to hypercalcaemia, or other provoking agents, e.g., alcohol. Recent evidence suggests that relative deficiency of CT may contribute to the bone disease seen in chronic renal failure, and possibly to other diseases. CT is secreted in excess by medullary carcinomas of the thyroid and can be used for the detection of pre-symptomatic disease in family members. There is also evidence that CT is secreted in excess in the various other tumours, e.g., bronchial carcinomas.

CT is used in man to reduce the excessive resorption and turnover of bone characteristic of Paget's disease and also to treat hypercalcaemia of malignancy.

Vitamin D (Vit D)

Animals derive their Vit D from the diet and from ultraviolet irradiation of 7-dehydrocholesterol in the skin. The essential steps in its subsequent metabolism are conversion to 25-hydroxycholecalciferol in the liver and the subsequent production of various dihydroxy metabolites in the kidney and elsewhere. The most important metabolite is probably $1,25(OH)_2D_3$, which has the major actions of the parent vitamin on intestine, bone and muscle. The intestinal absorption of Ca appears to involve both active transport and diffusion processes. Of the Ca regulating hormones, $1,25(OH)_2D_3$ is the most important in influencing intestinal absorption of calcium and phosphate. The precise mechanism of stimulation is unclear but seems to involve the synthesis of a calcium binding protein as well as a calcium-stimulated ATPase on the brush border surface of intestinal epithelial cells. The adjustment of Ca and phosphate absorption to body requirements and to changes in dietary intake seems to involve appropriate changes in the production of $1,25(OH)_2D_3$. One important feature of the control of Ca and phosphate absorption by $1,25(OH)_2D_3$ is that the adaption is relatively slow and takes several hours or days to come about, which is quite unlike the rapid responses of the kidney and bone to PTH and CT.

Intestinal absorption of Ca varies with age and sex and in various disease states, and some of these changes may reflect differences in either the production of $1,25(OH)_2D_3$, or in the target tissue responses to it. Although lack of vitamin D in man is associated with defective mineralisation of cartilage and bone, the question of whether Vit D or its metabolites act directly on skeletal tissues to increase mineralisation remains unsettled. It is possible that the effects of Vit D on bone mineralisation are secondary to changes in extracellular fluid concentrations of calcium and phosphate. The major effect of $1,25(OH)_2D_3$ on bone in culture is to increase resorption and this may contribute to making Ca and phosphate available for mineralisation. This effect of $1,25(OH)_2D_3$ on bone resorption probably occurs with physiological doses and $1,25(OH)_2D_3$ may be an important natural regulator of bone resorption.

Muscle weakness is also a well recognised feature of Vit D deficiency in man. It is possible that Vit D, or its metabolites, have direct effects on muscle function but little is known about this action.

The production of $1,25(OH)_2D_3$ by the kidney is closely regulated by a number of factors, including Vit D status, dietary calcium and phosphate, PTH, prolactin, growth hormone and oestrogens.

The factors which control Vit D metabolism in man require further elucidation but there is some evidence that the factors referred to above have some influence. Thus $1,25(OH)_2D_3$ levels are high in patients with hyperparathyroidism and low in hypoparathyroidism, indicating the possible influence of PTH or phosphate. Similarly, the low levels in chronic renal failure may indicate suppression by high plasma phosphate, or could be due to diminished renal mass. When the body's demand or requirements for Ca or P_i are increased, the production of $1,25(OH)_2D_3$ appears to be enhanced. For example, the increased intestinal absorption of Ca in growth, pregnancy and lactation may be mediated by increased production of $1,25(OH)_2D_3$, brought about by prolactin, growth hormone or oestrogens.

THE ROLE OF THE SKELETON

The structural organisation of bone is extremely complex and it has proved a difficult tissue to study biochemically.

In mature bone, three main cell types exist: osteoblasts responsible for bone formation; osteoclasts responsible for bone destruction; and osteocytes which are derived from osteoblasts and become trapped within the bone matrix as maturation proceeds. The osteocytes lie within a complex canalicular system and are probably responsible for many of the rapid ion fluxes that occur in bone. The origin, life span and fate of various cells in bone is gradually being elucidated Osteoprogenitor cells may arise from marrow stroma, whereas osteoclasts are derived, in part at least, from wandering mononuclear cells.

Tissue fluid surrounding the bone cells probably has a unique composition and is high in K^+ and contains particular plasma proteins in preference to others.

The mineral component of bone is predominantly hydroxyapatite. Since ionic exchange occurs between bone mineral and surrounding fluids it also contains other ions such as HCO_3, Mg^{2+}, Na^+, K^+, etc. Collagen is a major constituent of the organic matrix of bone and is largely responsible for its tensile strength. In addition the matrix contains several glycosaminoglycan components. The rate of deposition of bone matrix is controlled by hormonal factors (eg, PTH, GH), or by levels of Ca and phosphate, and by mechanical and electrical forces acting on bone.

Calcification is an important step in the transition between matrix production and the formation of mineralised bone. Mechanisms of calcification may differ according to the site at which it occurs. In general, the concentrations of Ca and phosphate in ECF are too low to initiate deposition of calcium and phosphate but can sustain crystal growth once it has started. In cartilage and bone, the first steps in calcification are now thought to take place in or around small membrane-bound vesicles found in the matrix. These vesicles probably arise from the plasma membranes of hypertrophic chondrocytes during the maturation of epiphysial cartilage, or from osteoblast membranes in osteoid tissue. The vesicle membranes are rich in alkaline phosphatase, an enzyme which has been known for many years to be associated with calcification.

Rickets is a term used to define failure of mineralisation of cartilage in long bones and is seen in growing children. Osteomalacia is a histological appearance of unmineralised bone matrix and has many causes apart from Vit D deficiency. For example, phosphate deprivation, or hypophosphatemia due to renal tubular disorders (eg, Vit D resistant rickets), may be associated with a defect in skeletal mineralisation

Calcification outside the skeleton (eg, in blood vessels, skin or muscles) occurs in many diseases and can sometimes be a serious clinical problem.

Resorption of bone is an essential part of the remodelling and growth process. Under physiological conditions, resorption is probably under the control of Vit D metabolites, PTH, thyroid hormones and steroids but will occur at a basal rate in the absence of these hormones. Many other factors have been found to stimulate bone resorption and these include Vit A, PG's, heparin and calcium ionophores. There are also peptide resorbing factors, such as OAF (osteoclast activating factor), produced by myeloma cells or activated lymphocytes.

Resorption of bone entails removal not only of the mineral components, but also of the matrix, and the latter probably requires the production of various degradative enzymes.

Bone resorption can be inhibited by CT and by certain drugs which include oestrogens, mithramycin and the diphosphonates.

INTEGRATION OF INDIVIDUAL ORGAN RESPONSES

It is helpful to draw the distinction between the way in which the plasma Ca is set at a particular value and the way in which the movements of calcium in and out of extracellular fluid are controlled. The plasma Ca is set close to a particular value in different individuals in normal and disease states. Deviations from this value are corrected by hormone-induced changes in the relative fluxes of calcium in and out of the ECF. Alterations in the flux rates are, therefore, monitored and adjusted by the changes in plasma Ca (or phosphate) concentrations they induce. This homeostatic system could operate with the plasma Ca set at any number of different values, with the relative rates of entry and exit of Ca to and from the ECF being altered as drift occurs from this set point.

Thus, in hyper- and hypoparathyroidism, the fluxes of Ca across the gut and in and out of bone may not be greatly different from normal and external Ca balance can be maintained (net intestinal absorption = urine loss) even though the plasma Ca is set at markedly different levels. The plasma Ca thus provides the point around which adjustments are made.

In considering homeostasis it is also helpful to distinguish between acute and chronic changes. When the system is disturbed, a steady state no longer exists and the response which occurs adjusts the system so that a new steady state comes into existence. The flux rates through individual organs and the level of plasma Ca may or may not be the same as previously, depending in part on whether the disturbance to the system is sustained or not.

CONTROL OF PLASMA CALCIUM

Acute Responses

Any deviation of plasma Ca^{2+} away from its normal value is rapidly corrected by alterations in the secretion of the regulating hormones, particularly PTH and, to a lesser extent, CT, in mammals. PTH and CT can be considered the fast acting (minutes to hours) component of the regulatory system, whereas Vit D is responsible for adaption in the longer term (hours to days). In experimental animals, removal of sources of PTH and CT (eg, thyroparathyroidectomy in dogs or rats) results in a slower return to normal of plasma Ca values in response to acute changes in plasma calcium. In man, the secretion of PTH responds within seconds to a change in plasma Ca^{2+} and in turn there is a rapid effect on renal tubular reabsorption and also on Ca movements in and out of bone.

The kidney is the key organ for determining the plasma concentrations of both Ca and phosphate. The most important hormonal influence on tubular reabsorption of these ions is PTH and there is little evidence that CT or Vit D metabolites at physiological concentrations have major effects on their renal handling.

Bone also plays a part in controlling plasma Ca by a buffering action, such that rises or falls in plasma Ca are partially compensated by increased net movements of Ca into or out of bone, respectively, without the intervention of any hormonal control factors.

One illustration of the contribution of bone to the control of plasma Ca is seen in young animals in whom injections of CT produce a rapid fall in plasma Ca due to inhibition of bone resorption. In mature animals in whom bone resorption is relatively slower, any inhibition by CT of bone resorption does not produce a large enough change in the net flux of Ca into the extracellular space to cause a significant fall in plasma Ca.

Chronic Responses

As indicated earlier, prolonged perturbations bring in contributions from changes in Vit D metabolism and adaptive responses in the intestine and bone. The responses to a reduction in dietary intake of Ca illustrates some of the changes involved. As intake is reduced, plasma Ca tends to fall and this increases the secretion of PTH. This will result in renal conservation of Ca and enhanced osteoclastic resorption of bone and an increase in $1,25(OH)_2D_3$ synthesis, which will augment the intestinal absorption of Ca and P_i and the resorption of these ions from bone. If the reduction in dietary Ca persists, these changes will act to restore the plasma Ca towards its previous value at the expense of a greater efficiency of intestinal Ca absorption and enhanced bone resorption. Bone formation rate will come to match that of the increased bone resorption rate by the coupling mechanism described below. In the new steady state, net Ca balance can again be maintained. Only when dietary deprivation is so severe that intake can no longer match output do the reserves of Ca in bone become utilised.

THE INFLUENCING HORMONES

Growth Hormone (GH)

Growth hormone is best known for its effects on the growth of cartilage, an effect which may be brought about indirectly by growth hormone-dependent production of somatamedins. GH also increases the renal tubular reabsorption of phosphate and thereby raises plasma P_i.

Thyroid Hormones

Thyroid hormones are essential for proper skeletal development, as indicated by the well known skeletal deformities of childhood hypothyroidism. In the adult, excess of triiodothyronine or thyroxine can be associated with hypercalciuria, hyperphosphataemia, raised alkaline phosphatase and occasionally hypercalcaemia.

In hyperthyroidism there is increased bone turnover, probably due to direct actions of the thyroid hormones on bone.

Insulin

Insulin probably has growth promoting properties on the skeleton in man, and insulin deficient diabetics appear to be prone to osteoporosis.

Adrenal Steroids

Glucocorticoids have complex effects on Ca metabolism and bone. In man their exogenous administration appears to suppress bone formation and may enhance bone resorption, and is therefore frequently associated with osteoporosis.

Sex Steroids

Characteristic growth abnormalities are associated with alterations of secretion of male or female sex steroids. In adults the effects of oestrogens are of particular interest because of the loss of bone that occurs in women after the menopause. Administration of exogenous oestrogen may slow down this loss.

Prostaglandins (PG's)

Their is considerable evidence that PG's, particularly of the E series, are potent bone resorbing agents. They may be important in pathological states in the production of the hypercalcaemia associated with cancers and in the bone resorption that accompanies rheumatoid arthritis or dental cysts.

COUPLING OF BONE FORMATION AND RESORPTION

There are additional control mechanisms that exist within bone itself and which must be of great importance in homeostasis.

In a variety of both physiological and pathological steady states, it is clear that there is a remarkably close correlation between rates of mineral deposition and mineral resorption from bone. Even though these individual rates may be altered many-fold, the net gains or losses of skeletal mass are minimised by the tight coupling that exists between these rates. This important feature of homeostatic adaption is often overlooked but it is of vital importance in designing therapeutic agents to treat bone disease. Thus, it is difficult to achieve a sustained dissociation between rates of mineralisation and resorption and this is why so many potential therapeutic agents have proved disappointing in trying to produce a sustained increase in bone mass in osteoporosis. Transient dissociations in these rates can occur, for example, during the acute loss of bone mineral that occurs in response to immobilisation but within a few weeks or months the rates come into phase again.

One of the more impressive examples of the coupling phenomenon comes from Paget's disease, in which a tight correlation is seen between plasma alkaline phosphatase which reflects bone formation and urinary total hydroxyproline (THP), which reflects the resorption of bone collagen. This phenomenon can also be demonstrated using radioisotopes of Ca to measure bone mineral accretion and resorption.

The mechanisms underlying this coupling are unknown but may involve cell-to-cell communication within bone as well as the influence of external hormones.

It is important to note that plasma Ca can be normal in the face of either low, normal, or high rates of bone turnover.

A further feature of bone is its ability to respond to mechanical deformation by changing its structure and shape to counteract the stress. One mechanism proposed to account for this phenomenon involves the generation of small piezo-electric currents within bone in response to stress. These currents are then thought to influence bone cell metabolism.

DISORDERS OF CALCIUM HOMEOSTASIS

A detailed description of disorders of Ca metabolism is beyond the scope of this review but there have been several books and reviews published on this topic recently. To a large extent the effects of excess, or deficiency, of PTH, CT or Vit D are those to be expected from the known physiological and pharmacological actions of these agents.

There are, however, many disorders which at present cannot be explained in terms of alterations in endocrine function. These include various inherited abnormalities of bone, such as osteogenesis imperfecta, hypophosphatasia, fibrogenesis imperfecta ossium, osteopetrosis, and various epiphysial, metaphysial and diaphysial dysplasias.

A particularly important group from the medical and socio-economic point of view are the disorders associated with osteoporosis.

Osteoporosis

Osteoporosis or osteopenia may be defined as a diminution in bone mass without there being any detectable change in the chemical composition of the bone in terms of its mineral versus matrix content. There are many conditions associated with osteoporosis but the most important from a public health standpoint is that which occurs after the menopause in women and that associated with old age. Major complications of osteoporosis are fractures, particularly of the wrist, the vertebrae and of the femoral neck.

From a theoretical point of view, it is clear that bone mass can diminish as a result of a number of different disturbances of Ca metabolism. They share the common feature that in order for bone mass to diminish, the rate of bone resorption must exceed that of bone formation. Even if the difference between these rates is very small, e.g., in the region of 1 mmol/day, significant bone mass can be lost if the difference is maintained over a prolonged period.

The prevention of bone loss and restoration of bone mass is one of the greatest therapeutic challenges in the field of skeletal physiology, Many approaches have been tried but few have met with much success. However, there is now good evidence that administration of oestrogens to post-menopausal women can reduce the rate of bone loss. Other modern drug regimes include the use of fluoride to stimulate bone formation and/or Vit D metabolites to enhance intestinal absorption. What is needed is an agent that can lead to a sustained increase in bone formation rate relative to bone resorption.

TABLE 1 Some Causes of Osteoporosis (Osteopenia or Thin Bones)

Primary: Old age
Post-menopause or post-oophorectomy
Idiopathic juvenile osteoporosis

Secondary: Dietary deficiency of calcium or malabsorption - Steatorrhea
- Partial gastrectomy
- Chronic liver disease

Endocrine - Hyperparathyroidism
- Hyperthyroidism
- Cushing's syndrome
- Hypogonadism

Metabolic - Vitamin D deficiency
- Pregnancy
- Osteogenesis imperfecta

Drugs - Corticosteroids
- Heparin

Immobilisation, generalised, e.g., space flight
localised, e.g., after fracture
paraplegia

Rheumatoid arthritis
Chronic renal failure or dialysis

DRUGS ACTING ON CALCIUM METABOLISM

A number of commonly used drugs have some influence on Ca metabolism. In some cases these effects can be turned to therapeutic advantage, e.g., the effect of thiazides to reduce urinary Ca, which is used in the treatment of renal stone formation. Some of these effects are illustrated in Fig. 2.

PHARMACOLOGY OF CALCIUM METABOLISM

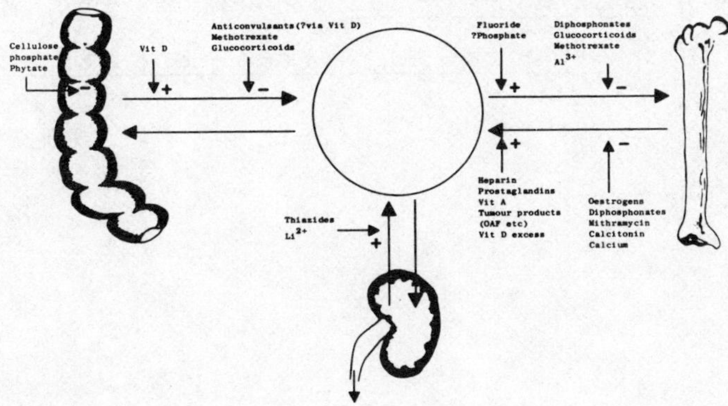

Fig. 2. Major sites of action of drugs influencing Ca metabolism

REFERENCES

Copp, D.H. & Talmage, R.V. (Ed) (1978) Endocrinology of Calcium Metabolism, Excerp. Medica, Amsterdam-Oxford

M.R. Haussler & T. McCain, Basic and clinical concepts related to Vit D metabolism and action, New Eng. J. Med. 297, 974 & 1041 (1977)

Krane, S.M. (1977) Paget's Disease, In: Clinical Orthopaedics & Related Research, vol 127

McKuisick, V.A. (1972) Heritable Disorders of Connective Tissue, 4th edn. St. Louis, Mosby

Massry, S.G. & Ritz, E. (Ed) (1977) Phosphate Metabolism, In: Advances in Exp. Medicine & Biology, vol 81, Plenum, New York

Morgan, D.B. (1973) Osteomalacia, Renal Osteodystrophy and Osteoporosis, Springfield, Thomas

Nordin, B.E.C. (1973) Bone and Stone Disease, Churchill Livingstone, Edinburgh

Norman, A.W. et al (Ed) (1977) Vitamin D, Biochemical, Chemical & Clinical Aspects, de Gruyter, Berlin, New York

Paterson, C.R. (1974) Metabolic Disorders of Bone, Oxford, Blackwell

Peacock, M. (Ed) (1977) The Clinical Uses of 1α-hydroxy vitamin D_3, Suppl. vol 7, Clinical Endocrinology

Potts, J.T. Jr. & Deftos, L.J. (1974) Duncan's Disease of Metabolism, 7th Edn. p 1225, Philadelphia, Saunders

L.G. Raisz & P.J. Bingham, Effects of hormones on bone development, Ann. Rev. Pharmacol. 12, 337 (1972)

R.G.G. Russell, Regulation of Calcium Metabolism, Ann. Clin. Biochem. 13, 518 (1976)

Vaughan, J.M. (1975) The Physiology of Bone, 2nd Edn. Oxford Univ. Press

Wynne-Davies, R. (1973) Heritable Disorders in Orthopaedic Practice, Blackwell, Oxford

Vitamin D and its Metabolites: Physiology, Biochemistry and Pharmacology

Hector F. DeLuca

Department of Biochemistry, College of Agricultural and Life Sciences, University of Wisconsin-Madison, Madison, Wisconsin 53706, U.S.A.

I. Introduction

Until the late 1960's it had been assumed that vitamin D must function directly without further modification (1). This view was further supported by the early work of Kodicek and his colleagues (1) with very low specific activity ^{14}C vitamin D in which they could detect no active metabolites of the vitamin. However, the synthesis of ^{3}H vitamin D of high specific activity in the author's laboratory (2) and the development of chromatographic methods adequate for the separation of vitamin D from its metabolites (3) ultimately led to the realization that vitamin D itself is not active directly but must be chemically converted to a hormone or hormones before it can act (4). Pursuit of this problem has led to the identification of a major endocrine system located in the kidney which converts the circulating form of vitamin D to an active hormone which then functions in regulating calcium and phosphorus metabolism (5). This paper will be devoted to a summary of this endocrine system and what is known concerning how the active hormone works in the target tissues.

II. Functions of Vitamin D

The basic function of vitamin D is to elevate plasma calcium and phosphorus concentrations to levels which will (A) support mineralization of newly forming bone (B) prevent hypocalcemic tetany and (C) prevent muscle weakness. In addition to these functions vitamin D may function directly on the bone cells carrying out the transfer of minerals for mineralization. Basically vitamin D through its active hormone(s) elevates plasma calcium and phosphrous concentrations by activating intestinal calcium and phosphorus transport reactions. These reactions appear to be active transport systems in which calcium and phosphate are independently transported across the epithelial membrane against electrochemical potential gradient (6,7). It is also known that vitamin D together with the parathyroid hormone is involved in the mobilization of calcium from the bone fluid compartment, (6,8,9). Phosphate is also mobilized from bone by this reaction since calcium phosphate hydroxyapatite is the ultimate source of bone calcium (10).

The function of vitamin D at the kidney is at the present time poorly understood (11). It does appear that vitamin D improves renal calcium reabsorption (11,12) although this conclusion is not shared by all investigators.

III. Metabolism of Vitamin D in Preparation for Function

In 1968 the first biologically active metabolite of vitamin D was isolated and the structure chemically determined to be 25-hydroxyvitamin D_3 (25-OH-D_3) (13). This metabolite appeared to act much more rapidly than vitamin D_3 in initiating intestinal calcium transport and the mobilization of calcium from bone (14). It also had activity in mineralization of bone superior to that of vitamin D. Although initially this appeared to be a good candidate for the active form of vitamin D it soon became clear that it was metabolized further before it could function (15). This led to the realization that the 25-OH-D_3 was further converted to a tissue active form. In 1971 this active form was isolated in pure form from the intestines of 1500 vitamin D-deficient chicks given a source of radioactive vitamin D of known specific activity (16). Its structure was conclusively demonstrated to be 1,25-dihydroxyvitamin D_3 (1,25-$(OH)_2D_3$) (16). Chemical synthesis of it (17) and of 1β,25-dihydroxyvitamin D_3 (18) conclusively removed any doubt that the active hormone is 1α,25-$(OH)_2D_3$. The known pathway, therefore, of vitamin D metabolism is shown in Figure 1. Not shown in this

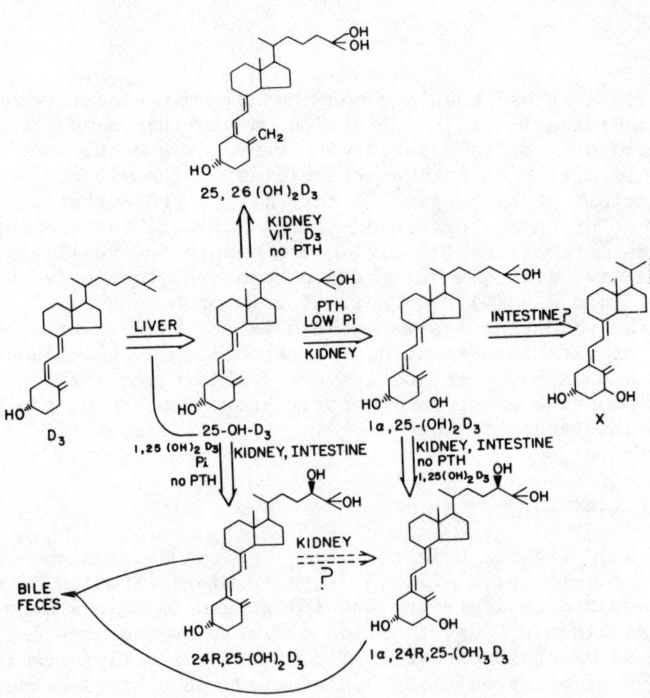

Fig. 1 Metabolism of Vitamin D

figure is the biogenesis of vitamin D_3 in skin from 7-dehydrocholesterol which exists in abundant quantities in the epidermis, a site to which 300 nm ultraviolet light easily penetrates. Work from two laboratories has conclusively demonstrated that the active substance produced in skin upon irradiation is vitamin D_3 (19,20). This reaction is not regulated and only limitation of ultraviolet exposure by either tanning of skin or pigmentation of skin appears to modify the reaction. Vitamin D_3 which is taken in either from this reaction or from diet rapidly accumulates in the liver with as much as 80% accumulating within an hour (21). In the liver it undergoes 25-hydroxylation in the endoplasmic reticulum by process supported by NADPH, molecular oxygen, and a cytosol protein (22). This reaction is carbon monoxide sensitive and appears to be cytochrome P_{450} dependent (Madhok and DeLuca, unpublished). It is not induced by phenobarbital or by the other known P_{450} inducing agents for liver microsomes. The 25-hydroxylation can also be found in intestine and in kidney, although the significance of the 25-hydroxylase in these tissues remains in some doubt (23,24). The 25-hydroxylase is feedback regulated by the product itself but this feedback regulation can be overcome by the administration of additional supplies of vitamin D_3 (25). 25-OH-D_3 does not act directly in any target tissue at physiologic concentration (26,27). Instead, it is further converted to an active hormone. The conversion to the active hormone occurs exlusively of the kidney and specifically mitochondria (28,29). In the kidney 25-OH-D_3 becomes hydroxylated in the 1α-position to produce the 1α,25-$(OH)_2D_3$. This system is a three component cytochrome P_{450} dependent reaction which is analagous to the adrenal steroidogensis system (30). It involves NADPH, a flavoprotein which is called renal ferredoxin reductase, an iron sulfur protein which is called renal ferredoxin and a cytochrome P_{450} which incorporates 1 atom of molecular oxygen into the 25-OH-D_3 substrate and the other into water (30). The 1,25-$(OH)_2D_3$ is then exported to the target tissues where it initiates the transport reactions described in section II above.

IV. Other Metabolism of Vitamin D

A major alternate to the 1-hydroxylation of 25-OH-D_3 is the 24R-hydroxylation to form 24R,25-dihydroxyvitamin D_3 (24,25-$(OH)_2D_3$) (31). In addition, the hormone 1α,25-$(OH)_2D_3$ can be 24R-hydroxylated to form 1α,24R,25-trihydroxyvitamin D_3 (1α,24R,25-$(OH)_3D_3$) (32). The 24R-hydroxylations take place primarily in the kidney and specifically in the mitochondria (33). However, recently, major 24R-hydroxylation has also been discovered in intestinal tissue (34) and also evidence presented that it is found in cartilage and bone tissue (35). It therefore appears that 24R-hydroxylation appears in all the target organs of 1,25-$(OH)_2D_3$ biogenesis. The nature of the 24R-hydroxylase is not thoroughly understood at the present time. It is less carbon monoxide sensitive than the 1-hydroxylase, is found in mitochondria in the case of kidney and appears to be dependent upon internally generated NADPH. Of great importance is the fact that this hydroxylase does not occur in the vitamin D-deficient animal (36). Instead this hydroxylase must be induced by the hormone 1α,25-$(OH)_2D_3$ (36). In man, nephrectomy markedly reduces the circulating 24R,25-$(OH)_2D_3$ but does not eliminate it (37), whereas nephrectomy does not reduce the circulating quantities of 1,24R,25-$(OH)_3D_3$ found in animals while markedly reducing the 24,25-$(OH)_2D_3$ found in the same animals (34).

There has been much discussion regarding the importance of 24R-hydroxylation in terms of specific biological activity. Reports have varied from a co-suppressant of parathyroid hormone secretion (38) to a mineralization form of the vitamin (39). So far evidence to support these concepts is tenuous and cannot be accepted at the present time. However, the question of whether there is a special function of 24R-hydroxylated vitamin D compounds must remain open. In animals it must be recognized that 24R-hydroxylation always diminishes and does not increase biological activity of the substrate; namely, 24,25-$(OH)_2D_3$ is less active than 25-OH-D_3 at least in birds and probably in mammals (40,41). 1,24R,25-$(OH)_3D_3$ is less active in all species examined than is 1,25-$(OH)_2D_3$ (42). Furthermore its onset of activity and its duration of activity is either less or identical with

$1,25\text{-}(OH)_2D_3$ (42). Therefore evidence for special function of 24R-hydroxylation is in considerable doubt and it is likely that this process is involved in the deactivation of $1,25\text{-}(OH)_2D_3$ and perhaps $25\text{-}OH\text{-}D_3$.

Another significant site of hydroxylation of the vitamin D molecules is the 26-position. $25,26\text{-}(OH)_2D_3$ was isolated and identified in 1969 (43) and chemically synthesized by several groups in the past few years. Recently we have been able to learn that the major site of 26-hydroxylation is in the kidney (44). The nature of this hydroxylase has not been thoroughly explored at the present time and is under investigation. However, the 26-hydroxylase is absent in vitamin D-deficient animals and must be induced by some form of vitamin D (44). $25,26\text{-}(OH)_2D_3$ has much less biological activity than $25\text{-}OH\text{-}D_3$ and so far its activity has been confined to intestinal calcium transport (43). Again, it is not clear whether this compound has any biological function or not. It is also possible that it is enroute to degradation of the vitamin D molecule. 26-hydroxylation like 24R-hydroxylation is regulated in a negative way. That is, under circumstances where the 1-hydroxylation is diminished by high calcium diets, by parathyroidectomy or high plasma phosphate concentrations, 24-hydroxylation and 26-hydroxylation are markedly stimulated.

Finally, mention should be made of the known fate of $1,25\text{-}(OH)_2D_3$. In experiments in which the recovery of radioactive $1,25\text{-}(OH)_2D_3$ was studied, it was learned that much of the 26,27-label was lost from the molecule (45). This was further confirmed when $1,25\text{-}(OH)_2[26,27\ ^{14}C]D_3$ was prepared and injected into animals. Approximately 25-30% of the C-14 appeared in the expired carbon dioxide indicating 30% of $1,25\text{-}(OH)_2D_3$ undergoes sidechain cleavage and oxidation. In the case of $25\text{-}OH\text{-}D_3$ only 7% is metabolized in this way and this 7% must course through the 1-hydroxylation pathway since nephrectomy prevents $^{14}CO_2$ production from 26,27 C-14 labeled $25\text{-}OH\text{-}D_3$. The side chain cleavage and oxidation reaction has been narrowed to the liver and intestine and is undergoing investigation (46). The product of this sidechain cleavage remains unknown.

V. Regulation of Vitamin D Metabolism

Since $1,25\text{-}(OH)_2D_3$ is made exclusively in the kidney and has its function in intestine and bone, it must be regarded as a hormone. In true hormonal fashion the biogenesis of $1,25\text{-}(OH)_2D_3$ is markedly regulated by the need for calcium. It now seems clear the slight hypocalcemia stimulates the parathyroid glands to secrete parathyroid hormone. This hormone then proceeds to the target tissue of the kidney and bone (47). In the kidney the parathyroid hormone stimulates production of $1,25\text{-}(OH)_2D_3$ in addition to stimulating renal reabsorption of calcium and the excretion of phosphorus. The $1,25\text{-}(OH)_2D_3$ then proceeds to the intestine where by itself and without futher action of the parathyroid hormones stimulates intestinal calcium transport and probably intestinal phosphate transport (9). At the bone, the parathyroid hormone and $1,25\text{-}(OH)_2D_3$ together mobilize calcium from the bone fluid compartment (9). These three sources of calcium then elevate plasma calcium to a point where the parathyroid glands are suppressed and this then suppresses the original signal which led to the production of $1,25\text{-}(OH)_2D_3$ in the first place.

On a minute to minute basis, the parathyroid hormone probably acts with endogenous $1,25\text{-}(OH)_2D_3$ to protect against hypocalcemia. In this response the parathyroid hormone would increase renal reabsorption of calcium together with endogenous levels of $1,25\text{-}(OH)_2D_3$ and would also stimulate the mobilization of calcium from bone again with endogenous levels of $1,25\text{-}(OH)_2D_3$. This would correct the hypocalcemia but the correction would be at the expense of bone. On a longer term basis continued excitation by hypocalcemia of the parathyroid glands would bring about stimulation of 1,25(OH)$_2$D$_3$ production a process which requires at least 2 or 3 hours (7). This system then brings into play the intestine the only organ which can sequester calcium from the environment. It is important to note that intestine does not bind parathyroid hormone (47) and is not a target tissue of parathyroid hormone action, at least as far as is known at the present time. Thus 1,25-(OH)$_2$D$_3$ is responsible for the utilization of calcium from the diet and hence from the environment. This is an important consideration in the disease process of such diseases as osteoporosis and renal osteodystrophy.

In addition to the need for calcium stimulating production of 1,25-(OH)$_2$D$_3$ it is also known that plasma phosphate plays an important role in regulating vitamin D metabolism (48); the exact site of phosphate intervention is not known but is likely at the 1-hydroxylase level and perhaps at the further metabolism and accumulation of 1,25-(OH)$_2$D$_3$ in the target tissues. It is well known that phosphate deprivation of animals markedly increases plasma levels of 1,25-(OH)$_2$D$_3$ (49). Some controversy exists as to whether this is the result of increased synthesis or decreased degradation. Certainly the 1-hydroxylase is elevated about 5-fold in the case of birds under phosphate deprivation (50). The increased 1,25-(OH)$_2$D$_3$ found in phosphate deprivation is probably an important physiologic mechanism which allows for greater utilization of phosphate from the small intestine which is a 1,25-(OH)$_2$D$_3$ dependent reaction (51).

Besides the need for calcium and phosphorus, other hormones appear to directly or indirectly affect the conversion of 25-OH-D$_3$ to the 1,25-(OH)$_2$D$_3$. In particular, considerable investigation has taken place in regard to the egg laying bird in which there is considerable mobilization of calcium from intestine and medullary bone at the time of egg shell formation. It was demonstrated that kidneys taken from egg laying birds have elevated 1-hydroxylase activity as compared to non-egg laying females or males of the same age and treatment (52). The injection of estradiol to the mature males markedly stimulates the 1-hydroxylase activity. Since the stimulation did not occur in immature birds, studies were then initiated in the castrate male birds which eliminated endogenous sex hormones. In this system testosterone was required for estradiol to stimulate the 25-OH-D$_3$ 1-hydroxylase (53). Progesterone could substitute for testosterone but all three sex hormones given together gave the highest 1-hydroxylase activity. Therefore at least in egg laying birds, the sex hormones also regulate the production of 1,25-(OH)$_2$D$_3$ directly or indirectly.

VIII. Subcellular Location and Mechanism of Action of 1,25-(OH)$_2$D$_3$

Throughout the above discussion it has been assumed that 1,25-(OH)$_2$D$_3$ is not metabolized further before it carries out its known functions in intestine and bone. Although this seems quite likely to be the case, it can not be entirely dismissed as being proven. Current evidence suggests that 1,25-(OH)$_2$D$_3$ is not metabolized further before it functions but further modifications of this molecule may be involved in specific target organ responses. Only continued investigation will permit final conclusions on this point. In any case, much work has been expended in attempting to understand how 1,25-(OH)$_2$D$_3$ functions in the target tissue. During the past several years we have devoted a great deal of attention to chemical synthesis of radioactive 1,25-(OH)$_2$D$_3$ of specific activity sufficient to carry out autoradiography with physiologic doses. Dr. Yamada in our group was successful in synthesizing 25-hydroxy-[23,24-^{3}H]vitamin D$_3$ (25-OH-[23,24-^{3}H]D$_3$) of about 80 Ci/mmol (55). This could be converted enzymatically to 1,25-(OH)$_2$[^{3}H]D$_3$ of the same specific activity. Utilizing this material we were able to carry out autoradiography experiments with physiologic doses of 1,25-$(OH)_2D_3$ (56). Specific nuclear location of 1,25-$(OH)_2D_3$ could be demonstrated at 2-1/2 hours post injection in the crypt cells and villus cells of small intestine, but not in submucosa, muscle tissue, liver and most segments of the kidney. The accumulation of nuclear tritium from 1,25-$(OH)_2[^3H]D_3$ reaches a maximum at four hours in the small intestine whereas the calcium transport response in the case of birds is found maximally stimulated at 9 hours. The nuclear location of 1,25-$(OH)_2D_3$ by autoradiographic means confirms the previous chemical demonstration by crude subcellular fractionation that 1,25-$(OH)_2D_3$ is found associated with nucleus and nuclear chromatin (58,59).

Although there is considerable controversy as to whether 1,25-$(OH)_2D_3$ functions in the small intestine by nuclear or non-nuclear mediated reaction it appears clear that at least a portion of 1,25-$(OH)_2D_3$ function must be mediated by nuclear events. It is therefore considered likely that 1,25-$(OH)_2D_3$ acts like other steroid hormones by interacting with a specific cytosol receptor which then interacts with the nucleus to bring about the production of mRNA which would then code specifically for calcium and phosphate transport proteins. Brumbaugh and Haussler were the first to demonstrate the existence of highly specific 1,25-$(OH)_2D_3$ binding protein in chick intestine which can be considered a candidate for the role of receptor (60). After carefully washing chick and rat intestinal cells to eliminate proteolytic activity, and the inclusion of dithiothreitol, as well as high salt concentration, we could confirm the presence of highly specific 1,25-$(OH)_2D_3$ binding proteins not only in the intestinal cytosol of rachitic chicks (61) but also in the intestine of rats (62) and in the bone of both chicks and rats (63). These receptors are very highly specific for the 1,25-$(OH)_2D_3$ molecule as studied in competitive binding experiments (74). Brumbaugh and Haussler have demonstrated that the chick intestinal receptor and 1,25-$(OH)_2D_3$ are transfered to the nucleus to bind specifically to nuclear chromatin (65).

The exact mechanism whereby vitamin D stimulates the mobilization of calcium from bone or the transport of calcium and/or phosphorus in the small intestine remains largely unknown. However, it seems likely that 1,25-$(OH)_2D_3$ combines with a cytosolic receptor which then interacts with the nucleus to produce mRNA, the mRNA coding for calcium and phosphorus transport proteins. The nature of the calcium and phosphorus transport proteins remains largely unknown although Wasserman and his colleagues have demonstrated the existence of a calcium binding protein of 24,000 mw in the case of birds which would possibly function in transport (66). However there has been considerable question raised by several investigators as to whether this substance can be considered the calcium carrier in the case of the vitamin D mechanism and powerful evidence arguing against this possibility have been accumulated (67). Likely the calcium binding protein of Wasserman and his colleagues is involved in some way but other factors that are induced by 1,25-$(OH)_2D_3$ must also be involved.

Calcium apparently enters the brush border membrane by a vitamin D-dependent process and calcium is then transferred either by vesicles or mitochondria to the basal lateral membrane where,in a sodium dependent process, (7) calcium is expelled giving rise to transcellular calcium transport. Much remains to be learned concerning this mechanism and this area represents a major area of investigation of the vitamin D hormonal system.

Summary

A new endocrine system has been discovered which utilizes the vitamin D molecule as a building block and produces a hormone 1,25-$(OH)_2D_3$ which in turn regulates calcium and phosphorus metabolism. Its biogenesis is also regulated in a feedback manner by serum calcium and phosphorus concentrations. Low serum calcium functions through the parathyroid gland and parathyroid hormone stimulates 1,25-$(OH)_2D_3$ production. Other regulators such as sex hormones have been proposed. The 1,25-$(OH)_2D_3$ appears to function in the small intestine in a manner analagous to other steroid hormones involving a receptor, nuclear interaction, and production of mRNAs. The mRNAs are believed to code for calcium and phosphorus transport proteins that function at the brush border membrane to permit entry of calcium and phosphorus into the absorbing cells. Major areas of investigation now include the synthesis of analogs of 1,25-$(OH)_2D_3$ the mechanism of regulation of vitamin D metabolism and the mechanism of action of 1,25-$(OH)_2D_3$ at the target tissue sites. The findings in the vitamin D endocrine system and particularly the new active metabolites including 1,25-$(OH)_2D_3$ are now being rapidly developed for treatment of calcium and bone disorders.

REFERENCES

1. E. Kodicek, Metabolic studies on vitamin D in "Ciba Foundation Symposium on Bone Structure and Metabolism." (G.W.E. Wolstenholme and C.M. O'Connor, eds.) Little Brown & Co., Boston, pp. 161-174 (1956).
2. P. F. Neville and H. F. DeLuca, The synthesis of [1,2-^{3}H]vitamin D_3 and the tissue localization of 0.25 μg (10 IU) dose per rat. Biochemistry 5, 2201 (1966).
3. A. W. Norman and H. F. DeLuca, Chromatographic separation of mixtures of vitamin D_2, ergosterol, and tachysterol$_2$. Anal. Chem. 35, 1247 (1963).
4. J. Lund and H. F. DeLuca, Biologically active metabolite of vitamin D_3 from bone, liver and blood serum. J. Lipid Res. 7, 739 (1966).
5. H. F. DeLuca, Vitamin D: the vitamin and the hormone. Fed. Proc. 33, 2211 (1974).
6. H. F. DeLuca, Vitamin D in "The Fat-Soluble Vitamins", (H. F. DeLuca, ed.) Vol. 2 of Handbook of Lipid Research, Plenum Press, New York, Chap. 2, pp. 69-132 (1978).
7. H. F. DeLuca, Vitamin D and calcium transport in "Calcium Transport and Cell Function", (A. Scarpa and E. Carafoli, eds.) Vol. 307 of Annals of the New York Academy of Sciences, New York Acad. Sci. New York, pp. 356-376 (1978).
8. H. Rasmussen, H. F. DeLuca, C. Arnaud, C. Hawker, and M. von Stedingk, The relationship between vitamin D and parathyroid hormone. J. Clin. Invest. 42, 1940 (1963).
9. M. Garabedian, Y. Tanaka, M. F. Holick, and H. F. DeLuca, Response of Intestinal calcium transport and bone calcium mobilization to 1,25-dihydroxyvitamin D_3 in thyroparathyroidectomized rats. Endocrinology, 94, 1022 (1974).
10. L. Castillo, Y. Tanaka, and H. F. DeLuca, The mobilization of bone mineral by 1,25-dihydroxyvitamin D_3 in hypophosphatemic rats. Endocrinology, 97, 995 (1975).
11. R. A. L. Sutton, Effects of vitamin D on renal tubular calcium transport, in "Proceedings of the VII International Congress of Nephrology" (R. Barcelo, M. Bergeron, S. Carriere, J. H. Dirks, K. Drummond, R. D. Guttmann, G. Lemieux, J-G Mongeau and J. F. Seely, eds.) S. Karger, Basel, pp. 463-468 (1978).
12. T. H. Steele, J. E. Engle, Y. Tanaka, R. S. Lorenc, K. L. Dudgeon, and H. F. DeLuca, Phosphatemic action of 1,25-dihydroxyvitamin D_3. Am. J. Physiol., 229, 489 (1975).
13. J. W. Blunt and H. F. DeLuca, The synthesis of 25-hydroxycholecalciferol. A biologically active metabolite of vitamin D_3. Biochemistry 8, 671 (1969).
14. J. W. Blunt, Y. Tanaka, and H. F. DeLuca, The biological activity of 25-hydroxycholecalciferol, a metabolite of vitamin D_3. Proc. Nat. Acad. Sci. USA 61, 1503 (1968).

15. R. J. Cousins, H. F. DeLuca, and R. W. Gray, Metabolism of 25-hydroxycholecalciferol in target and nontarget tissues, Biochemistry 9, 3649 (1970).
16. M. F. Holick, H. K. Schnoes, H. F. DeLuca, T. Suda, and R. J. Cousins, Isolation and identification of 1,25-dihydroxycholecalciferol. A metabolite of vitamin D active in intestine, Biochemistry, 10, 2799 (1971).
17. E. J. Semmler, M. F. Holick, H. K. Schnoes, and H. F. DeLuca, The synthesis of 1α,25-dihydroxycholecalciferol--a metabolically active form of vitamin D_3, Tetrahedron Letters, 40, 4147 (1972).
18. H. E. Paaren, H. K. Schnoes, and H. F. DeLuca, Synthesis of 1β-hydroxyvitamin D_3 and 1β,25-dihydroxyvitamin D_3, Chem. Commun. in press (1977)
19. R. P. Esvelt, H. K. Schnoes and H. F. DeLuca, Vitamin D_3 from rat skins irradiated in vitro with ultraviolet light, Arch. Biochem. Biophys., in press, (1978).
20. M. F. Holick, J. E. Frommer, S. C. McNeill, N. M. Richtand, J. W. Henley, and J. T. Potts, Jr., Photometabolism of 7-dehydrocholesterol to previtamin D_3 in skin, Biochem. Biophys. Res. Commun. 76, 107, (1977).
21. E. B. Olson, Jr., J. C. Knutson, M. H. Bhattacharyya, and H. F. DeLuca, The Effect of hepatectomy on the synthesis of 25-hydroxyvitamin D_3, J. Clin. Invest. 57, 1213 (1976).
22. M. Bhattacharyya and H. F. DeLuca, Subcellular location of rat liver calciferol 25-hydroxylase, Arch Biochem. Biophys. 160, 58 (1974).
23. G. Tucker, III, R. E. Gagnon, and M. R. Haussler, Vitamin D_3-25-Hydroxylase: tissue occurrence and apparent lack of regulation, Arch. Biochem. Biophys. 155, 47 (1973).
24. M. H. Bhattacharyya and H. F. DeLuca, The regulation of calciferol-25-hydroxylase in the chick. Biochem. Biophys. Res. Commun. 59, 734 (1974).
25. M. H. Bhattacharyya and H. F. DeLuca, The regulation of rat liver calciferol-25-hydroxylase, J. Biol. Chem. 248, 2969 (1973).
26. I. T. Boyle, L. Miravet, R. W. Gray, M. F. Holick, and H. F. DeLuca, The response of intestinal calcium transport to 25-hydroxy and 1,25-dihydroxy vitamin D in nephrectomized rats, Endocrinology 90, 605 (1972).
27. M. F. Holick, M. Garabedian and H. F. DeLuca, 1,25-Dihydroxycholecalciferol: metabolite of vitamin D_3 active on bone in anephric rats. Science 176, 1146 (1972).
28. D. Fraser and E. Kodicek, Unique biosynthesis by kidney of a biologically active vitamin D metabolite. Nature 228, 764 (1970).
29. R. Gray, I. Boyle, and H. F. DeLuca, Vitamin D metabolism: the role of kidney tissue. Science 172, 1232 (1971).
30. J. C. Ghazarian, C. R. Jefcoate, J. C. Knutson, W. H. Orme-Johnson and H. F. DeLuca, Mitochondrial cytochrome P_{450}: a component of chick kidney 25-hydroxycholecalciferol-1α-hydroxylase. J. Biol. Chem. 249, 3026 (1974).
31. M. F. Holick, H. K. Schnoes, H. F. DeLuca, R. W. Gray, I. T. Boyle, and T. Suda, Isolation and identification of 24,25-dihydroxycholecalciferol: a metabolite of vitamin D_3 made in the kidney. Biochemistry 11, 4251 (1972).
32. M. F. Holick, A. Kleiner-Bossaller, H. K. Schnoes, P. M. Kasten, I. T. Boyle, and H. F. DeLuca, 1,24,25-Trihydroxyvitamin D_3: a metabolite of vitamin D_3 effective on intestine. J. Biol. Chem. 248, 6691 (1973).
33. J. C. Knuston and H. F. DeLuca, 25-Hydroxyvitamin D_3-24-hydroxylase: subcellular location and properties. Biochemistry 13, 1543 (1974).
34. R. Kumar, H. K. Schnoes, and H. F. DeLuca, Rat intestinal 25-hydroxyvitamin D_3-and 1α,25-dihydroxyvitamin D_3-24-hydroxylase, J. Biol. Chem., in press (1978).
35. M. Garabedian, M. T. Corvol, M. Baily du Bois, M. Lieberherr and S. Balsan. The biological activity of 24,25-dihydroxycholecalciferol on cultured chondrocytes and its in vitro production in cartilage and calvarium. in "Proc. Sixth Parathyroid Conference" Vancouver, Canada, Excerpta Medica, in press (1977).

36. Y. Tanaka, R. S. Lorenc, and H. F. DeLuca, The role of 1,25-dihydroxyvitamin D_3 and parathyroid hormone in the regulation of chick renal 25-hydroxyvitamin D_3-24-hydroxylase, Arch. Biochem. Biophys. 171, 521 (1975).
37. H. F. DeLuca, 24-Hydroxylation of the vitamin D metabolites: its site and physiologic significance in man, in "Proceedings of VII International Congress of Nephrology" (R. Barcelo, M. Bergeron, S. Carriere, J. H. Dirks, K. Drummond, R. D. Guttmann, G. Lemieux, J-G. Mongeau, and J. F. Seely, eds.) S. Karger, Basel, Switzerland, pp. 447-454 (1978).
37. H. L. Henry, A. N. Taylor and A. W. Norman, Response of chick parathyroid glands to the vitamin D metabolites 1,25-dihydroxyvitamin D_3 and 24,25-dihydroxyvitamin D_3, J. Nutr. 107, 1918 (1977).
38. P. Bordier, A. Ryckwaert, P. Marie, L. Miravet, A. Norman and H. Rasmussen, Vitamin D metabolites and bone mineralization in man, in "Vitamin D: Biochemical, Chemical and Clinical Aspects Related to Calcium Metabolism" (A. W. Norman, K. Schaefer, J. W. Coburn, H. F. DeLuca, D. Fraser, H. G. Grigoleit and D. v. Herrath, eds.) Walter de Gruyter, Inc., Berlin, pp. 897-911 (1977).
40. M. F. Holick, L. A. Baxter, P. K. Schraufrogel, T. E. Tavela, and H. F. DeLuca, Metabolism and biological activity of 24,25-dihydroxyvitamin D_3 in the chick. J. Biol. Chem. 251, 397 (1976).
41. Y. Tanaka, H. F. DeLuca, N. Ikekawa, M. Morisaki, and N. Koizumi, Determination of stereochemical configuration of the 24-hydroxyl group of 24,25-dihydroxyvitamin D_3 and its biological importance. Arch. Biochem. Biophys. 170, 620 (1975).
42. L. Castillo, Y. Tanaka, H. F. DeLuca and N. Ikekawa, On the physiological role of 1,24,25-trihydroxyvitamin D_3, Mineral & Elect. Metab. in press (1978).
43. T. Suda, H. F. DeLuca, H. K. Schnoes, Y. Tanaka, and M. F. Holick, 25,26-Dihydroxycholecalciferol, a metabolite of vitamin D_3 with intestinal calcium transport activity, Biochemistry, 9, 4776 (1970).
44. Y. Tanaka, R. M. Shepard, H. F. DeLuca, and H. K. Schnoes, The 26-hydroxylaton of 25-hydroxyvitamin D_3 in vitro by chick renal homogenates, Biochem. Biophys. Res. Commun. in press (1978).
45. R. Kumar, D. Harnden, and H. F. DeLuca, Metabolism of 1,25-dihydroxyvitamin D_3: evidence for side-chain oxidation. Biochemistry 15, 2420 (1976).
46. R. Kumar and H. F. DeLuca, Side chain oxidation of 1,25-dihydroxyvitamin D_3 in the rat: effect of removal of the intestine. Biochem. Biophys. Res. Commun. 76, 253 (1977).
47. J. E. Zull and D. W. Repke, The tissue localization of tritiated parathyroid hormone in thyroparathyroidectomized rats. J. Biol. Chem. 247, 2195 (1972).
48. Y. Tanaka and H. F. DeLuca, The control of 25-hydroxyvitamin D metabolism by inorganic phosphorus. Arch. Biochem. Biophys. 154, 566 (1973).
49. M. R. Hughes, P. F. Brumbaugh, M. R. Haussler, J. E. Wergedal and D. J. Baylink, Regulation of serum 1α,25-dihydroxyvitamin D_3 by calcium and phosphate in the rat. Science, 190, 578 (1975).
50. L. A. Baxter and H. F. DeLuca, Stimulation of 25-hydroxyvitamin D_3-1α-hydroxylase by phosphate depletion. J. Biol. Chem. 251, 3158 (1976).
51. T. C. Chen, L. Castillo, M. Korycka-Dahl, and H. F. DeLuca, Role of vitamin D metabolites in phosphate transport of rat intestine. J. Nutr. 104, 1056 (1974)
52. Y. Tanaka, L. Castillo, and H. F. DeLuca, Control of the renal vitamin D hydroxylaes in birds by the sex hormoens. Proc. Nat. Acad. Sci. USA 73, 2701 (1976).
53. L. Castillo, Y. Tanaka, H. F. DeLuca, and M. L. Sunde, The stimulation of 25-hydroxyvitamin D_3-1α-hydroxylase by estrogen. Arch. Biochem. Biophys. 179, 211 (1977).
54. A. D. Kenny, Vitamin D metabolism: physiological regulation in egg-laying Japanese quail. Am. J. Physiol. 230, 1609 (1976).

55. S. Yamada, H. K. Schnoes, and H. F. DeLuca, Synthesis of 25-hydroxy-[23,24-^{3}H] vitamin D_3. Anal. Biochem. in press (1978).
56. M. Zile, E. C. Bunge, L. Barsness, S. Yamada, H. K. Schnoes, and H. F. DeLuca, Localization of 1,25-dihydroxyvitamin D_3 in intestinal nuclei in vivo. Arch. Biochem. Biophys. in press (1978).
57. M. R. Hughes and M. R. Haussler, 1,25-Dihydroxyvitamin D_3 receptors in parathyroid glands. Preliminary characterization of cytoplasmic and nuclear binding components. J. Biol. Chem. 253, 1065 (1978).
58. T. C. Chen, J. C. Weber, and H. F. DeLuca, On the subcellular location of vitamin D metabolites in intestine. J. Biol. Chem. 245, 3776 (1970).
59. M. R. Haussler, J. F. Myrtle and A. W. Norman, The association of a metabolite of vitamin D_3 with intestinal mucosa chromatin in vivo. J. Biol. Chem. 243, 4055 (1968).
60. P. F. Brumbaugh and M. R. Haussler, Nuclear and cytoplasmic binding components for vitamin D metabolites. Life Sciences 16, 353 (1975).
61. B. E. Kream, R. D. Reynolds, J. C. Knutson, J. A. Eisman, and H. F. DeLuca, Intestinal cytosol binders of 1,25-dihydroxyvitamin D_3 and 25-hydroxyvitamin D_3. Arch. Biochem. Biophys. 176, 779 (1976).
62. B. E. Kream, S. Yamada, H. K. Schnoes, and H. F. DeLuca, Specific cytosol binding protein for 1,25-dihydroxyvitamin D_3 in rat intestine. J. Biol. Chem. 252, 4501 (1977).
63. B. E. Kream, M. Jose, S. Yamada, and H. F. DeLuca, A specific high-affinity binding macromolecule for 1,25-dihydroxyvitamin D_3 in fetal bone. Science 197, 1086 (1977).
64. B. E. Kream, M. J. L. Jose, and H. F. DeLuca, The chick intestinal cytosol binding protein for 1,25-dihydroxyvitamin D_3: a study of analog binding. Arch. Biochem. Biophys. 179, 462 (1977).
65. P. F. Brumbaugh and M. R. Haussler, 1α,25-Dihydroxycholecalciferol receptors in intestine. II. Temperature-dependent transfer of the hormone to chromatin via a specific cytosol receptor. J. Biol. Chem. 249, 1258 (1974).
66. R. H. Wasserman, and J. J. Feher, Vitamin D-dependent calcium-binding proteins, in "Calcium Binding Proteins and Calcium Functions" (R. H. Wasserman, R. A. Corradino, E. Carafoli, R. H. Kretsinger, D. H. MacLennan, and S. L. Siegel, eds.) Elsevier North-Holland, Inc, pp. 392-302 (1977).

ACKNOWLEDGMENTS

This work was supported by National Institutes of Health Program-Project grant #AM-14881, the U.S. Energy Research and Development Administration contract #EY-76-S-02-1668 and the Harry Steenbock Research Fund.

Clinical Pharmacology and Application of the Active Forms of Vitamin D

Jack W. Coburn and Arnold S. Brickman

Medical and Research Services, VA Wadsworth Hospital Center (Los Angeles) and VA Hospital (Sepulveda) and the Department of Medicine, UCLA School of Medicine, Los Angeles, California, U.S.A.

The vast accumulation of new information about the metabolism and mechanisms of action of vitamin D over the past decade has been followed closely by the pharmacologic evaluation of the newly identified vitamin D sterols in man and their application to the management of certain disease states. Overwhelming evidence suggests that vitamin D_3, itself, is a precursor to the steroid hormone, 1,25-dihydroxy-vitamin D_3 {$1,25(OH)_2D_3$}; the latter is generated exclusively in the kidney in response to a number of physiologic stimuli which act to maintain calcium and phosphorus homeostasis (1). It seems probable that $1,25(OH)_2D_3$ exerts its actions on target tissues in a manner analogous to those of other steroid hormones (1). The present discussion reviews: 1) the pharmacologic effects of $1,25(OH)_2D_3$ in man, 2) certain disease states which arise due to alterations in vitamin D metabolism, 3) the therapeutic uses of active forms of vitamin D and synthetic analogues of $1,25(OH)_2D_3$, which do not require renal 1-hydroxylation before acting, and 4) considers the biologic actions of 24,25-dihydroxy-vitamin D {$24,25(OH)_2D_3$}, another naturally occurring vitamin D sterol.

PHYSIOLOGIC AND PHARMACOLOGIC ACTIONS OF 1,25-DIHYDROXY-VITAMIN D IN NORMAL MAN

Research on the effect of vitamin D carried out 30-40 years ago suggested that there is little biologic effect in normal man unless very large quantities were given. Knowledge that vitamin D must undergo two-step bioconversion first to 25-hydroxy-vitamin D_3 {$25(OH)D_3$} and then to $1,25(OH)_2D_3$ and data indicating that this bioconversion is under close endocrine control provide an explanation for the relatively low biologic activity of vitamin D_3 in these early experiments, Newer observations made after the administration of $1,25(OH)_2D_3$ in both the vitamin D-deficient or replete state have yielded new information about the physiologic and pharmacologic actions of vitamin D sterols.

When $1,25(OH)_2D_3$ is given orally to normal, vitamin D-replete volunteers who are ingesting their usual diet, the first effect is an increase in urinary calcium (Ca) apparent after 6-8 hours (2). In some but not all subjects, small but significant increments in serum Ca can be identified. Serum levels of immunoreactive parathyroid hormone (iPTH) decrease in some but not all patients; however, there is a decrease in the excretion of urinary cyclic AMP, an observation consistent with suppression of parathyroid hormone (PTH) secretion (Fig. 1). However, it is not yet certain whether PTH suppression occurs indirectly as a consequence of an increased

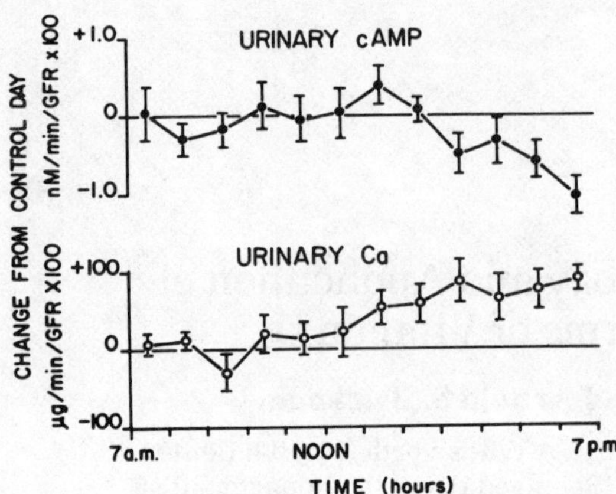

Figure 1. Changes in urinary Ca and urinary cyclic AMP (cAMP) in normal subjects given $1,25(OH)_2D_3$, 2.7 μg, at 7 a.m. Values represent the differences between observations on experimental and treatment day (mean ± SE). Reproduced from (2) with permission.

serum Ca or arises due to a direct effect of $1,25(OH)_2D_3$ on the parathyroid glands (3).

Measurements of intestinal absorption of ^{47}Ca, studied 2 to 7 days after the administration of $1,25(OH)_2D_3$, reveal a substantial augmentation of absorption (4). The administration of $1,25(OH)_2D_3$, 1-2 μg/day, increases intestinal absorption of Ca to values that are well above normal. It has been possible to establish a dose-response curve for changes in ^{47}Ca absorption in normal human subjects given oral $1,25(OH)_2D_3$ and $25(OH)D_3$ (Fig. 2).

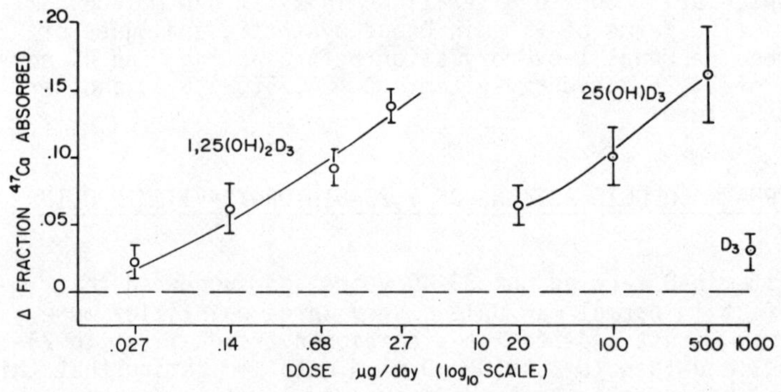

Figure 2. Changes in ^{47}Ca absorption in normal subjects given a vitamin D sterol for 7-9 days. Only a single dose of vitamin D was studied. (From 5,6)

The identification of an effect of $1,25(OH)_2D_3$ on bone resorption is far less certain in normal man. Fasting urinary Ca levels, an index of bone resorption, are increased only slightly; thus, the major increment in urinary Ca occurs during the post-prandial state, which suggests that increased intestinal absorption accounts for the hypercalciuria.

Metabolic balance studies indicate that net absorption of phosphorus (Pi) is also increased substantially by $1,25(OH)_2D_3$ (7). The effect of vitamin D compounds on intestinal absorption of magnesium (Mg) is not clearly defined: Metabolic balance studies reveal little or no effect of $1,25(OH)_2D_3$ on net Mg absorption. Observations that urinary excretion of Mg is not augmented by $1,25(OH)_2D_3$ support the validity of the metabolic balance results (8). On the other hand, measurements of Mg transport in perfused segments of human intestine suggest that active vitamin D sterols can substantially stimulate the transport of Mg (9). Such divergent results could arise if net absorption of Mg by one segment of the intestine was stimulated by $1,25(OH)_2D_3$ while its overall action of the entire gut is so small that a local action is obscured during metabolic balance studies.

Studies in experimental animals suggest that $1,25(OH)_2D_3$ has an effect on renal tubular handling of Pi and/or Ca (10). However, such effects have not been clearly identified in man. Changes in urinary Pi have not been observed unless they are accompanied by increased intestinal phosphorus absorption; moreover, effects of $1,25(OH)_2D_3$ on renal phosphorus handling could arise due to the suppression of PTH secretion secondary to increased Ca absorption (7). Hypercalciuria is common when $1,25(OH)_2D_3$ is given; however, there is no proof that this arises from a direct action of $1,25(OH)_2D_3$ on the renal tubule. Hypercalciuria could arise due to an increased filtered load of Ca and/or reduced renal tubular reabsorption of Ca arising from PTH suppression. The possibility that $1,25(OH)_2D_3$, itself, acts to increase urinary Ca is suggested from observations in the vitamin D-replete rat (11).

Vitamin D has an important effect on the maintenance of normal muscle metabolism and to stimulate bone mineralization. These effects have generally been observed only in vitamin D-deficient man or animals, and there are no data indicating that "physiologic" quantities of $1,25(OH)_2D_3$ have effects on muscle or stimulate bone formation in normal man. Such effects may only be detectable when vitamin D or $1,25(OH)_2D_3$ is given to a vitamin D-deficient man or animal.

Observations carried out in normal man and in patients with disorders of calcium metabolism provide some insight into the duration of action of $1,25(OH)_2D_3$. Thus, when treatment with $1,25(OH)_2D_3$ is discontinued in subjects receiving a constant diet, the decrement in urinary Ca toward the control rate provides an indication of the rate of decay of the biologic response to the sterol. In subjects receiving $1,25(OH)_2D_3$, urinary Ca fell on the first day treatment was stopped, and the decrement approximated a 1-component, exponential decay, with a half-life (t 1/2) for disappearance of the effect on urinary calcium of 1.5 to 2.7 days (Fig. 3) (5). This relatively short half-time for the disappearance of a biologic effect may indicate a therapeutic advantage of $1,25(OH)_2D_3$ over other vitamin D-sterols.

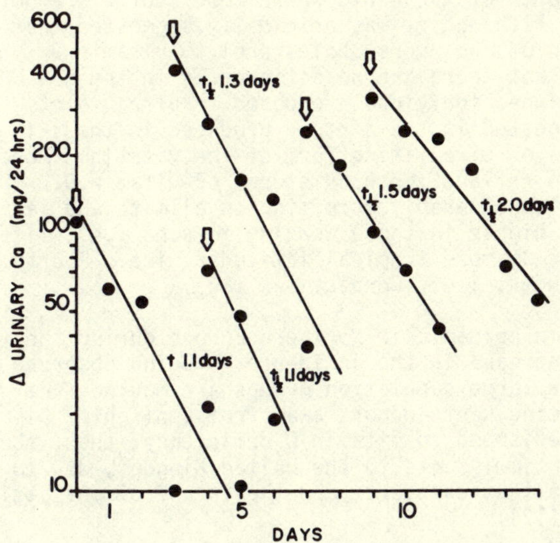

Figure 3. Dissipation of biologic effect of $1,25(OH)_2D_3$, measured by a return of urinary Ca to pre-treatment levels in 5 normal subjects. Arrows indicate the last day of treatment with the sterol. Regression lines are calculated from least squares. Adapted from (5).

CLINICAL CONDITIONS WITH REDUCED ACTION OF VITAMIN D

It is convenient to classify the clinical disorders with reduced action of vitamin D according to the affected step in vitamin D bioconversion metabolism (Table 1). Such a classification is arbitrary, particularly with certain disorders about which knowledge is fragmentary.

TABLE 1

CONDITIONS ASSOCIATED WITH DEFICIENT VITAMIN D ACTION

1. Reduced Availability of Vitamin D_2 or D_3
   a) Inadequate sunlight exposure
   b) Nutritional rickets or osteomalacia
      ? chappati diet
   c) Intestinal malabsorption of vitamin D_2 or D_3
      - malabsorption syndromes
      - subtotal gastrectomy
      - ileal bypass surgery
2. Reduced Availability of 25-OH-D_3
   a) Liver disease
   b) Anticonvulsant therapy:
      - phenobarbital, phenytoin, glutethemide
   c) Reduced enterohepatic circulation, malabsorption syndrome
   d) Nephrotic syndrome
3. Reduced Availability of 1,25$(OH)_2D_3$ (proven or probable)
   a) Vitamin D-dependency rickets
   b) Renal insufficiency
   c) Hypo- & pseudohypoparathyroidism
   d) Fanconi syndrome
   e) Osteomalacia associated with mesenchymal tumor
   f) Itai-itai disease (cadmium toxicity)
   g) Diphosphonate treatment
4. End-Organ Unresponsiveness
   a) Vitamin D-dependency rickets, Type II
   b) Steroid osteopenia
5. Disorders of Possible Relation to Vitamin D
   a) Diabetes mellitus
   b) Neonatal hypocalcemia
   c) Magnesium depletion
   d) Steroid-induced osteopenia
   e) Osteoporosis
   f) X-linked, hypophosphatemic rickets
   g) Fibrogenesis imperfecta ossium

Reduced Availability of Vitamin D_3 (Cholecalciferol)

Inadequate exposure to the appropriate wavelengths of ultraviolet light is a major cause of vitamin D deficiency in many parts of the world where food stuffs are not regularly supplemented with vitamin D. Although it was originally suggested that skin pigmentation retards the conversion of 7-dihydro-cholesterol to vitamin D_3, direct experimental evidence indicates that there are no differences in the generation of vitamin D_3 by dark or light-skinned individuals exposed to ultraviolet light. The vitamin D that is either ingested in the diet or produced in the skin is converted readily to 25$(OH)D_3$, the major circulating form of the vitamin. Moreover, plasma levels of 25(OH)D appear to reflect the body stores of vitamin D. Under steady-state conditions they vary considerably according to climate and latitude; thus, blood levels of 25(OH)D are higher in those working in sun, i.e., lifeguards, and blood levels are much higher at more tropical latitudes, i.e., Puerto Rico, compared to more northernly latitudes, i.e., Connecticut (12).

Historically, vitamin D-deficiency became epidemic in Northern Europe during the industrial revolution, and, a similar increase in the incidence is being observed in "developing" tropical countries where large population groups are moving from the country to cities and infants are being kept indoors away from sunlight. Dietary factors and cultural habits can predispose to vitamin D deficiency; thus, the incidence of rickets is high among Asian immigrants to the United Kingdom, due to the low intake of vitamin D, minimal sunlight exposure, and the intake of chapatti floor which reduces calcium absorption (13).

In the U.S., where foods are supplemented with vitamin D, nutritional rickets and osteomalacia are rare and are nearly limited to malabsorption syndromes, characterized by impaired fat absorption, and other disorders of vitamin D bioconversion. The malabsorption syndromes include sprue, celiac disease, regional enteritis, and intestinal by-pass surgery. The fact that $1\alpha(OH)D_3$, an analogue of $1,25(OH)_2D_3$, is highly effective in increasing Ca absorption and serum Ca levels in patients with celiac disease (14) indicates that impaired intestinal absorption of Ca is probably related to abnormal absorption of vitamin D rather than from an inability of the intestine to transport Ca.

Reduced Availability of 25-Hydroxy-Vitamin D_3

25-hydroxy-vitamin D_3, the circulating form of vitamin D, is generated in the liver from either endogenous or exogeneous vitamin D_3. Hepatocellular disease and abnormalities of the biliary system, which can impair fat absorption, may be associated with low plasma levels of 25(OH)D. The hepatic disorders most commonly associated with skeletal disease and osteomalacia are various disorders of the biliary tract, and in particular, biliary cirrhosis. Both defective hepatocellular function and defective enterohepatic circulation of both vitamin D_3 and $25(OH)D_3$ (15) contribute to the pathogenesis of this problem.

Despite the use of the anticonvulsant drugs, phenobarbital and phenytoin, for many years, a high incidence of osteomalacia in patients receiving these drugs was only noted within the last decade. Considerable data indicate that the induction of a hepatic microsomal enzyme system by these drugs leads to increased degradation of D_3 and $25(OH)D_3$ and causes the "vitamin D deficiency" (16). As many as 15-25% of institutionalized patients with severe and recurrent seizure disorders who receive these drugs may have evidence of osteomalacia. Plasma levels of 25(OH)D have been found to be low, but a puzzling finding has been the normal plasma levels of $1,25(OH)_2D_3$ (17). The mechanism whereby osteomalacia occurs despite the presence of normal levels of $1,25(OH)_2D$ is uncertain. The administration of vitamin D_2 or D_3, 800-1,600 IU/day (20-40 μg) can prevent or treat the disorder; substantially lower amounts of $25(OH)D_3$ have also been effective.

Recent reports indicate that significant urinary losses of 25(OH)D occur in association with the massive proteinuria and increased excretion of vitamin D-binding protein that exist in the nephrotic syndrome (18). Although the clinical significance of this abnormality is not certain, this defect of vitamin D metabolism probably accounts for the intestinal malabsorption of Ca, hypocalciuria and low levels of 25(OH)D observed in patients with the nephrotic syndrome.

Reduced Availability of 1,25-Dihydroxy-Vitamin D_3

The most closely regulated step in vitamin D metabolism is the production of $1,25(OH)_2D_3$. A variety of disorders, including destructive kidney disease (chronic renal failure), disorders of tubular function (i.e., Fanconi syndrome, cadmium toxicity), and certain hormonal disorders, characterized by an absence of trophic stimulation of the 25-hydroxy-D_3-1-hydroxylase, can lead to a deficiency of $1,25(OH)_2D_3$. An hereditary disorder, "vitamin D-dependency rickets", which responds to pharmacologic but not "replacement" doses of vitamin D_3, responds favorably to 0.5 to 1.0 μg/day of $1,25(OH)_2D_3$ (19). The latter condition is believed to arise due to a congenital absence of renal $25(OH)D_3$-1-hydroxylase.

Despite a large number of disorders that are characterized by low plasma levels of $1,25(OH)_2D$, osteomalacia is not uniformly present. Patients with hypoparathyroidism, pseudohypoparathyroidism and renal failure may have low or even undetectable levels of $1,25(OH)_2D_3$ (17) and yet not exhibit osteomalacia. Certain features of conditions with low plasma levels of $1,25(OH)_2D$ and other conditions associated with osteomalacia are shown in Table 2. It would seem that a low level of 1,25-$(OH)_2D$ and hypophosphatemia must both be present before rickets or osteomalacia occurs.

TABLE 2

BIOCHEMICAL FEATURES AND VITAMIN D STATUS IN DISORDERS WITH ALTERED VITAMIN D METABOLISM AND PRESENTING WITH OR WITHOUT OSTEOMALACIA

	Serum Levels					Skeletal Features	
	25(OH)D	$1,25(OH)_2D$	iPTH	Pi	Ca	OM	OF
Vitamin D deficiency	↓↓	↓↓	↑↑	↓↓	↓↓	+++	++
Chronic renal insufficiency	N	↓↓	↑↑↑	↑↑(N)	↓	+	+++
Mesangial tumor with osteomalacia	N	↓↓	N	↓↓	N	+++	±
Hypoparathyroidism	N	↓↓	↓↓	↑	↓↓	0	0
Pseudohypoparathyroidism	N	↓↓	↑↑	↑	↓↓	0	++
Sex-linked hypo-phosphatemia (VDRR)	N	N	↓(N)	↓↓	N	+++	±
Vitamin D-dependency I	N	↓↓	↑↑	↓↓	↓↓	+++	++
Vitamin D-dependency II	N	↑↑	↑↑	↓↓	↓	+++	+
Anticonvulsant osteomalacia	↓↓	N	↑	↓	↓	++	+
Phosphate depletion	N	↑	↓	↓	N	+++	0

iPTH - immunoreactive PTH; OM - osteomalacia (mineralizing defect); OF - osteitis fibrosa (fibrosteoclasia)

In hypo- and pseudohypoparathyroidism, plasma levels of $1,25(OH)_2D$ are low, probably due to the lack of trophic stimulus of PTH; the hyperphosphatemia may also inhibit generation of $1,25(OH)_2D_3$. From a therapeutic standpoint, $1,25(OH)_2D_3$ or its synthetic analogue, $1\alpha(OH)D_3$, have been very useful in the management of patients with hypocalcemia due to hypoparathyroidism or pseudohypoparathyroidism. Quantities of $1,25(OH)_2D_3$ of 0.5-2.0 μg/day can increase serum Ca to normal in these patients (20) (Fig. 4). Unpublished observations from our laboratory suggest that the dose of $1\alpha(OH)D_3$ required to stimulate Ca absorption may be 3 to 6 times greater than that of $1,25(OH)_2D_3$ in patients with hypo- and pseudohypoparathyroidism. These conditions are characterized by lack of effective PTH action, raising the possibility that PTH could play a role in the 25-hydroxylation of $1\alpha(OH)D_3$ to 1,25-$(OH)_2D_3$. Whether PTH may also stimulate the 25-hydroxylation of D_3 to $25(OH)D_3$ is not certain at this time.

Chronic renal failure is a condition where alterations in vitamin D metabolism have been implicated as a cause of altered Ca and Pi homeostasis. Since the kidney produces $1,25(OH)_2D_3$ from its precursor, it has been attractive to speculate that impaired renal production of this sterol accounts for impaired Ca absorption and osteomalacia in uremia. Certain observations, reviewed in detail elsewhere (21) are consistent with this premise; thus, nephrectomized rats fail to produce $1,25(OH)_2D_3$, a decrease in the intestinal Ca transport correlates with reduced intestinal localization of $1,25(OH)_2D_3$ in a model of renal failure in the chick, and there is absent generation of $1,25(OH)_2D_3$ in uremic humans given ^3H-labeled $25(OH)D_3$. Moreover, there is recovery of the ability to generate $1,25(OH)_2D_3$ from ^3H-$25(OH)D_3$

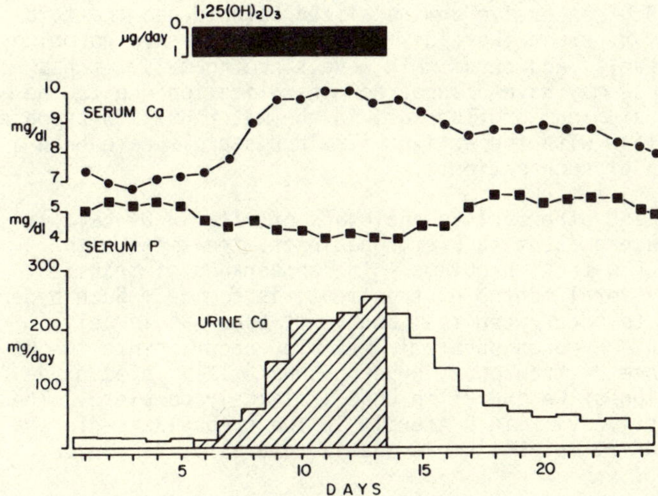

Figure 4. Changes in serum Ca and P and urinary Ca in a patient with surgical hypoparathyroidism given $1,25(OH)_2D_3$, 1.0 μg/day by mouth for 8 days. The rapid response and quick dissipation of effect are apparent.

following successful renal transplantation. Each laboratory carrying out measurements of plasma $1,25(OH)_2D$ has found very low or undetectable levels in patients with renal insufficiency.

The administration of $1,25(OH)_2D_3$, 0.25-1.0 μg/day, raises the absorption of Ca in uremic patients to normal or even supranormal values, and serum Ca increases particularly if it is low (4). Such observations have prompted the therapeutic evaluation of $1,25(OH)_2D_3$ and other active vitamin D sterols for the management of Ca and Pi abnormalities in uremia. Although one might expect that osteomalacic skeletal lesions might be the most sensitive to such treatment, the skeletal disease and the biochemical features which show the most dramatic improvement toward normal are the features of secondary hyperparathyroidism. In such patients, the administration of $1,25(OH)_2D_3$, 0.5-1.5 μg/day, lowers elevated alakaline phosphatase values and returns serum iPTH toward normal. Bone biopsy findings of osteitis fibrosa or fibroosteoclasia are improved. In such cases, the fall in serum iPTH is usually inversely related to a rise in serum Ca; and such results do not indicate that $1,25(OH)_2D_3$ has direct effect to inhibit PTH secretion. However, it is possible that $1,25(OH)_2D_3$ may sensitize the parathyroid glands so they may respond to any given change in Ca level in extracellular fluid. Thus, one can achieve much greater suppression of secondary hyperparathyroidism following the administration of $1,25(OH)_2D_3$ than occurs when serum Ca is increased to hypercalcemic levels by hemodialysis using a high dialysate Ca level. In many cases, treatment leads to an improvement of subperiosteal resorption as seen in X-ray. The patients commonly note a striking improvement in their symptoms, with reversal of bone pain and improved muscle strength (22). Such observations lend support to the view that altered vitamin D metabolism plays a major role in the pathogenesis of secondary hyperparathyroidism in uremia. In the United States, osteomalacia is infrequent in uremic patients; the presence of normal or elevated levels of serum Pi in uremic patients may be a factor that contributes to this low prevalence. Thus, the increase in serum Pi level may "protect" these patients from the mineralizing defect of osteomalacia.

When uremic patients do exhibit a mineralizing defect, treatment with $1,25(OH)_2D_3$ led to a favorable response with a substantial decrease in the volume of bone occupied by unmineralized osteoid. Another group of uremic patients have been identified with an unusual type of mineralization defect. The bone of these patients is characterized by the presence of excessive unmineralized osteoid and the total absence of evidence of secondary hyperparathyroidism. They usually have normal or even mildly elevated serum Ca levels, and serum iPTH levels are normal. It has been suggested that these patients may have disordered mineralization due to the presence of substances, such as aluminum or fluoride, in the water used to prepare dialysate (23). Therapeutic trials with the active vitamin D sterols have been a useful aid in the identification of such patients.

Although $1,25(OH)_2D_3$, $1\alpha(OH)D_3$, and other active analogues of vitamin D_3 have improved the severe skeletal disease present in uremic patients, treatment with these patent sterols has not been without problems. The appearance of transient hypercalcemia, sometimes after several months of treatment, is common. Such hypercalcemia is particularly likely to occur when the "abnormal" bone has largely "completed" its remineralization. Hyperphosphatemia may also occur; since the active vitamin D sterols can enhance Pi transport, hyperphosphatemia is also prone to occur when the rapid deposition of Ca and Pi in bone is largely complete. The value of $1,25(OH)_2D_3$ or other active vitamin D sterols in the prophylaxis of renal osteodystrophy and altered Ca and Pi metabolism in uremia remains to be proven.

Altered vitamin D metabolism may occur in other disorders affecting the kidney. The Fanconi syndrome, which is characterized by proximal renal tubular dysfunction with impaired absorption of amino acids, Pi, bicarbonate, glucose, and uric acid, is commonly accompanied by hypocalcemia and osteomalacia. In an animal model of the Fanconi syndrome produced by the administration of maleic acid, Brewer et al. (24) suggested that altered mitochondrial function in tubular cells may give rise to the altered tubular transport and impaired conversion of $25(OH)D_3$ to $1,25(OH)_2D_3$. In children with the Fanconi syndrome, Brewer et al. (25) further found a failure to generate radiolabeled $1,25(OH)_2D_3$ from administered ^3H-labeled $25(OH)D_3$. Such observations raise the possibility that the abnormal metabolism of Ca, Pi and bone seen in the Fanconi syndrome may arise because of defective biconversion of vitamin D.

The co-existence of osteomalacia and severe hypophosphatemia with certain mesenchymal tumors has long been puzzling; even more surprising has been the reversal of renal Pi wasting and improvement of bone disease following surgical extirpation of the tumor. Drezner and Feinglas (26) reported low plasma levels of $1,25(OH)_2D$ in a patient with this syndrome; treatment with $1,25(OH)_2D_3$, 1-3.0 μg/day, reduced urinary Pi losses and improved the hypophosphatemia. They suggest that the tumor impairs the generation of $1,25(OH)_2D_3$. Other studies (27) suggest the tumor may generate a factor which impairs renal Pi transport leading to the "phosphate diabetes"; the abnormal vitamin D metabolism could be a secondary event.

"Itai-itai", a disorder characterized by hypocalcemia and osteomalacia, is endemic to certain areas of Japan; its appearance seems to be related to environmental exposure to cadmium. Experimental cadmium exposure impairs intestinal Ca absorption and reduces the deposition of Ca in bone (28). A relationship between this disorder and altered biotransformation of vitamin D has not yet been established.

Steroid Osteopenia. The possibility that treatment with large doses of glucocorticoids may block the action or metabolism of vitamin D and produce the osteopenia or osteoporosis of steroid therapy has long been suspected. The observations of Klein et al. (29) that treatment with $1,25(OH)_2D_3$, 0.4 μg/day, corrects the malabsorption of Ca produced by steroid treatment suggests that altered biotransformation of vitamin D may contribute to abnormal Ca metabolism that occurs with steroid treatment. It is possible that treatment with appropriate vitamin D analogues may be useful in the treatment and/or prevention of steroid-induced osteopenia.

There are a number of clinical disorders where the relationship to vitamin D metabolism or action is not established. The syndrome, "X-linked, hypophosphatemic rickets" or Vitamin D-resistant Rickets, has shown no improvement following treatment with $1,25(OH)_2D_3$, $1\alpha(OH)D_3$, or $25(OH)D_3$. The patients respond with an increase in Ca absorption, but the renal phosphate wasting is not improved.

The possibility has been raised that osteoporosis, either post-menopausal, senile, or idiopathic, may arise from altered vitamin D metabolism (30). Data from Scandinavia suggest that treatment with an active vitamin D sterol may slow the progressive bone loss, as measured by photon absorptiometry (31). However, results from metabolic balance techniques (32) have failed to reveal a positive Ca balance during treatment of osteoporosis with $1,25(OH)_2D_3$. Nonetheless, Riggs and Gallagher (30) suggest that minor abnormalities of vitamin D and estrogen metabolism may exist in osteoporosis. Whether treatment with one of the active vitamin D sterols will be beneficial for osteoporosis remains uncertain.

OTHER ANALOGUES OF VITAMIN D

Several vitamin D sterols other than $1,25(OH)_2D_3$ have a "hydroxyl" group that is analogous to that in the 1α-position. Thus, $1\alpha(OH)D_3$, 5,6-trans-D_3, 25(OH)-5,6-trans-D_3 and dihydrotachysterol (DHT) are synthetic sterols which act without 1α-hydroxylation. The "A" ring is rotated 180° in the last 3 sterols, leaving the 3-hydroxyl group, typical of the basic vitamin D structure, in a "pseudo-1-α" configuration. Of these compounds, DHT has been used extensively; it is converted to 25-OH-DHT before acting, and doses of 0.125 to 1.0 mg are usually needed for a therapeutic effect. There is 25-hydroxylation of $1\alpha(OH)D_3$ and 5,6-trans-D_3 before these sterols exert their biological response. $1\alpha(OH)D_3$, the most potent of the synthetic sterols, is available in Europe and the U.K., and proceedings of a symposium on its use has recently appeared (33). Whether these sterols will have any advantages over $1,25(OH)_2D_3$ remains, at this time, unknown.

$25(OH)D_3$, the circulating form of D , may have biologic properties in promoting bone mineralization that are distinct from those of $1,25(OH)_2D_3$ (34). Whether this action is specific for $25(OH)D_3$ or occurs because the patients could produce both $1,25(OH)_2D_3$ and $24,25(OH)_2D_3$ is not certain. Further studies are needed to clarify whether 25-OH-D_3 may possess unique actions or exerts special actions because it may form $1,25(OH)_2D_3$, $24,25(OH)_2D_3$ or $1,24,25(OH)_3D_3$.

24,25-DIHYDROXY-VITAMIN D_3

This sterol, generated in the kidney and also in the intestine, normally circulates with a plamsa level that is one-tenth $25(OH)D_3$ but 100 times higher than $1,25(OH)_2D_3$. $24,25(OH)_2D_3$ has been considered of little biologic importance, since it is less active than $1,25(OH)_2D_3$ in stimulating intestinal Ca absorption or in mobilizing bone Ca in animals; moreover, these effects are abolished by nephrectomy (35). Thus, $24,25(OH)_2D_3$ has been considered a metabolic degradation product of D_3 which requires 1-hydroxylation to 1,24,25-trihydroxy-vitamin D_3 to exert a biological action (35).

However, there may be qualitative differences between actions of various vitamin D sterols: thus, $24,25(OH)_2D_3$ increases serum Ca in vitamin D-deficient animals with low serum Ca but causes a decrease in serum Ca in normocalcemic animals (36). Bordier et al. (34) suggested that 25(OH)D was more active in inducing bone mineralization than either $1,25(OH)_2D_3$ or $24,25(OH)_2D_3$, given alone. However, combined therapy with the last 2 sterols improved mineralization to normal. Henry et al. (37) reported that $1,25(OH)_2D_3$ requires the presence of $24,25(OH)_2D_3$ to cause regression of hyperplastic parathyroid glands in vitamin D depleted chicks despite a similar increment in serum Ca in chicks given 1,25$(OH)_2D_3$, alone. Also, 24,25-$(OH)_2D_3$ increased Ca absorption without changing urinary Ca in normal man, observations different from those noted with 1,25$(OH)_2D_3$ treatment (38). Canterbury et al. noted a decrease in plasma iPTH levels after a bolus injection of 24,25$(OH)_2D_3$, while 1,25$(OH)_2D_3$ led to a transient increase in iPTH levels (39).

In patients with advanced renal failure, we found a fall in serum Ca and rise in alkaline phosphatase after treatment with 24,25$(OH)_2D_3$, 2.0 μg/day and 4.0 μg/day. Values returned to the pre-treatment levels 2 weeks after the sterol was discontinued. Intestinal absorption of ^{47}Ca was unaffected by 24,25$(OH)_2D_3$, and serum iPTH rose as serum Ca fell. Such observations and those reported by Bordier et al. (34) are consistent with the stimulation of osteoblastic activity, accounting for a higher alkaline phosphatase and Ca deposition in the bone, which could lead to a lower serum Ca. The lack of decrease in serum iPTH in response to 24,25$(OH)_2D_3$ differs from the results of Canterbury et al. (39). Uremic patients almost certainly have low 1,25$(OH)_2D$ levels, and both 1,25$(OH)_2D_3$ and 24,25$(OH)_2D_3$ may be required to suppress PTH secretion. The role of 24,25$(OH)_2D_3$ in normal physiology is uncertain, but we would speculate that 1,25$(OH)_2D_3$ and 24,25$(OH)_2D_3$ may interact in Ca, Pi and bone homeostasis (Fig. 5). 1,25$(OH)_2D_3$ is a potent stimulus for the production of 24,25$(OH)_2D_3$; thus, PTH may stimulate the production of 1,25$(OH)_2D_3$, which in turn stimulates production of 24,25$(OH)_2D_3$. The two sterols together may cause feedback suppression of PTH secretion. In bone, most data indicates that 1,25$(OH)_2D_3$ promotes bone resorption, particularly in the presence of PTH; if 24,25$(OH)_2D_3$ can stimulate bone formation this would provide a mechanism for the well-known coupling of the resorption and formation processes in bone remodeling. This suggestion is highly speculative, and further data are needed to clarify a possible role of 24,25$(OH)_2D_3$ in Ca and Pi homeostasis.

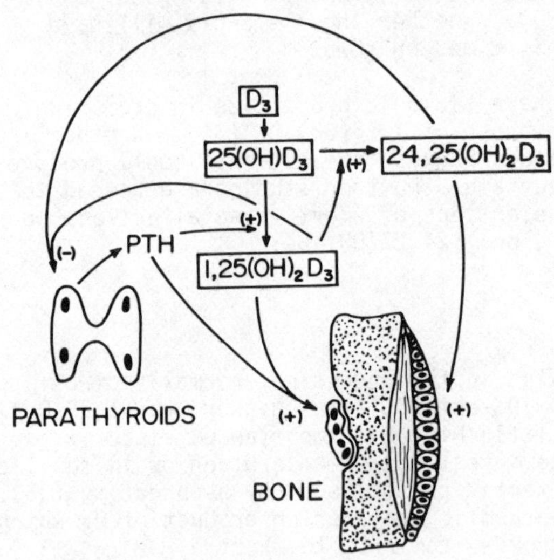

Figure 5. Schematic showing how 24,25-$(OH)_2D_3$, which is stimulated by 1,25-$(OH)_2D_3$, may interact with the latter to suppress PTH secretion and also to lead to coupled resorption {PTH + 1,25-$(OH)_2D_3$} and formation {24,25$(OH)_2D_3$} of bone.

SUMMARY

The identification and synthesis of the active form of vitamin D have provided valuable tools for the investigation of the role of vitamin D in normal Ca homeostasis in man. Insight into the pathogenesis of certain disorders of Ca metabolism has been obtained, and, these active vitamin D sterols may have therapeutic advantages over their precursor, D_3, particularly regarding the predictability of response and rapid reversibility of an effect, if toxicity occurs.

ACKNOLWEDGEMENTS

Supported, in part, by USPHS Grant AM-14750 and by VA Research Funds. We are deeply appreciative for the valuable assistance provided by Patti Kentor in the preparation of this manuscript.

REFERENCES

1. M.R. Haussler and T. McCain. N. Engl. J. Med. 297:974-983 and 1041-1050, 1977.

2. F. Llach, J.W. Coburn, A.S. Brickman, et al. J. Clin. Endocrinol. Metab. 44: 1054-1060, 1977.

3. B.S. Chertow, D.J. Baylink, J.F. Wergedal, et al. J. Clin. Invest. 56:668-678, 1975.

4. A.S. Brickman, J.W. Coburn, S.G. Massry, et al. Ann. Int. Med. 80:161, 1974.

5. A.S. Brickman, J.W. Coburn, G.R. Friedman, et al. J. Clin. Invest. 57:1540-1547, 1976.

6. I.H. Colodro, A.S. Brickman, J.W. Coburn, et al. Metabolism 27:745-753, 1978.

7. A.S. Brickman, D.L. Hartenbower, A.W. Norman, et al. Am. J. Clin. Nutr. 30: 1064, 1977.

8. J.W. Coburn, D.L. Hartenbower, A.S. Brickman, et al. In: Calcium Metabolism in Renal Failure and Nephrolithiasis, edited by David, D.S., New York, John Wiley & Sons, 1977, p. 77.

9. P.G. Brannan, P. Vergne-Marini, C.Y. Pak, et al. J. Clin. Invest. 57:1412-1418, 1976.

10. J.W. Coburn, R.I. Saltzman and S.G. Massry. In: Methods in Pharmacology, 4A, edited by Martinez-Maldanado, Plenum Press, New York, pp. 227-267, 1976.

11. J.P. Bonjour, U. Treschsel, H. Fleisch, et al. Am. J. Physiol. 229:402-408, 1975.

12. R. Neer, M. Clark, V. Friedman, et al. In: Vitamin D. Biochemical, Chemical and Clinical Aspects Related to Calcium Metabolism, edited by A.W. Norman, K. Schaefer, J.W. Coburn, H.F. DeLuca, D. Fraser, H.G. Grigoleit, D. van Herrath. de Gruyter, Berlin, 1977, pp. 595-606.

13. K.M. Goel, R.W. Logan, G.C. Arneil, et al. Lancet 1:1141-1148, 1976.

14. P. Bordier, M.M. Pechet, R. Hesse, et al. N. Engl. J. Med. 291:866-871, 1974.

15. S.B. Arnaud, R.S. Goldsmith, P.W. Lambert, et al. Proc. Soc. Exp. Biol. Med. 149:570-572, 1975.

16. T.J. Hahn and L.V. Avioli. Arch. Int. Med. 135:997-1000, 1975.

17. M.R. Haussler, M.R. Hughes, T.A. McCain, et al. Calci. Tis. Res. 22:1-18, 1977.

18. H. Schmidt-Gayk, C. Grawunder, W. Tschöpe, et al. Lancet 2:105-108, 1977.

19. D. Fraser, S.W. Kooh, H.P. Kind, et al. N. Engl. J. Med. 289:817-822, 1973.

20. J.W. Coburn, A.S. Brickman, K. Kurokawa, et al. Endocrinology 1973, edited by S. Taylor, W. Heineman, pp. 385-392, 1974.

21. J.W. Coburn, D.L. Hartenbower and A.S. Brickman. Am. J. Clin. Nutr. 29:1283, 1976.

22. J.W. Coburn, A.S. Brickman, D.J. Sherrard, et al. In: Vitamin D. Biochemical, Chemical and Clinical Aspects Related to Calcium Metabolism. op cit pp. 657-666.

23. M.K. Ward, T.G. Feest, H.A. Ellis, et al. Lancet 1:841-845, 1978.

24. E.D. Brewer, H.C. Tsai, K.S. Szeto, et al. Kidney Internat. 12:244-252, 1977.

25. E.D. Brewer, H. Tsai, R.C. Morris Jr. In: Vitamin D. Biochemical, Chemical and Clinical Aspects Related to Calcium Metabolism. op cit pp. 937-949.

26. M.K. Drezner and M.N. Feinglas. J. Clin. Invest. 60:1046-1053, 1977.

27. L.C. Aschinberg, L.M. Solomon, P.M. Zeis, et al. J. Pediatr. 91:56-60, 1977.

28. M. Ando, Y. Sayato, M. Tonomura, et al. Toxicol. Appl. Pharm. 39:321-327, 1977.

29. R.G. Klein, S.B. Arnaud, J.C. Gallagher, et al. J. Clin. Invest. 60:253-259, 1977.

30. B.L. Riggs and J.C. Gallagher. In: Vitamin D. Biochemical, Chemical and Clinical Aspects Related to Calcium Metabolism. op cit pp. 639-648.

31. B. Lund, R.B. Anderson, M.S. Christensen, et al. In: Vitamin D. Biochemical, Chemical and Clinical Aspects Related to Calcium Metabolism. op cit pp. 399-406.

32. M. Davies, E.B. Mawer and P.H. Adams. J. Clin. Endocrinol. Metab. 45:199-208, 1977.

33. M. Peacock, editor. Clin. Endocrinol. 7 (suppl):1S-246S, 1977.

34. P. Bordier, H. Rasmussen, P. Marie, et al. J. Clin. Endocrinol. Metab. 46: 284-294, 1978.

35. I.T. Boyce, J.L. Omdahl, R.W. Gray, et al. J. Biol. Chem. 248:4174-4180, 1973.

36. L. Miravet, J. Redel, M.L. Queille, et al. Calcif. Tis. Res. 21:145-152, 1976.

37. H. Henry, A. Taylor and A.W. Norman. J. Nutr. 107:1918-1926, 1977.

38. J.A. Kanis, T. Cundy, M. Bartlett, et al. Brit. Med. J. 1:1382-1386, 1978.

39. J.M. Canterbury, S. Lerman, A.J. Claffin, et al. J. Clin. Invest. 61:1375-1383, 1978.

40. Y. Tanaka and H.F. DeLuca. Science (Wash., D.C.) 183:1198, 1974.

Biosynthesis, Metabolism, and Mode of Action of Parathyroid Hormone

John T. Potts, Jr.

Department of Medicine, Harvard Medical School, and Endocrine Unit, Massachusetts General Hospital, Boston, Massachusetts 02114, U.S.A.

INTRODUCTION

A clearer understanding of the physiological role of parathyroid hormone in calcium homeostasis and bone metabolism should ultimately emerge from a detailed knowledge of the biosynthesis, secretion, metabolism, and mode of action of the hormone. The biological role of parathyroid hormone in calcium and phosphate metabolism and the exchanges of these mineral ions with bone is obviously linked to an understanding of initial and sequential consequences of the interaction of the hormone with receptors in bone and kidney. The overall homeostatic function of the hormone is also, however, determined by the rate of delivery of the active molecular species of the hormone with these receptors; this latter process is regulated by the production (biosynthesis and secretion) and the modification or clearance of the secreted peptide from blood and tissue fluids (metabolism).

The hormone functions in the body to maintain the concentration of soluble calcium and phosphate in the extracellular fluid within the desired range via its direct actions in promoting mobilization of calcium and phosphate from bone mineral and by altering reabsorption of calcium and phosphate from renal tubular fluid (1). An additional, physiologically important, action of the hormone is to promote increased intestinal absorption of calcium by increasing the conversion of 25-hydroxyvitamin D to the active form, 1,25-dihydroxyvitamin D; the latter stimulates intestinal calcium transport (2). Extracellular fluid calcium in turn, by negative feed-back control on the parathyroid glands, regulates the release of hormone. Increased calcium suppresses and decreased calcium stimulates hormone secretion and eventually to corresponding changes in hormone biosynthesis (3).

The structure of the 84-amino acid parathyroid hormone molecule has been determined for the bovine, porcine and human species. The human sequence has been recently completed by Dr. Henry Keutmann in the Endocrine Unit, Massachusetts General Hospital, in collaboration with Dr. Jeffrey O'Riordan and colleagues at Middlesex Hospital in London. Figure 1 illustrates the structural similarities and differences among the hormone molecules from the three species; the porcine and bovine each differ from the human at eleven sequence positions, most markedly through the middle region of the molecule.

It is the purpose of this review to focus on recent advances in the biosynthesis, metabolism, and mode of action of the hormone.

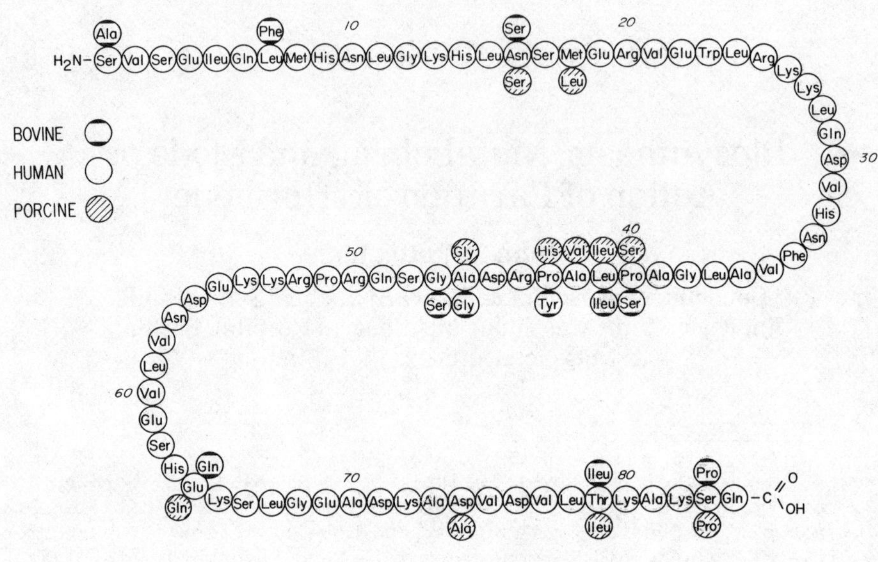

Fig. 1. The amino acid sequence of parathyroid hormone

BIOSYNTHESIS

Studies *in vitro* of the biosynthesis of parathyroid hormone have revealed that it arises by way of two sequential enzymic cleavages from a larger protein precursor. Pre-proparathyroid hormone, a polypeptide of 115 amino acids, is first modified to an intermediate proparathyroid hormone consisting of 90 amino acids. The prohormone then undergoes final cleavage to the hormone. Current evidence suggests that the initial product, pre-proparathyroid hormone, belongs to a class of biosynthetic precursors common to all proteins destined for export from their cells of origin and that the hydrophobic amino-terminal leader sequence, specific for the precursor, functions in the transport of the newly-synthesized hormone into the intra-cisternal space of the endoplasmic reticulum. Enzymic conversion of pre-proparathyroid hormone to proparathyroid hormone takes place within seconds of its synthesis. This process ensures the accurate segregation of the hormone into the secretory pathway. The cellular functions of the prohormone are unknown. Its conversion to parathyroid hormone occurs in the Golgi apparatus of the cell 12-15 minutes after its initial synthesis in the rough endoplasmic reticulum. It is speculated that the prohormone may be involved in the transport of the hormonal polypeptide to the Golgi and may function in some way in the formation of the secretion granules (Fig. 2).

The existence of a biosynthetic pathway that involves the formation of parathyroid hormone by cleavages of larger precursors raises the possibility that the cleavage steps may be regulated events and may serve as control points wherein the cells can vary the amounts of hormone produced in response to extracellular demands.

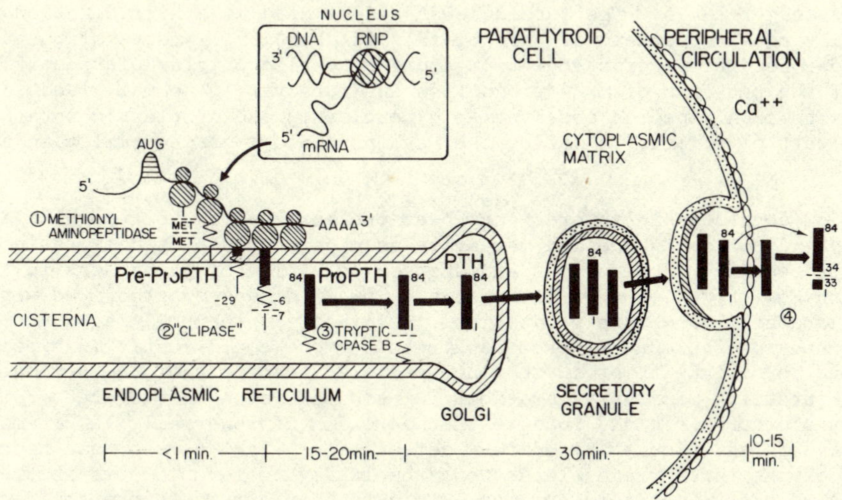

Fig. 2. The proposed intracellular pathway of the biosynthesis of parathyroid hormone

In Fig. 2, pre-proparathyroid hormone (Pre-ProPTH), the initial product of synthesis on the ribosomes, is converted into proparathyroid hormone (ProPTH) by removal of ① the amino-terminal methionyl residues and ② the amino-terminal sequence (-29 to -7) of 23 amino acids during and/or within seconds after synthesis, respectively. The conversion of Pre-ProPTH probably occurs during transport of the polypeptide into the cisterna of the rough endoplasmic reticulum. By 20 minutes after synthesis, ProPTH reaches the Golgi region and is converted into PTH by ③ removal of the amino-terminal hexapeptide. PTH is stored in the secretory granule until released into the circulation in response to a fall in the blood concentration of calcium. The time needed for these events is given below the schema.

One mechanism by which amounts of hormone available for secretion may be regulated over a relatively short time is the modulation of an intracellular degradative pathway; studies of hormone production *in vitro* have identified a calcium-sensitive degradative pathway.

Some information is available concerning effects of changes in extracellular calcium concentration on biosynthesis *per se*. Certain observations suggest that regulation of biosynthesis may occur at the transcriptional level rather than at the more rapid translational level. It has not yet been possible to measure directly rates of mRNA formation (or turnover) in parathyroid glands. Recent progress, however, in studies of the chemistry and biology of the specific DNA's and RNA's (genes and messages, respectively) that code for parathyroid hormone and other hormones and cellular protein products, provide great promise as tools uniquely useful for the study of transcriptional and translational control.

Kronenberg and colleagues, in a collaborative study between our laboratory and that of Alexander Rich at Massachusetts Institute of Technology, have successfully prepared the DNA complementary in sequence to the parathyroid hormone mRNA (cDNA). Inasmuch as cDNA molecules, prepared with radioactive nucleotides, will associate tightly and specifically with the corresponding mRNA, methods have been developed (hybridization assays) to permit quantitative measurements of specific secretory protein messenger RNA's. The labeled cDNA can be used as a hybridization probe to specifically measure amounts of hormone mRNA in cellular extracts of parathyroid glands under conditions of stimulation and suppression of glandular activity. Details of the patterns of mRNA production and turnover in normal glands can be compared with mRNA kinetics observed in hyperplastic and neoplastic parathyroid glands as part of efforts to define the pathophysiology of abnormal glandular activity.

A subject of considerable interest has been whether any of the recognized disorders of parathyroid hormone secretion may arise as a consequence of defects in the cellular processing of precursors and whether detection of precursors in the circulation may prove to be of diagnostic value in diseases associated with abnormal parathyroid hormone production. A failure in the cellular conversion of pre-proparathyroid hormone to proparathyroid hormone could result in hypoparathyroidism due to an impairment in the intracellular transport of the hormone into the secretory pathway. Similarly, a defect in the conversion of proparathyroid hormone to parathyroid hormone could lead to the secretion of abnormally large amounts of prohormone. In addition to hereditary defects, one might expect that neoplastic disorders of the parathyroid glands would be likely to favor errors in the processing of biosynthetic precursors. To date, however, attempts to identify release of precursors from both normal and neoplastic parathyroid tissues both *in vivo* and *in vitro* have not succeeded. For a recent review of this topic plus a current list of primary references of multiple laboratory groups, see Ref. (4).

METABOLISM

It is now clear that circulating endogenous parathyroid hormone is heterogeneous; in addition to the intact hormonal peptide, at least one principal fragment of the hormone devoid of the biologically-active region at the amino terminus is present in blood at concentrations that are considerably greater than that of the intact hormone. The metabolic fate of parathyroid hormone once the hormone enters the general circulation has proven to be much more complex than was previously appreciated. A highly-specific, high-capacity, but saturable enzymic activity involved in proteolysis and peripheral metabolism of parathyroid hormone has been identified. Present evidence points toward the liver and the kidney as organs principally concerned with the post-secretory modification of parathyroid hormone. Other evidence has led to the suggestion that fragments of parathyroid hormone may be secreted from the gland as well as the intact hormone. For a recent review of this topic plus a current list of primary references of multiple laboratory groups, see Ref. (5).

It is at present uncertain to what degree peripheral metabolism versus gland secretion accounts for the heterogeneity of circulating immunoreactive parathyroid hormone. What seems clear is that the principal circulating form of immunoreactive hormone is biologically inert since it consists of the middle and carboxyl two-thirds of the hormone rather than the intact native peptide, the biologically-active form of the molecule. These findings concerning the heterogeneity of circulating hormone are of great fundamental interest but, at the same time, pose considerable difficulties in the application of immunoassays to a quantitative description of the normal rates of secretion and turnover of biologically-active parathyroid hormone in man and in other animal species.

Since present measurements of immunoreactive hormone reflect primarily the rate of production and turnover of the inactive carboxyl fragments rather than of the active hormone, detailed analysis of the process of hormone metabolism and the nature of the heterogeneity of circulating hormone is clearly essential for proper interpretation and improved specificity of immunoassays. Our recent efforts to understand the heterogeneity of circulating hormone led us to development of several useful methods. Immunoassays made specific for detection of limited regions of the hormone sequence were extensively developed in our laboratory. These assays were made possible, in part, because of the multiple fragments of hormone synthesized. Application of these assays to measurements of circulating hormone were combined with an alternate technique which we developed, microsequence analyses of fragments of radioiodinated hormone generated by peripheral metabolism after injection into the circulation. Because of the extraordinary sensitivity of the latter technique, chemical identification of the nature of hormone fragments is possible using concentrations of hormone that are in the physiologic range. This method will also be extensively applied in newer studies proposed in the present application as the equivalent of a specific enzyme assay for the enzymes involved *in vivo* in proteolytic modification of hormone; that is, the chemical nature of fragments generated in *in vitro* studies can be compared for their authenticity to those resulting from hormonal metabolism *in vivo*.

Recognition of the sites of cleavage within parathyroid hormone that occur during peripheral metabolism has led us to the recent synthesis of a fragment, bPTH-(28-48) (Fig. 3), that spans the region involved in proteolytic cleavage of parathyroid

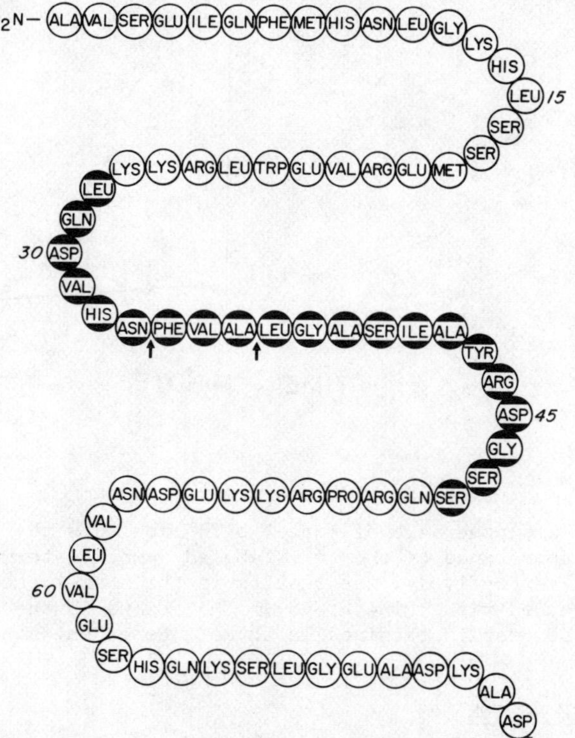

Fig. 3. The amino acid sequence of bovine parathyroid hormone

In this figure, shaded residues depict the sequence of the synthetic hormone fragment bPTH-(28-48). Arrows indicate the principal sites of cleavage of intact hormone *in vivo*.

hormone, thereby providing a fragment that is a competitive inhibitor of the peripheral metabolism of parathyroid hormone (Fig. 4). This approach represents an important step in understanding the metabolic significance of the peripheral metabolism of the hormone in the overall expression and regulation of parathyroid hormone's homeostatic effects on calcium and phosphate transport. The overall results from multiple laboratories and methods now available show considerable promise for current efforts to more accurately measure biologically-active hormone in tissue fluids, through development of more specific assays, alternate immunological, as well as bioreceptor, assays, using chemically-synthesized fragments of the hormone. These developments are critical to linking biosynthesis to secretion and turnover of hormone in tissue fluids in order to understand hormone action quantitatively by evaluation of the minute-to-minute concentration of biologically-active hormone.

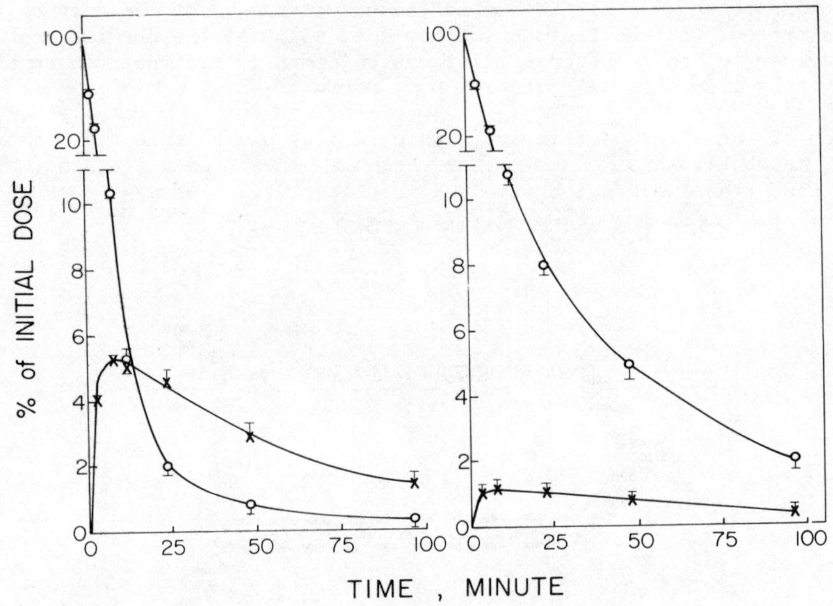

Fig. 4. Cleavage chromatogram

Figure 4 shows the disappearance of intact bPTH labeled with ^{125}I (o) and the appearance and disappearance of the ^{125}I-labeled carboxyl-terminal fragments (x). A and B represent, respectively, the results in the sham-treated control rats and in the rats treated with bPTH-(28-48). Each value is the mean ± standard error of the mean of results obtained in three independent studies.

MODE OF ACTION

Much progress has also been made in understanding the mode of action of parathyroid hormone at the cellular level in target tissues, a final process whose definition is essential in order to fully describe the overall role of parathyroid hormone in calcium and phosphate homeostasis. Multiple approaches are now available to critically define the essential details of hormone/receptor interaction, cyclic AMP production, and phosphokinase activity, the processes speculated to be the initial steps in the hormone-driven cascade of specific cellular responses and mineral ion transport. For a recent review of this topic plus a current list of primary references of multiple laboratory groups, see Ref. (6). Initial progress has been made by our group and by others in identification of hormone-specific interactions between parathyroid hormone and binding sites in membrane preparations from kidneys and bone cells, quantitative definition of cyclic AMP production by natural and synthetic fragments of hormone in these membrane and cell systems, analysis of the role of guanyl nucleotides as a cofactor in adenylate cyclase stimulation, and identification of hormone-specific phosphorylation of cell proteins in renal cells. Systematic structure/activity studies using many synthetic fragments and analogs which are analyzed in various bioassays, has led our group to the design of analogs of greater potency than that of the naturally-occurring hormone and inhibitors of parathyroid hormone/receptor interaction, a critical tool in dissecting initial steps of hormone effects in target tissues. Analogs of the sequence bPTH-(3-34) lacking agonist activity have proved to be, in *in vitro* systems, pure competitive inhibitors of hormone/receptor interaction (Fig. 5).

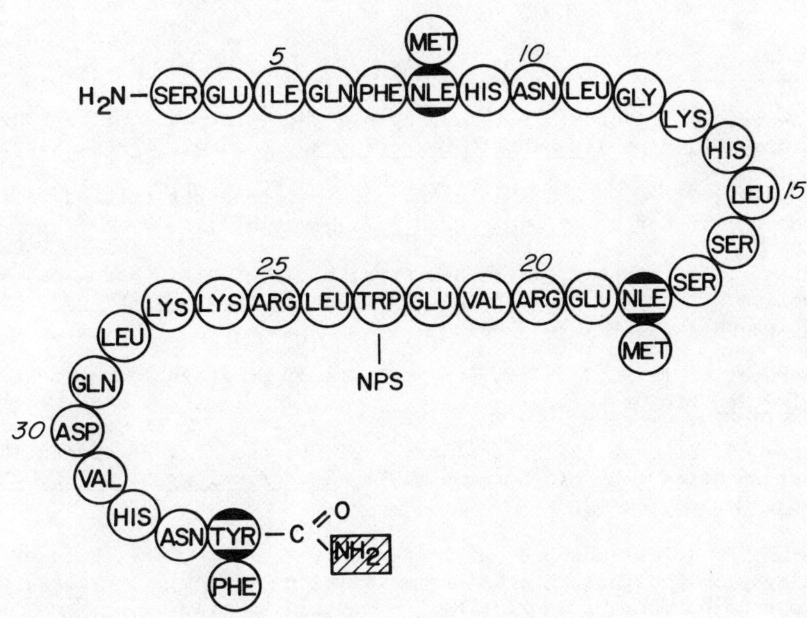

Fig. 5. The amino acid sequence of the synthetic hormone antagonist [Nle-8,Nle-18,Tyr-34]bPTH-(3-34)amide

Shaded residues depict positions of substitution for the native sequence of bPTH-(3-34). In addition, the carboxyl-terminal carboxyl function has been modified to a carboxamide ($CONH_2$).

The sixty or more chemically-synthesized fragments and analogs of parathyroid hormone that express in varying degrees agonist or antagonist activity of parathyroid hormone have been tested in existing receptor binding assays for hormone as well as in conventional adenylate cyclase assays to test the kinetics and specificity, quantitative cyclic AMP production, and/or evidence of phosphokinase activity, as well as to analyze control points in initial steps of hormone action such as receptor availability ("down regulation"), and to acquire more detailed knowledge of the role of guanyl nucleotides and other cofactors of PTH on adenylate cyclase activity. A careful examination of similarities or differences in responses seen with different PTH peptides *in vivo* versus those seen *in vitro* with emphasis on similarities and/or differences between those effects ascribable to skeletal versus renal actions of parathyroid hormone should provide the necessary physiological and pharmacological correlations with which to examine critically whether the receptor binding studied in different *in vitro* systems meets physiological criteria.

What is needed at present is a careful delineation of the specificity and affinity of parathyroid hormone/receptor interactions, examination of the similarities or differences, if any, in receptor characteristics in bone and kidney cells, and further exploration of cellular regulation of hormone/receptor interaction and the possible participation of phosphokinase in the initial stages of hormone action. It would be particularly useful to analyze quantitatively the presumed sequential correlations among hormone binding to receptors, cyclic AMP production, and phosphokinase activity.

REFERENCES

1. J. F. Habener, J. T. Potts, Jr., Parathyroid physiology and primary hyperparathyroidism (in press) Metabolic Bone Disease, Academic Press, New York.

2. M. R. Haussler, T. A. McCain, Basic and clinical concepts related to vitamin D metabolism and action, N Engl J Med 297, 974 (1977).

3. J. F. Habener, J. T. Potts, Jr., Chemistry, biosynthesis, secretion, and metabolism of parathyroid hormone (1976) Handbook of Physiology, American Physiological Society, Washington.

4. J. F. Habener, J. T. Potts, Jr., Biosynthesis of parathyroid hormone, N Engl J Med (in press).

5. G. V. Segre, P. D'Amour, M. Rosenblatt, J. T. Potts, Jr., Heterogeneity and metabolism of parathyroid hormone (1978) Endocrinology of Calcium Metabolism, Excerpta Medica, Amsterdam.

6. D. Goltzman, E. N. Callahan, G. W. Tregear, J. T. Potts, Jr., Role of 5'-guanylylimidodiphosphate in the activation of adenylyl cyclase by parathyroid hormone, Endocrine Res Commun, 3, 407 (1976).

Calcitonin: From Physiology to Therapeutic Uses

Gérard Milhaud

Department of Biophysics, Faculty of Medicine Saint-Antoine, 75012 Paris, France

An overall review of the field of calcitonin today can no longer be limited to the control of calcium metabolism. The story of calcitonin started when Harold Copp had to postulate the existence of a hypocalcemic factor to interpret perfusion studies performed on the thyroparathyroid apparatus of the dog, with blood of a high and low calcium concentration (1). From then on, calcitonin was considered as a calcium regulating hormone and the hypocalcemic effect elicited by this principle provided not only an easy way of detection but also the basis for an accurate bioassay. However two of the basic assumptions made at this early time proved later to need revision.

The first one concerned the source of calcitonin, which is not the parathyroids as originally thought by Copp, but the thyroid as demonstrated by Foster and coll. who perfused separately the thyroid and parathyroid glands in the goat (2). Real progress started when Hirsch and Munson discovered a calcium and phosphate lowering principle in the thyroid of the rat, which they named thyrocalcitonin to stress the thyroid origin (3). The thyroid proves therefore to be a double endocrine gland in mammals : calcitonin is secreted by the C-cells, which are scattered through the gland usually in a parafollicular position (4). C-cells originate from the neural crest and belong to the diffuse endocrine system or to the clear cell system. The function of the diffuse endocrine system is to maintain regulations and modulations at a fundamental level. Secretory cells of this group may well escape the classical approach used in endocrinology, which consists in producing a disorder by removing the endocrine and than in correcting it by the administration of the corresponding extract. As compared to classical endocrines, the function of the diffuse endocrine system is indeed difficult to demonstrate, as in addition to the difficulty of removing selectively scattered hormone producing cells, the consequences of the hormonal suppression may not be quite obvious. The fact that imbalances in calcitonin secretion do not elicit striking changes is not a convincing argument to consider the hormone as vestigial in mammals.

The second assumption concerns the precision of the regulation of plasma calcium level, which has been said to be unexceeded by any other electrolyte. In fact, calcium and phosphate levels fluctuate according to a circadian rythm and in health the levels are constant only if blood is sampled at the same sideral time (5). The homeostatic control does not simply involve feedback loops with the two antagonistic hormones like parathyroid hormone and calcitonin but includes anticipatory mechanism able to cope with the events happening in the "milieu extérieur" (6).

I shall attemps to give a review of the role of calcitonin as a calcium and phosphate regulating hormone in fish, as a lipid regulating hormone in the newborn rat and eventually as a hormone in man, a work carried out first at the Pasteur Institute and later at the Saint-Antoine Faculty of Medicine over the last thirteen years.

Calcitonin, a major gill hormone

The discovery of the production of very large amounts of calcitonin by the ultimobranchial gland of fishes (7) raised the question of the physiological role of this hormone in fish. By analogy with the hypocalcemic and hypophosphatemic effect in mammals, large does of hog and salmon calcitonin were injected into elasmobranch and teleost fish but the attemps to lower plasma calcium failed. These negative results could not be due to a species specificity as salmon calcitonin itself was inactive in the salmon. They have led some investigators to consider calcitonin as a vestigial hormone in fishes.

We did not find this explanation satisfactory and we made the hypothesis that other target organs than bone should be considered. The ultimobranchial body is embedded in the gill, an anatomical situation which leads one to investigate the gill as a target organ. The ionic composition of the "milieu intérieur" of the fish is very different from that of sea water, with a very high calcium content and a very low phosphate content. Homeostatic mechanisms are obviously required in order to maintain the constancy of fish plasma. The gill represents the major organ for ionic, water and respiratory exchange with the environment.

We used an isolated gill preparation under simulated in vivo conditions for the investigation of hormonal actions on a simplified system (8). Isolated salmon gills were suspended in a bath of aerated seawater containing either 45Ca Cl_2 or (32P) phosphate and perfused with a fish Ringer's solution. Fractions of the perfusate were collected at regular time intervals and their volume and radioactivity were measured. Three parameters were studied : the rate of flow of perfusate, the calcium influx and the phosphate influx from sea water through the gill. Hormonal effects were produced by injecting norepinephrine or calcitonin as pulses into the inlet cannula of the perfusion device. Norepinephrine increased the rate of flow of perfusate and the influx of calcium and phosphate (Fig. 1 and 2). Only the increased water flow was mediated by β-adrenergic receptors, as this effect could be prevented by β-blocking agents such as propranolol. The enhanced calcium influx induced by norepinephrine was not sensitive to propranolol.

The dissociation of the two responses by propranolol suggests that they result from a direct effect on gill epithelium rather than from a hemodynamic effect. Calcitonin elicits an effect opposite to that of catecholamine on the three parameters. It decreases the influx of water, calcium (Fig. 3) and phosphate from seawater and these effects are not sensitive to propranolol. Calcitonin can prevent excesscalcium entry from the calcium-rich seawater. This effect involves calcitonin concentration of the order of 10^{-7} M. In vivo, circulating calcitonin levels are of the order of 10^{-9} M. The difference between in vivo and in vitro may seem rather large, but can well be explained by the response to pulse flow compared to continuous perfusion. This was found to be the case using the same concentration as in vivo (9). Both calcitonin and norepinephrine therefore contribute to the homeostasis of the "milieu intérieur" of the fish and calcitonin appears as hormone for maintaining the constancy of the ionic composition of the plasma.

As spawning involves dramatic environmental metabolic changes, difference is sensitivity to calcitonin and norepinephrine was investigated between gills from fishes in pre- and post-spawning conditions. The ED_{50} is higher by almost two orders of magnitude in postspawning conditions, both for water and calcium influx, at a time when the physiological catecholamine concentration found in teleost blood is markedly increased (10). After spawning the gills are most sensitive to calcitonin, corresponding either to a decreased production or secretion of calcitonin or to an increased receptor sensitivity. It is known that fish experimentally returned from

freshwater to seawater will die : this could be explained if calcitonin production could not be raised rapidly enough to control the increased calcium influx due to the excess catecholamine output.
It is remarkable that calcitonin plays an analogous role in fish and mammals protecting the "milieu intérieur" from an increase in calcium. In mammals, the primary non dietary source of calcium is bone. Calcitonin by inhibiting bone resorption and enhancing bone formation affects the bone blood equilibrium. In fish, the source of calcium is seawater and calcitonin decrease the influx through the gill.
The antagonism between the adrenergic system and calcitonin may not be present in the fish only. In mammals, it could play a role in controlling the catecholamine production linked to stress. In man, an imbalance due to excess catecholamine output, which could not be matched by an adequate calcitonin production, could well lead to ischemic heart disease. Two observations are in favour of this hypothesis, both in rat and man. In the rat, catecholamines increase the plasma calcium level in the absence of the thyroid and parathyroid glands, an effect which can be suppressed by exogenous calcitonin administration (11) (12).
In man, isoproterenol, a β-adrenergic agonist, increases and propranolol, a β-blocker, decreases significantly the circulating calcitonin level without the induction of changes in serum calcium (13).

As aging is accompanied by a high rate of osteoporosis, atherosclerosis and ischemic heart disease, an inadequate calcitonin production could well be a common denominator in the pathogenesis of these disorders.

Calcitonin and triglycerides level

In the suckled new born rat, calcitonin administration is able to decrease the level of triglycerides in the plasma (14). This effect results mainly from an inhibition of gastric emptying. As circulating calcitonin levels are always higher in suckled than in fasted animals, maternal milk is likely to be the stimulus for calcitonin secretion. Rat milk is rich in triglycerides ; their role as a calcitonin secretagogue was tested using soybean oil as a substitute whose triglyceride composition is quite close to that of rat milk. One hour after the administration of soybean oil by stomach gavage, the circulating calcitonin level was significantly increased in the 2-h-old rat. These observations suggest that calcitonin is participating in the regulation of nutrient absorption in the suckled new born animal. A regulatory loop is likely to involve gastrointestinal hormones. In the 2-h-old fasted rat, the thyroid gland responds to a parenteral calcium load with a significant rise in circulating calcitonin level, yet neither pentagastrin nor glucagon are able to stimulate calcitonin secretion.
This leaves two messengers for consideration as calcitonin secretagogue, namely pancreozymin-cholecystokinin and the gastric inhibitory peptide. Thus the unexpected role played by calcitonin in the lipid metabolism of the newborn may continue later in life. We will discuss later the prevention if immune arteriosclerosis by calcitonin on an experimental model.

Calcitonin in man

The discovery of the existence of calcitonin in the human thyroid (15) suggested that the hormone should have physiological and pathological implications relevant to man as well as therapeutic applications. This is not the place to discuss the role of calcitonin in human pathology and I will restrict myself to the physiological involvement of the hormone and to it's therapeutic use. Calcitonin as a hormone produced by the diffuse endocrine system is likely to have pleiotropic actions, extending beyond the regulation of calcium metabolism

Osteoporoses.- These represent the most common metabolic bone diseases. The various

types of osteoporosis differ in cause, in clinical kinetic profiles and even in roentgenological appearance, but all of them result in a loss of bone mass, the remaining bone tissue being normally mineralized.
Affected bone segments have an increased porosity and are brittle. Albright described osteoporosis as a disease of "too little bone tissue" in contrast to osteomalacia, a disease of "too little bone mineral". Among the different types of osteoporosis, the most common are postmenopausal and senile osteoporoses. They relate to aging and are being found to an increasingly large number of human beings due to the extension of life expectancy and to improvements in diagnostic procedures.
The largest group of patients consiswomen in the postmenopausal state and the true incidence can be estimated to affect in France 5,5 millions, 15 % of them suffering from painful deformity of fractures, leading eventually to immobilization, another cause for osteoporosis. After the age of fifty, fractures of the proximal femur are 8 times more frequent in women than in men. This is not only due to the less dense skeleton in women, but also to a sharp loss of bone around the age of 50 to 55. In men, this does not happen to the sextent and the onset of senile osteoporosis is delayed by a decade.
The management of age related osteoporoses is a difficult task. The definition of the therapeutic goal is already the subject of a basic controversy : Is bone loss irreversible ? Is aging irreversible ? Can bone loss be prevented ? Can bone loss be arrested or can the process be reversed ? If the aging of bone can be stopped or slowed down, this would mean that it is possible to have the patient's biological time stopping or running slower whereas the flow of the sideral time remains unchanged. Obviously the answer to these fundamental questions requires the availability of convincing methods for assessing the efficacy of long term treatments. Clinical examination, kinetic data, balance studies, histological examination, bone mineral density and a sophisticated statistical analysis of the results have to be performed, if poorly grounded controversy is to be avoided.

The various osteoporoses being metabolic bone diseases, kinetic studies are needed to quantitate the main pathways of calcium metabolism and to point to the mechanism of bone loss. Since the calcium balance is the difference between bone formation and bone resorption, the result relating to osteoporotic patients is that excess bone resorption produces a negative calcium balance and the loss of bone. Urine calcium excretion (Vu) is increased particularly during the acute stages of the disease and the intestinal absorption of calcium is decreased. Therefore four out five of the main routes of calcium metabolism are disturbed in senile and post-menopausal osteoporosis (Fig. 4).

The effects of calcitonin on calcium metabolism are exactly a miror image of the metabolic disturbances found in osteoporosis as they include the inhibition of bone resorption, the decrease of urine calcium excretion, the increase of bone formation and intestinal absorption (16). The following hypothesis could account for the pathogenesis of bone loss :

Insufficient calcitonin production
↓
Increased bone resorption
↓
Negative calcium balance
↓
Bone loss

An aternative view consisting in an excess production of parathyroid hormone, though less likely, will be discussed later.
This hypothesis leads us tcnsider senile and postmenopausal osteoporoses as disorders resulting from a calcitonin deficiency and to propose a long range treatment

with substitutive doses of calcitonin, provided the sensitivity of the hormone receptors is still present. The replacement dose can be assessed using the quantitative relationship between the hypocalcemic effect (ΔCa) elicited by the hormone and bone resorption (17).

$$(\Delta Ca) = (Ca)\ \frac{Vo-}{P}$$

where (ΔCa) = hypocalcemic effect produced 1 hour after calcitonin injection, in mg %
(Ca) = serum calcium level, in mg %
P = rapidly exchangeable compartment of the calcium pool in mg
Vo- = bone resorption in mg/hour

The first step consists in calculating the maximal hypocalcemic response by determining plasma calcium level and the size of the calcium pool. Then increasing doses of calcitonin are administered and the hypocalcemic response is measured. The maximal hypocalcemia is elicited by the complete inhibition of bone resorption. Experimentally, one M.R.C. unit of porcine calcitonin is sufficient for the complete arrest of bone resorption in man. It is of interest to point to the calcitonin content of the normal human thyroid, which is of the order of 0.2 M.R.C. unit and to the total amount of the circulating hormone, which is of the order 0.015 M.R.C. unit. Calcitonin and 1,25-dihydroxycholecalciferol are the hormones possesing the highest specific biological activity.

Calcitonin was injected at low doses for several months to 42 patients suffering from several types of osteoporosis (senile and postmenopausal, 28 patients ; cortisone induced, 7 patients ; idiopathic male, 4 patients ; juvenile, 3 patients) (18). The effectiveness of the treatment was assessed using three types of informations :
- *Clinical results* : relative to pain, 3 patients out of 4 responded well to treatment, the beginning of the clinical improvement starting from less than one month to 4 months. The best results were obtained in postmenopausal and senile osteoporosis.
- *Kinetic and balance data* : In 20 patients repeated kinetic analysis of calcium metabolism was performed. The calcium balance (Δ) became positive in 9 patients, was in equilibrium in 2 patients and remained negative in 9 patients. The time course of overall balance changes showed, after 3 months of treatment, a $\Delta = +33$ mg/day, and after 6 months a $\Delta = +160$ mg/day ($P < 0.05$). These effects were the consequence of a decrease in $Vo\_$ during the treatment and of an increased absorption of calcium by the gastrointestinal tract.
- *Histological examination* : In 16 patients repeated histological examinations were performed on iliac crest biopsy specimens. Bone mass was decreased in 13 patients before the treatment : it increased in 10 patients (two third of the cases), remained unchanged in 3 patients and decreased slightly in 2 others without reaching abnormal values. At the histological level, calcitonin administration acts on both bone resorption and formation : osteoclastic activity is inhibited, in addition osteogenesis may be enhanced, osteoid material is deposited and bone mass increases.

The efficacy of the long term treatment of postmenopausal osteoporosis was further established by principal component analysis (19). The classical statistical methods are inapropriate because : a control group of patients is hardly available is hardly for an experimental approach extending over months and including major risks, like fractures ; the disease is progressive with periods of regression, which could make the efficacy of the treatment difficult to demonstrate.

It is nevertheless theoretically possible to establish the efficacy of a treatment without a control group if there is a clinical improvement in the disease each time the treatment is instituted and if a relapse occurs after withdrawal of the treatment. Five parameters of calcium kinetic were used : bone formation, the

exchangeable calcium pool, the daily decay constant of serum calcium radioactivity, urine calcium, the calcium balance ; pain was quantitated.
In addition, the incidence of fractures before and during treatment has been noted. The results of the study showed :
a) a symptomatiincreasing with the length of treatment.
b) an improvement of calcium balance in a high percentage of patients ($P < 0.02$)
c) the absence of any additional fractures in all patients with complicated se- postmenopausal osteoporosis.

The improvement was due to an initial increase in bone accretion accompanied by a longlasting decrease in bone resorption ($P < 0.05$), urine calcium ($P < 0.025$) and an increase in calcium absorption during digestion ($P < 0.05$) (Fig. 4). A break in the treatment generally produced a relapse.

The principal componentanalysis enabled the following assessments to be made :
1) when no calcitonin treatment was given, the degree of pain was related to the biological parameters of the disease.
2) The therapeutic protocol does not require long lasting calcitonin preparations as the principal components analysis establishes that 1 M.R.C. unit of porcine calcitonin administered three times a week for three months, followed by an intermission of two months, gave therapeutic results which compared favourably to the daily injection. In addition the side effects mentioned in the published reports relating to the administration of large doses like transient flushing, nausea, diarrhea are not observed with the physiologic rdoses, which have the further advantage of having practically no antigenicity. We have been unable to detect circulating antibodiagainst porcine calcitonin if administered at low doses without gelatin even after five or more years of treatment, whereas antibodies to salmon and porcine calcitonin may arise in as many as half of the patients treated with large doses. We also chechat the low doses successfully avoida parathyroid response.

This successfull approach was based on the results of kinetic analysis in osteoporotic patients and on the mode of action of calcitonin. Although hormone were utilized, the treatment could not be stricto sensu a replacement one, as no direct evidence of the hormonal imbalance of the osteoporotic patient was available. In theory two disturbances could account for the kinetic data, namely of excess production of parathyroid hormone or a calcitonin insufficiency. Direct measurements of circulating levels should help selecting the proper hypothesis provided sensitive and specific radioimmunoassays would be available for C- and N-terminal fragments of human parathyroid hormone and for decreased calcitonin levels. Before discussing the results of the radioimmunoassays used, it may be appropriate to comment on the difficulties inherent to the radioimmunoassays of these two hormones. For parathyroid hormone, a radioimmunoassay specific of the biologically active fragment of the hormone is necessary as it's known thanks to the pioneering work of Potts and coll. (20) that the hormone is present in the peripheral circulation in two forms, a short lived biologically active N-terminal fragment and a metabolically inert C-terminal fragment with a much longer half-life. We have developed a sensitive homologous assay for the active fragment of the human parathyroid hormone in plasma according to the sequence proposed by Niall (21).

As for calcitonin, though only the intact unoxydized molecule is biologically active, the interpretation of radioimmunoassay data is complicated by the restricted size of the antigenic sites of the molecule as compared to the complete hormonal sequence. Furthermore, we and others have shown that calcitonin in plasma is heterogenous due to the presence of "big calcitonins" which react differently with various antibodies used for radioimmunoassay of the hormone (22). It follows that measurement of circulating levels of calcitonin should be performed using antibodies directed towards different regions of the molecule. Calcitonin determinations should therefore not be based on a single antibody but must include the use of

several radioimmunoassays specific for different regions of the human molecule and sensitive enough to measure decreased circulating levels. We have applied these assays to patients suffering from osteoporosis as evidenced by clinical, biochemical and roentgenological criteria. Preliminary investigations showed that such patients tended to have lower circulating calcitonin levels than normal subjects. This finding is observed both in male and female atients. In a second investigation, postmenopausal osteoporosis was selected as a well defined and very common condition. For age mached patients and control subjects, values of calcitonin were obtained by two or three differents antibodies as well as of N- and C- terminal fragments of parathyroid hormone. A constant finding was the decrease in circulating calcitonin level in osteoporotic patients irrespective of the radioimmunoassay system used (Fig. 5) while values of both N- and C- terminal fragments were normal in the two groups (Fig. 6). This observation is the first substantiation of osteoporosis as a calcitonin deficiency disorder.
Calcitonin treatment can therefore be considered as a replacement therapy.
Results obtained in postmenopausal osteoporotics have led us to study the relationship between sexual hormones and calcitonin secretion.

Calcitonin and sex hormones

Loss of ovarian function is promptly followed by increased resorption of bone and loss of bone mass leading to osteoporosis. Albright was the first to suggest a relationship between loss of ovarian function and osteoporosis (1940) and this hypothesis has led to advocate the treatment with oestrogens (23). Since we have just shown that osteoporosis is associated with a calcitonin deficiency, we have reinvestigated the effects of castration and substitutive oestrogen therapy in an experimental model. We measured the circulating levels of calcitonin in the female rat at different ages, during the oestrial cycle, and assessed the effect of ovariectomy.

Calcitonin levels are higher in sexually mature animals than in immature rats. Castration elicits a dramatic fall in calcitonin levels only in the sexually mature animal (Fig. 7), indicating an unsuspected relationship beween the ovary and the C-cells of the thyroid. Administration of oestrogen and progesterone alone or in combination did not reestablish normal calcitonin levels (Fig. 8). Therefore if calcitonin secretion in the female rat is in part under ovarian control, sexual steroid hormones may not be involved. Work is in progress to assess if this regulation is a direct one (ovarian factors) or an indirect one via the hypothalamus and or the hypophysis. The probable regulation of calcitonin secretion in the female by the ovary may play a role in the protection of the maternal skeleton by this hormone during pregnancy and lactation and may explain the high incidence of postmenopausal osteoporosis, provided that the physiological cessation of ovarian endocrine functions should include not only the arrest of the secretion of oestrogen and progesterone but also of a "calcitonin releasing factor". If these results can be extrapolated to man, the loss of ovarian function is followed by a decreased secretion of calcitonin whereas parathyroid hormone secretion is maintained. This hormonal imbalance may lead to an increased bone resorption and to the loss of bone. The ideal treatment of postmenopausal osteoporosis seems to consist in correcting this imbalance by raising the calcitonin level, a goal which oestrogen therapy may not be able to reach.

Calcitonin and arteriosclerosis

A common finding in elderly people consists in the association of spine demimeneralisation associated with arterial calcification : on roentgenographic examinations, the vertebrae may be so radiolucent that it seeems as if the calcified aorta were supporting the body.

The question arises as to whether these calcium movements in the two systems are associated with one other and share a common pathogenesis.
We have observed a decrease of circulating calcitonin level in postmenopausal osteoporotic women, which is consistent with the main disturbances of calcium metabolism in this condition. The deposition of calcium salts in cells and matrix macromolecules of the aorta is one of the important biochemical alterations linked to aging and arteriosclerosis. Two types of calcium deposition are observed in the large blood vessels : a diffuse one on the elastic fibers and a localised one with crystalline or amorphous calcium salt accumulation in cells and atherosclerotic plaques. Diffuse deposition of calcium in elastic fibers is a continuous process, increasing steadily during the whole life span. As very little is known with respect to the role of endocrine factors in the regulation of calcium metabolism in the artrial wall, the effect of calcitonin administration on the deposition of calcium salts in cells and matric macromolecules of the aorta was investigated on an animal model (24). Typical sclerotic calcified lesions can induced in the rabbit by immunisation with K-elastin in complete Freund's adjuvant. These lesions include fragmentation of the elastic fibers and lead to an increase in the solubility of structural glycoproteins, collagen as well as polymeric collagen. In animals treated simultaneously with calcitonin, the morphological and most of the biochemical modifications produced by K-elastin immunisation were either not observed or lessened. Elastic lamellae were much less fragmented and calcium deposition in the media was markedly decreased. The solubility of structural glycoproteins collagen and polymeric glycogen was normal. These results indicate that calcitonin at low doses exerts a protective action against the morphological and the biochemical modifications produced by immun-arteriosclerosis.
If this observation can apply to man, calcitonin well be one of the factors influencing the alterations of the elastic fibers during arteriosclerosis and aging. Therefore a decrease in calcitonin production may lead to osteoporosis as well as to the deposition of calcium salts in the cells and interfibrillar matrix of the arterial wall and be the common denominator of the two disorders.

CONCLUDING REMARKS

At a pharmacological meeting, it may be appropriate to put calcitonin in perspective, at a time when the formerly so creative therapeutical chemistry is leading to steadily decreasing yields. If medicines are now a day coping with most of the acute disorders, which had very bad prognosis in the past, they are not able to manage the disorders linked to aging. In this respect, another approach should be designed if this type of disorder is to be treated. Calcitonin is precisely a successful example of the new approach, in which physiological messengers are isolated on the basis of their biological activity.
The mode of action has to be assessed and human disorders linked to the inadequate secretion of the messenger have to be recognized. Eventually, the management of these disorders by replacement therapy will have to be carefully evaluated.
This means that in designing new medicines, a different attitude may be required. The architecture of new drugs should no more rely on the knowledge of the empiric rules linking pharmacological activity to chemical structure and be restricted to the possible substitutions of the organic compound to interact with the cell receptors. Instead, the logic of the body should be revealed by isolating and assessing the mode of action of the compounds produced in order to fulfill the functions of the living organisms. This approach (which is lacking any a priori assumptions) requires a high degree of integration between biochemists, organic chemists, endocrinologists, pharmacologists and last but not least clinicians. The difficulty of the task may account for the fact that the progress of modern medical practice still rely on the pasteurian revolution. It is obvious that the evolution of the concepts of molecular biology has not yet had sufficient practical impact on the

management of human disorders. Even when an important potentially valuable natural agent has been identified, a great deal of sophisticated work is required to develop a well-tolerated and clinically effective preparation. A fatal drawback may be waiting at every crossroad, but because of their role as circulatory mediators the hormones offer better therapeutic prospects than other classes of naturally occuring drugs.

The study of calcitonin has been particularly rewarding, opening a new chapter in human pathology and making promises for the management of severe disorders linked to aging.

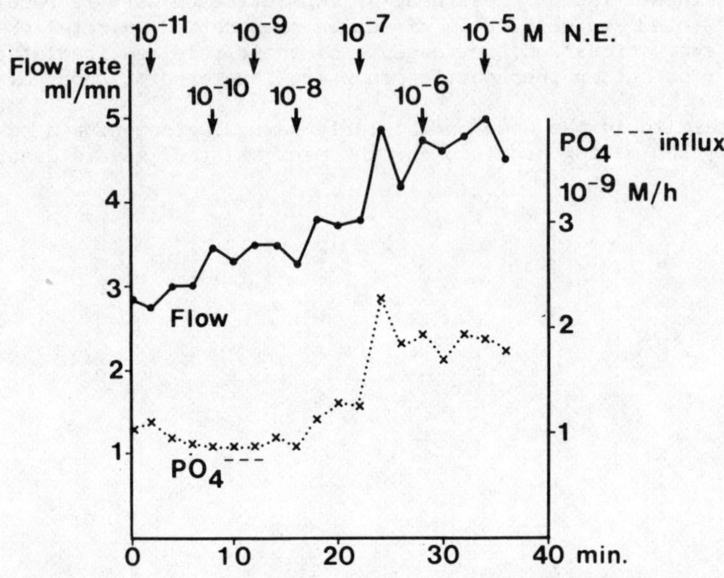

Fig. 1 : Effect of norepinephrine (N.E.) on the rate of flow and phosphate influx through the gill. Flow rats is measured directly ; phosphate influx is expressed per Kg of fish/hr multiplied by 8 (number of gill arches). Female Oncorhynchus keta.

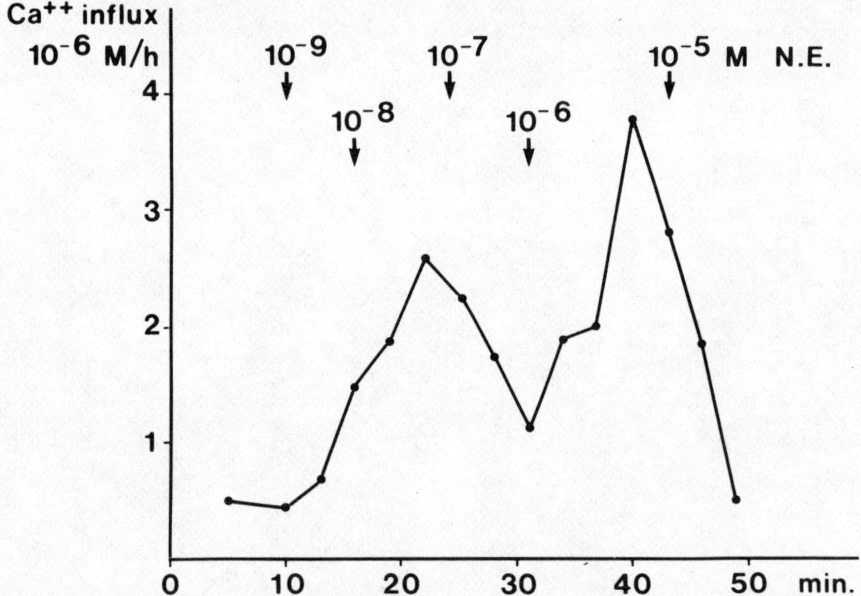

Fig. 2 : Effect of norepinephrine (N.E.) on calcium influx through the gill expressed per Kg of fish/hr multiplied by the number of gills. Female Oncorhynchus keta.

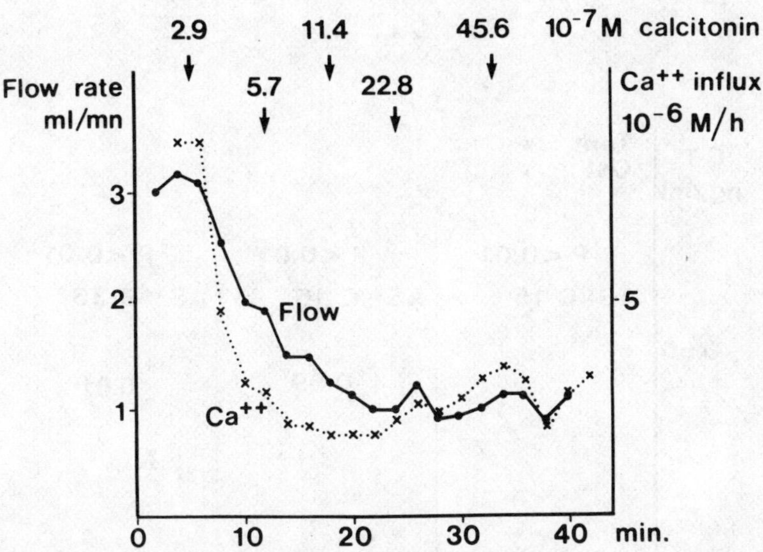

Fig. 3 : Effect of calcitonin on the rate of flow and calcium influx through the gill. Male Oncorhynchus keta.

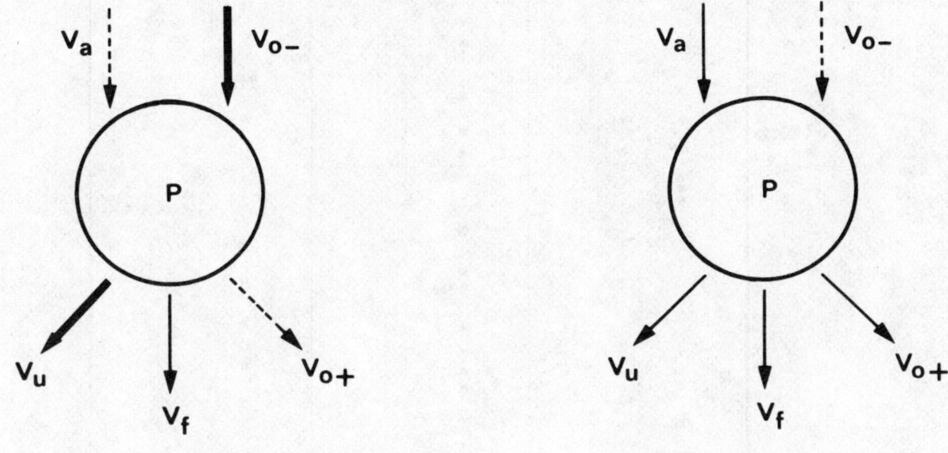

Fig. 4 : A) Disturbances of calcium metabolism in postmenopausal & senile osteoporosis.

B) Correction by replacement therapy with calcitonin

Process —— normal
▬▬ increased
----- decreased

Va calcium absorbed during digestion
Vo + bone formation
Vo - bone resorption
Vu urine calcium
V_f endogenous faecal calcium

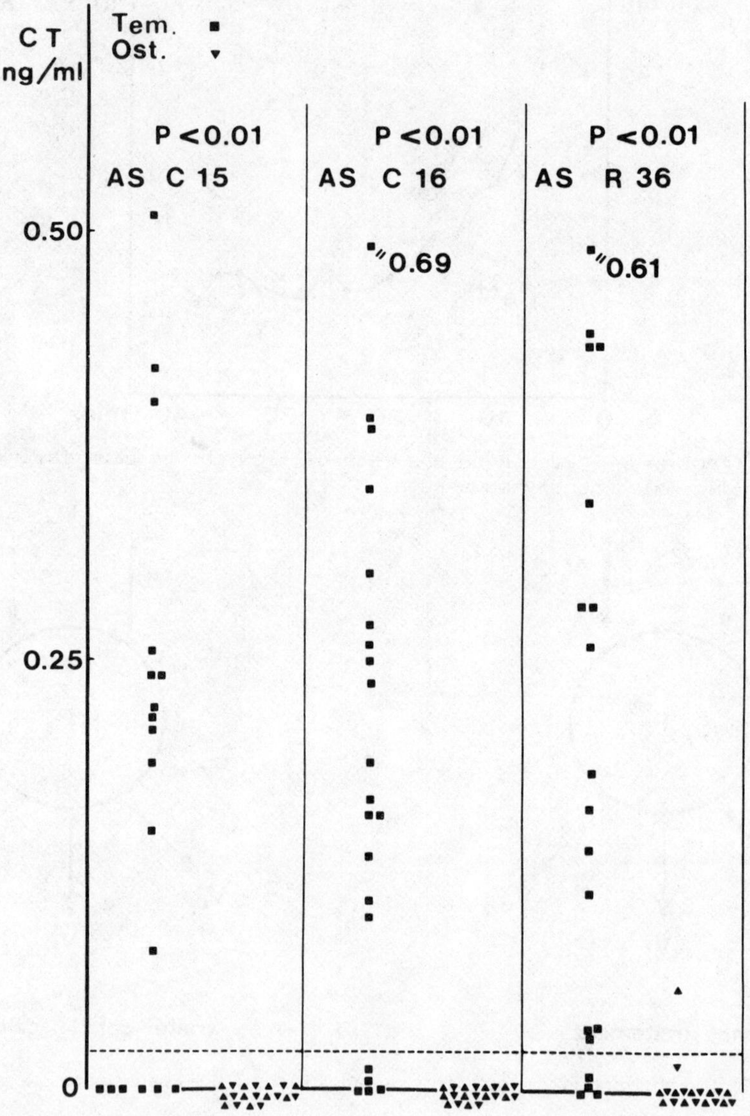

CT ng/ml

Tem. ■
Ost. ▼

P < 0.01 AS C 15

P < 0.01 AS C 16

P < 0.01 AS R 36

0.50

0.25

0

″0.69

″0.61

Fig. 5 : Decrease of circulating calcitonin levels in osteoporotic patients ▼ as compared to age matched control subjects ■, measured by 3 antibodies.

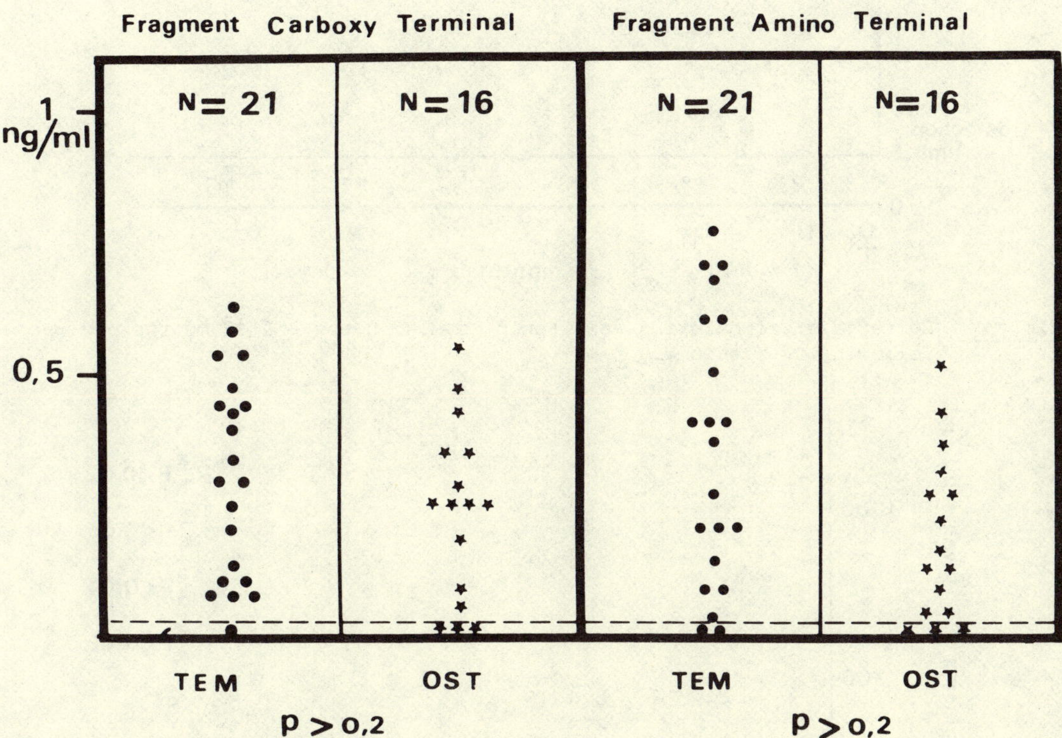

Fig. 6 : Levels of circulating N- & C- terminal parathyroid hormone fragments in osteoporotic patients.

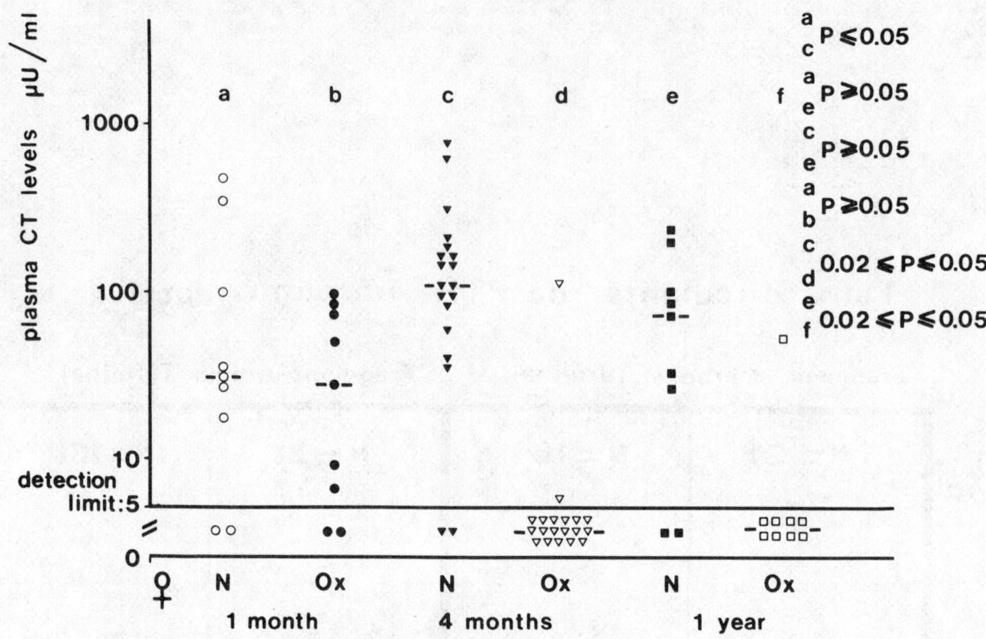

Fig. 7 : Decrease of circulating calcitonin levels in the 4 & 12 months old rat following ovariectomy.

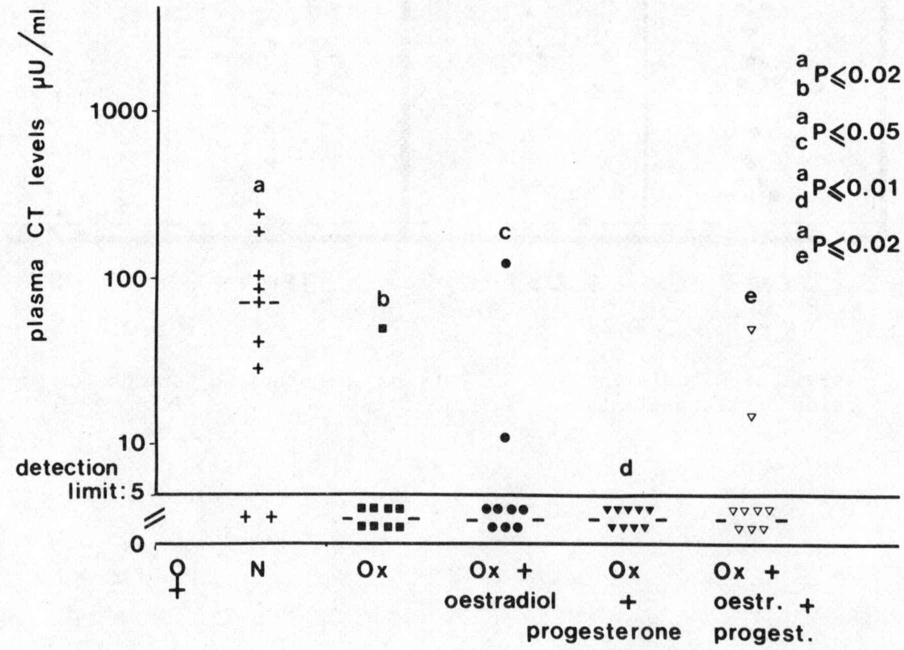

Fig. 8 : Failure of oestrogen and progesterone, alone or in combination, to raise the level of circulating calcitonin in the ovariectomized rat.

BIBLIOGRAPHY

1) D.H. COPP, E.C. CAMERON, B.A. CHENEY, A.G.F. DAVIDSON & K.G. HENZE, "Evidence for calcitonin -a new hormon from the parathyroid that lowers blood calcium". Endocrinology 70, 638 (1962).

2) G.V. FOSTER, A. BAGHDIANTZ, M.A. KUMAR, E. SLACK, H.A. SOLIMAN, I. Mac INTYRE, "Thyroid origin of calcitonin". Nature (London), 202, 1303 (1964).

3) P.F. HIRSCH, E.F. VOELKEL & P.L. MUNSON, "Thyrocalcitonin : hypocalcemic hypophosphatemic principle of the thyroid gland". Science 146, 412 (1964).

4) G. BUSSOLATI & A.G.E. PEARSE, "Immunofluorescent localisation of calcitonin in the "C-cells" of pig & dog thyroid". J. Endocr. 37, 205 (1967).

5) G. MILHAUD, A.M. PERAULT-STAUB & J.F. STAUB, "Diurnal variation of plasma calcium & calcitonin function in the rat". J. Physiol. 222, 559 (1972).

6) A.M. PERAULT-STAUB, J.F. STAUB & G. MILHAUD, "A new concept of plasma calcium homeostasis in the rat". Endocrinology 95, 458 (1974).

7) D.H. COPP, D.W. COCKROFT & Y. KUEH, "Calcitonin from ultimobranchial of dog, fish & chicken". Science 158, 924 (1967).

8) G. MILHAUD, J.C. RANKIN, L. BOLIS & A.A. BENSON, "Calcitonin : its hormonal action on the gill". Proc. Natl. Acad. Sci. U.S.A. 74, 4693 (1977).

9) C. MILET, J. PEIGNOUX & E. MARTELLY, "Personal communication".

10) M. FONTAINE, M. MAZEAUD & F. MAZEAUD, "L'adrénalinémie du *Salmo salar L* à quelques étapes de son cycle vital et de ses migrations". C. R. Hebd. Séances Acad. Sci. 256, 4562 (1963).

11) W.A. HSU & C.W. COOPER, "Hypercalcemic effect of catecholamines and its prevention by thyrocalcitonin". Calc. Tiss. Res. 19, 125 (1975).

12) D. DARMAUN, "Effets antagonistes des catécholamines et de la calcitonine sur le métabolisme calcique du rat ". D.E.A. Endocr. (1977).

13) N.M. VORA, G.A. WILLIAMS, G.K. HARGIS, E.N. BOWSER, W. KAWAHARA, B.L. JACKSON, W.J. HENDERSON & S.C. KUKREJA,
"J. Clin. Endocrinol. Metab. 46, 567 (1978).

14) J.M. GAREL & A. JULLIENNE,
"Plasma calcitonin levels in pregnant and newborn rats".
J. Endocr. 75, 373 (1977).

15) G. MILHAUD, M.S. MOUKHTAR, J. BOURICHON & A.M. PERAULT,
"Existence et activité de la thyrocalcitonine chez l'homme".
C.R. Hebd. Séances Acad. Sci. 261, 4513 (1965).

16) G. MILHAUD,
"Place de la calcitonine en pathologie et en thérapeutique".
Actualités pharmacologiques 30ème Série, Masson (Paris), 7 (1977).

17) G. MILHAUD,
"Utilization of thyrocalcitonin in man in normal & pathological conditions in Parathyroid hormone and thyrocalcitonin".
Ed. R.V. TALMAGE & L.F. BELANGER, Excerpta Medica Foundation (Amsterdam) 86 (1968).

18) H. BLOCH-MICHEL, J. CAYLA, F. DELBARRE, G. MILHAUD, J.C. RENIER, G. VIGNON, C. BREGEON, G. COUTRIS, D. PANSU, J. RONDIER & P. WALTZING,
"Etude coopérative sur le traitement au long cours de l'ostéoporose par la thyrocalcitonine à faible dose".
Rev. Rhum. 41, 93 (1974).

19) G. MILHAUD, J.N. TALBOT & G. COUTRIS,
"Calcitonin treatment of postmenopausal osteoporosis. Evaluation of efficacy by principal components analysis".
Biomedecine 22, 223 (1975).

20) G.V. SEGRE, H.D. NIALL, J.F. HABENER & J.T. POTTS,
"Metabolism of Parathyroid hormone : Physiology and clinical significance".
Am. J. Med. , 56, 774 (1974)

21) H.D. NIALL, R. SAUER, J. JACOBS, H.T. KEUTMANN, G.V. SEGRE, J.L.H. O'RIORDAN, G.D. AURBACH & J.T. POTTS,
"The aminoacid sequence of the aminoterminal 37 residues of human parathyroid hormone".
Proc. Natl. Acad. Sci. USA, 71, 384 (1974).

22) A. JULLIENNE, C. CALMETTES, D. RAULAIS, G. MILHAUD & M.S. MOUKHTAR,
"Immunochemical characterization of calcitonin in human disorders" in Endocrinology of calcium metabolism".
Ed. Excerpta Medica (Amsterdam) 55 (1978).

23) E.C. REIFENSTEIN & F. ALBRIGHT,
"The metabolic effect of steroid hormones in osteoporosis".
J. Clin. Inv., 26, 24 (1947).

24) L. ROBERT, D. BRECHEMIER, G. GODEAU, M.L. LABAT & G. MILHAUD,
"Prevention of experimental immun-arteriosclerosis by calcitonin".
Biochem. Pharm. 26, 2129 (1977).

Diphosphonates

Herbert Fleisch

Department of Pathophysiology, University, 3010 Bern, Switzerland

INTRODUCTION

The concept that diphosphonates could be used in certain diseases involving calcium metabolism came from earlier studies on inorganic pyrophosphate (PPi) (for literature see 1-3). This compound had been found to inhibit *in vitro* the formation as well as the aggregation of hydroxyapatite and calcium oxalate crystals. Besides, PPi also inhibits the rate of dissolution of already formed crystals. Both these effects on formation and dissolution are related to the marked adsorption of PPi onto the crystal surface. When given *in vivo* to animals, parenterally administered PPi is able to prevent various types of experimental soft tissue calcification.

Since PPi was found to be present in various biological fluids such as plasma, urine, saliva and synovial fluid as well as in mineralized tissues such as bone and teeth, it has been suggested that this compound plays a role in the protection of soft tissues from calcification and in the regulation of both the formation and the destruction of hard tissues. The local concentration would be under the control of alkaline phosphatase, which is known to possess a pyrophosphatase activity, and other pyrophosphatases. It is possible that PPi also plays a role in certain diseases. Thus, it has been found to be increased both in plasma and urine in hypophosphatasia, a condition characterized by a lack of alkaline phosphatase and a deficient mineralization, and to be decreased in patients with urinary stones.

However, PPi and condensed phosphates were unlikely to be useful clinically since they are not absorbed in the intestinal tract when given orally and, when given parenterally, they are very quickly destroyed enzymatically. Thus it was attempted to find substances related in structure to PPi and with similar physical-chemical effects but which were not destroyed *in vivo*.

EXPERIMENTAL INVESTIGATIONS

The diphosphonates, compounds characterized by two P-C bonds instead of the P-O-P bond of PPi, seem to fulfil this requirement. Many diphosphonates have been tested in various systems up to now, but only three have been used as yet in man and will be discussed here. Their structure is the following, all three being of the geminal type (P-C-P bond):

O	O	O	O
‖	‖	‖	‖
OH - P - OH	OH - P - OH	OH - P - OH	OH - P - OH
|	|	|	|
O	OH - C - CH_3	Cl - C - Cl	NH_3 - CH_2 - CH_2 - C - OH
|	|	|	|
OH - P - OH	OH - P - OH	OH - P - OH	OH - P - OH
‖	‖	‖	‖
O	O	O	O
inorganic pyrophosphate acid	ethane-1-hydroxy-1,1-diphosphonic acid (EHDP)	dichloro-methane diphosphonic acid (Cl_2MDP)	3-amino-1-hydroxy-propane-1,1-diphosphonic acid (AHPDP)

Physical Chemical Effect

Many of the diphosphonates were found to have physical chemical effects very similar to those of PPi. Thus *in vitro* they also inhibit the crystallization (4,5) and the aggregation (6) of calcium phosphate and calcium oxalate (7) crystals and slow down the dissolution of hydroxyapatite (8). Like PPi, the diphosphonates are strongly bound onto the crystal surface, with a relation between the binding and the effect (9).

Effect of Calcification In Vivo

Again like PPi the diphosphonates are very efficient in inhibiting experimental soft tissue calcification in animals such as the arterial and renal calcification induced by vitamin D (5,10), the skin calcification induced by DHT and the periarticular calcification that accompanies experimental arthritis. However, unlike PPi, which is effective only when administered parenterally, the diphosphonates are active also when given orally. There is a close correlation between the ability of individual diphosphonates to inhibit crystal growth *in vitro* and soft tissue calcification *in vivo*. The most efficient compounds are those possessing a P-C-P bond (geminal diphosphonates) while compounds containing P-C-C-P (vicinal diphosphonates) or single C-P bonds are generally ineffective. In the dental field EHDP has been found, when applied topically, to reduce the development of dental calculus in rats (11,12).

Unlike PPi, certain diphosphonates, when given at high doses, also inhibit the mineralization of normally calcifying tissues. Thus, from a dose of about 5 mg/kg/day s.c., EHDP causes the appearance of unmineralized osteoid in the bone of several species (13-15). In growing animals there is also an inhibition of cartilage mineralization so that the epiphyseal plate becomes wider and the X-ray picture somewhat resembles that of classical vitamin D-deficient rickets (16). Histologically, however, the two conditions differ fundamentally, at least in the chick, since the cartilage from vitamin D deficiency shows an accumulation of proliferating cells while that of animals treated with EHDP consists mainly of hypertrophic cells (17). The inhibition of the mineralization of osteoid appears at a lower dose than that of cartilage. Both effects are reversible when the administration of the compound is discontinued. Finally EHDP inhibits also the mineralization of dentine (18). The simplest explanation for these effects is that EHDP prevents the formation of apatite crystals at the site of mineralization by a direct physico-chemical action. However, other mechanisms, especially at the cellular level, cannot be excluded. It is of clinical interest that other diphosphonates have a much smaller inhibitory activity on the calcification of cartilage and bone than EHDP. This is especially true for Cl_2MDP, although it has a very strong effect on ectopic calcification. The cause of this discrepancy is not yet known.

Effect on Bone Resorption

It has been found by means of a variety of experimental designs that diphosphonates are also very powerful inhibitors of bone resorption, in contrast to pyrophosphate, for which no such effect could be shown. In culture, diphosphonates lower the resorption in mouse calvaria both when endogenous resorption is measured and when the bone destruction is induced by various agents such as PTH, lipopolysaccharides, cyclic AMP or prostaglandins (8,19). Interestingly, Cl_2MDP is always nearly an order of magnitude more potent than EHDP, with AHPDP lying in between these two. The inhibition of resorption can also be demonstrated *in vivo*. Thus, Cl_2MDP given to growing rats induces a morphological picture of the metaphysis characteristic of a complete inhibition of chondroclastic and osteoclastic resorption (16). In this system AHPDP seems to be the most powerful, followed by Cl_2MDP, which is again more active than EHDP. When Cl_2MDP is given to newborn mice, resorption is impaired so strongly that the bones resemble those of the grey-lethal strain of congenitally osteopetrotic mice, and the animals develop the symptoms of this disease (20). The effect of the diphosphonates on bone resorption is also shown by the decrease in the excretion of hydroxyproline, especially of the dialysable fraction (21) and by using ^{45}Ca kinetic techniques (22). The latter show a log dose effect on the resorption, Cl_2MDP producing a detectable effect at a dose as low as 0.04 mg/kg/day s.c. and being again about ten times more potent than EHDP.

The mechanism by which the diphosphonates inhibit bone resorption is not as yet completely understood. It seems likely however that the effect is obtained by more than one mechanism. Indeed the originally invoked physical chemical inhibition of the dissolution of the apatite appears to be not the only mechanism. Thus, for example, it could not explain the greater effect of Cl_2MDP *in vivo*, while *in vitro* EHDP is the better inhibitor of apatite dissolution. Recent results have shown that the diphosphonates can enter cells and induce various cellular effects which might well be relevant with respect to bone resorption (23). Interestingly there is now a certain correlation between many of these cellular effects and the inhibition of resorption. Thus, Cl_2MDP as well as AHPDP inhibits the growth of calvaria cells in cultures, while EHDP has only little or no effect (23). Furthermore Cl_2MDP and to a lesser extent EHDP added to culture medium alters the morphology of the osteoclasts of calvaria, while little or no effect is obtained on other cells. When injected into animals, both diphosphonates increase the number of osteoclasts as well as the number of nuclei per cell and induce various cytological alterations (16,24).

The diphosphonates have also been found to have potent effects on carbohydrate metabolism. Thus in cultured calvaria cells Cl_2MDP and to some extent EHDP induce a decrease of glucose consumption and of lactate production (23). The latter is also seen in culture in whole calvaria. Since the local acid production is thought to be one of the mechanisms by which the cells induce a resorption of bone, the decrease in lactic acid production might well explain, at least partly, the inhibiting effect of some of the diphosphonates, such as Cl_2MDP. However, this mechanism cannot be applied to all compounds of this class. Thus AHPDP and long-chain diphosphonates are very effective inhibitors of bone resorption but either have no influence or, on the contrary, increase lactic acid production.

Finally, the diphosphonates have also been found to exert some other cellular effects. Thus, they inhibit both *in vitro* and *in vivo* various lysosomal enzymes (25) and inhibit the release of calcium from mitochondria (26). Whether these effects are of importance in explaining their action on bone resorption is not yet known.

In the light of these various effects the question can be raised as to why the diphosphonates act *in vivo* only on bone and not on other tissues. The answer probably lies in the fact that these compounds, because of their high affinity to bone, are cleared very quickly from blood (27). Thus soft tissues are exposed to them for only a very short time. On the contrary this is not the case for bone cells, especially those involved in bone destruction. Indeed, the diphosphonates will be released again from the bone when the apatite crystals are dissolved such that the cells in the relevant region will be exposed to very high concentrations.

Effect on Experimental Osteoporosis

The results on bone resorption raise the possibility that the diphosphonates might be used in the prevention or the treatment of osteoporosis. It seems that the former possibility has a good chance of being achieved. Indeed Cl_2MDP and to a lesser degree EHDP has been found to prevent either completely or partially the osteoporosis induced by immobilization after nerve section (28,29). Interestingly polyphosphates, calcitonin and fluoride as well as inorganic phosphate have no effect, the diphosphonates being thus the only compounds as yet known to be active in this model. EHDP also partially prevents the osteoporosis induced by heparin or coumarine, while Cl_2MDP prevents that induced by cortisol (30). However, the loss of bone induced by a low calcium diet is not influenced (31).

Unfortunately until now there has been no good experimental model to test the effect of the drug on an already established osteoporosis. Therefore the question of a possible curative administration cannot be answered. However, the fact that the administration of diphosphonates and especially Cl_2MDP induces an increased net calcium balance as well as an increased mass of bone in normal animals would suggest that an effect could be hoped for.

CLINICAL APPLICATION OF DIPHOSPHONATES

Bone Scanning Agents

The strong affinity of hydroxyapatite to pyrophosphate, polyphosphates and diphosphonates (9) and the ability of these compounds to pass rapidly from blood into bone (27) have been used for their application in clinical medicine as bone scanning agents. The compounds can be linked in the presence of stannous ions or a reducing agent to ^{99m}technetium, a gamma-emiting isotope with a short half-life. Instead of going to the soft tissues the technetium will now follow the pyrophosphate or diphosphonate and be taken up very quickly in the calcified tissues (32-34). These compounds, which have now nearly completely replaced radioactive strontium or ^{18}F, have been found to be very useful in the detection of changes in the bone turnover such as in bone metastases, Paget's disease and other bone diseases. They have furthermore been found to localize in areas of tissues necrosis and thus have been used for the localization of heart infarctions.

Dental Calculus

EHDP has been found, when applied topically, to diminish the development of dental calculus (12,35). This finding is potentially important for dental health, since plaque formation and calculus are thought to predispose to periodontal disease and tooth loss. Lately a tooth-paste containing EHDP has been put on the market in some countries.

Ectopic Calcification

Small numbers of patients with ectopic calcifications in such diseases as sclerodorma, dermatomyositis and calcinosis universalis have been given EHDP. Unfortunately no clear-cut results have emerged. While some reports suggest a regression of the lesions under EHDP, others have found no effect. It is difficult to assess benefit in diseases occurring rarely and where spontaneous exacerbations and remissions occur.

The disease most extensively studied has been myositis ossificans progressiva (36-38). Unfortunately no double-blind study has been performed, so that the results are again difficult to interpret. If all the results are considered, it seems that EHDP given at oral doses of 10 mg/kg/day or above does slow down the formation of ectopic bone, but does not inhibit it completely, so that the disease is still progressing although probably at a slower rate. The occurrence of new bouts with swelling, pain and tenderness does not seem to be inhibited, but under treatment the lesions most often do disappear again without calcifying. Whether the treatment with EHDP allows for the surgical removal of ectopic bone has not been settled. While results in adults have been encouraging, those in children were less so. It has to be noted that the dose of EHDP which seems to be effective also leads to an inhibition of the calcification of normal bone, resulting at least in children in an X-ray picture similar to those of rickets and fractures. Thus, if EHDP is given it should be limited in time and should be administered only for short periods when a new bout of the disease occurs.

Finally, in a double-blind study performed in patients with spinal cord injury (39) EHDP was shown to diminish the appearance of heterotopic ossification. A double-blind study has also been performed in ectopic calcification of the hip after total hip replacement (40). While the calcification was diminished during the three months of EHDP administration it reappeared after discontinuation of the drug, suggesting, as in myositis ossificans, that EHDP did not block the production of ectopic matrix but only its mineralization. However, despite the late calcification, the mobility of the hip was clinically significantly better in the patients receiving EHDP, suggesting that there might be some advantage in using this compound in this condition, especially in those patients who require a second operation due to ossification after the first intervention.

Urinary Stones

In view both of the effect of the diphosphonates on crystal growth and aggregation and of the fact that they are absorbed in the intestine and excreted in the urine with a clearance greater than that of inulin, it appeared possible that these compounds could inhibit stone formation. About 2 % of the ingested EHDP is absorbed, half of it being excreted in the urine, so that with a dose of about 1,000 mg EHDP/day a concentration in urine can be obtained which will have inhibitory effects on crystal formation. By administering the diphosphonate four times a day, a constant excretion within 24 hours could be obtained (41). In a pilot study the administration of 1,100 mg of EHDP per day given in four divided doses showed a diminution of the recurrence of calcium stones (41). Interestingly this inhibitory effect does not stop after the EHDP administration is discontinued, although only undetectable amounts of diphosphonate then come out in the urine. This discrepancy is perhaps explained by the fact that the diphosphonate led to an increased excretion of pyrophosphate which was sustained after stopping the treatment (41). Although this preliminary result suggests that diphosphonates might be useful in the prevention of urinary stones, it must be stressed that the dose used was in the range which decreases bone turnover and induces an inhibition of mineralization. Considering these adverse effects on the skeleton, it seems doubtful that the possible benefit in stone disease is large enough to permit an application, except perhaps in some rare cases. However, the concept of the increase of endogenous or exogenous inhibitors in urine looks promising. One approach might be to find a phosphonate which is excreted in the urine but has fewer effects on bone.

Paget's Disease

The most extensive studies have been performed on Paget's disease, a condition characterized by an increased turnover of bone. EHDP has been found extremely effective at reversing the rise in alkaline phosphatase and in urinary hydroxyproline, the two parameters characteristic of the increased bone turnover (42-48). The effect is dose-dependent, normal values being obtained in many patients within three months of treatment. The powerful effect on turnover is also shown by morphological studies (42-44,48). Thus, both formation and resorption assessed by morphometry are decreased, the number of osteoclasts and osteoblasts is reduced, and the bone under EHDP returns to a normal lamellar type. EHDP also corrects the abnormality of calcium kinetics (42-44) and that seen in the skeleton scan (45-48). However, the X-ray picture does not seem to be improved, at least not during the treatment period (44,46,48). Finally, the subjective symptoms are also improved (45-48). Similar results have lately also been found with AHPDP (49).

The first studies were done with a relatively high dose, 10-20 mg of EHDP/day p.o. Under this treatment the inhibitory effect on normal calcification seen in animals was apparent. Thus the rate of calcification dropped and there was a clear-cut increase of the surface and the thickness of the osteoid (42-44,48). This inhibition of mineralization, which was reversible when treatment is stopped, was accompanied clinically by the occurrence of bone pains and fractures (45,46). However, all these various side effects were not seen under an oral dose of 5 mg/kg/day, although the bone turnover is still decreased (48). Thus the latter is the dose which is most often recommended today. If a larger amount is nevertheless administered, the treatment should not be longer than three months.

Interestingly in contrast to calcitonin, where the parameters of increased bone turnover return quickly to pre-treatment levels as soon as the administration of the hormone is stopped, this is not the case with diphosphonates. Indeed, many patients keep their turnover low over years after discontinuation of the treatment. Others do show some recurrence, yet are sensitive to a second treatment (42,44,46-48). This and the fact that the diphosphonates can be given orally and for relatively short periods make them probably more attractive than calcitonin in the treatment of this disease. Just recently EHDP has been put on the market in the U.S.

Side Effects of Diphosphonates

As in animals the studies in man have revealed little in the way of side effects. The two main ones are the inhibition of mineralization of the skeleton and an increase in plasma phosphate.

The inhibition of bone calcification has been found at doses at and above 10 mg/kg/day orally. The effect was seen in patients with osteoporosis (50) and in patients undergoing total hip replacement (40) as well as in both normal and pathological bone in patients with Paget's disease (42-44,48). Furthermore there is also an inhibition of the mineralization of cartilage in children, although this tissue seems to be more resistant than the osteoid. In both adults and children the inhibition of mineralization has led to fractures. After periods of treatment of more than a year, the children also developed a proximal muscular weakness leading to an abnormal gait. Both the radiological and the clinical defects are however regressive when therapy is discontinued.

In both normal persons and in patients EHDP causes a conspicuous rise in plasma phosphate, often to high levels (42-44,50-52). This is most clearly seen with a dose of 20 mg EHDP/kg/day. The change in plasma phosphate is associated with a change in the renal handling of phosphate, namely an increase in the tubular maximum rate of reabsorption of phosphate (51,52). The mechanism of this change is as yet unclear. It is interesting that it has not been found in any other animal species studied so far.

PROSPECTS

From both the experiments in animals and the results obtained in man, it appears that diphosphonates have, beside their use in nuclear medicine as ^{99m}technetium complexes, a future in the treatment of various bone diseases. The use in ectopic calcification has been hindered until now by the fact that EHDP also inhibits the calcification of normal tissues. However, from animal studies it seems that Cl_2MDP, while being very effective on ectopic calcification, has a much smaller effect on bone and cartilage calcification, so that this compound might have some application in these disorders. This difference between Cl_2MDP and EHDP, and the fact that Cl_2MDP is more effective in inhibiting bone resorption, will make it the favourite compound in Paget's disease and other diseases with increased bone resorption as well as in diseases such as osteoporosis where an uncoupling between bone formation and destruction is sought. The first clinical trials along this line are currently underway. Although the animal experiments and the first clinical data with AHPDP seem promising, until a toxicological study and more data are available, a full appreciation will not be possible.

Since small changes in the structure of the diphosphonates have been found to have profound effects on their action *in vivo*, it can be hoped that in the future other diphosphonates will be found with more specific and more powerful effects on the various processes involved in calcium metabolism. It appears that with the discovery of this class of compounds a new path in the treatment of calcium disorders has been opened.

ACKNOWLEDGEMENTS

This work has been supported by the Swiss National Science Foundation (3.725.76), by the Procter and Gamble Company, USA, and by the Ausbildungs- und Förderungsfonds der Arbeitsgemeinschaft für Osteosynthese (AO), Chur, Switzerland.

REFERENCES

1. Fleisch, H. and Russell, R.G.G., Experimental and clinical studies with pyrophosphate and diphosphonates, *Calcium Metabolism in Renal Failure and Nephrolithiasis*, Wiley, New York, 293 (1977).
2. Russell, R.G.G. and Fleisch, H., Pyrophosphate and diphosphonate, *The Biochemistry and Physiology of Bone*, Academic Press, New York, 61 (1976).
3. Fleisch, H. and Russell, R.G.G., Pyrophosphate and polyphosphate, *Pharmacology of the Endocrine System and Related Drugs*, Pergamon Press, Oxford, 61 (1970).
4. Francis, M.D., The inhibition of calcium hydroxyapatite crystal growth by polyphosphonates and polyphosphates, *Calc.Tiss.Res.*3, 151 (1969).
5. Fleisch, H., Russell, R.G.G., Bisaz, S., Mühlbauer, R.C., and Williams, D.A., The inhibitory effect of phosphonates on the formation of calcium phosphate crystals in vitro and on aortic and kidney calcification in vivo, *Europ.J.Clin.Invest.* 1, 12 (1970).
6. Hansen, N.M., Felix, R., Bisaz, S., and Fleisch, H., Aggregation

of hydroxyapatite crystals, Biochim.Biophys.Acta 451, 560 (1976).

7. Fraser, D., Russell, R.G.G., Pohler, O., Robertson, W.G., and Fleisch, H., The influence of disodium ethane-1-hydroxy-1,1-diphosphonate (EHDP) on the development of experimentally induced urinary stones in rats, Clin.Sci. 42, 197 (1972).
8. Russell, R.G.G., Mühlbauer, R.C., Bisaz, S., Williams, D.A., and Fleisch, H., The influence of pyrophosphate, condensed phosphates, phosphonates and other phosphate compounds on the dissolution of hydroxyapatite in vitro and on bone resorption induced by parathyroid hormone in tissue culture and in thyroparathyroidectomized rats, Calc.Tiss.Res. 6, 183 (1970).
9. Jung, A., Bisaz, S., and Fleisch, H., The binding of pyrophosphate and two diphosphonates by hydroxyapatite crystals, Calc. Tiss.Res. 11, 269 (1973).
10. Rosenblum, L.Y., Flora, L., and Eisenstein, R., The effect of disodium ethane-1-hydroxy-1,1-diphosphonate (EHDP) on a rabbit model of athero-arteriosclerosis, Atheroscl. 22, 411 (1975)
11. Briner, W.W., Francis, M.D., and Widder, J.S., The control of dental calculus in experimental animals, Int.Dent.J. 21, 61 (1971).
12. Mühlemann, H.R., Bowles, D., Schatt, A., and Bernimoulin, J.-P., Effect of diphosphonate on human supragingival calculus, Helv. Odont.Acta 14, 31 (1970).
13. Jowsey, J., Holley, K.E., and Linman, J.W., Effect of sodium etidronate in adult cats, J.Lab.Clin.Med. 76, 126 (1970).
14. King, W.R., Francis, M.D., and Michael, W.R., Effect of disodium ethane-1-hydroxy-1,1-diphosphonate on bone formation, Clin. Orthop.Rel.Res. 78, 251 (1971).
15. Russell, R.G.G., Kislig, A.-M., Casey, P.A., Fleisch, H., Thornton, J., Schenk, R., and Williams, D.A., Effect of diphosphonates and calcitonin on the chemistry and quantitative histology of rat bone, Calc.Tiss.Res. 11, 179 (1973).
16. Schenk, R., Merz, W.A., Mühlbauer, R., Russell, R.G.G., and Fleisch, H., Effect of ethane-1-hydroxy-1,1-diphosphonate (EHDP) and dichloromethylene diphosphonate (Cl_2MDP) on the calcification and resorption of cartilage and bone in the tibial epiphysis and metaphysis of rats, Calc.Tiss.Res. 11, 196 (1973).
17. Bisaz, S., Schenk, R., Kunin, A.S., Russell, R.G.G., Mühlbauer, R., and Fleisch, H., The comparative effects of vitamin D deficiency and ethane-1-hydroxy-1,1-diphosphonate administration on the histology and glycolysis of chick epiphyseal and articular cartilage, Calc.Tiss.Res. 19, 139 (1975).
18. Larsson, A., The short-term effects of high doses of ethylene-1-hydroxy-1,1-diphosphonate upon early dentin formation, Calc.Tiss. Res. 16, 109 (1974).
19. Reynolds, J.J., Minkin, C., Morgan, D.B., Spycher, D., and Fleisch, H., The effect of two diphosphonates on the resorption of mouse calvaria in vitro, Calc.Tiss.Res. 10, 302 (1972).
20. Reynolds, J.J., Murphy, H., Mühlbauer, R.C., Morgan, D.B., and Fleisch, H., Inhibition by diphosphonates of bone resorption in mice and comparison with grey-lethal osteoporosis, Calc.Tiss.Res. 12, 59 (1973).
21. Goulding, A. and McChesney, R., Comparison of effects of parathyroidectomy and ethane-1-hydroxy-1,1-diphosphonate (EHDP) administration upon bone synthesis of hydroxyproline and urinary excretion of hydroxyproline in rats, Calc.Tiss.Res. 23, 115 (1977).

22. Gasser, A.B., Morgan, D.B., Fleisch, H.A., and Richelle, L.J., The influence of two diphosphonates on calcium metabolism in the rat, Clin.Sci. 43, 31 (1972).
23. Fast, D.K., Felix, R., Dowse, C., Neuman, W.F., and Fleisch, H., The effects of diphosphonates on the growth and glycolysis of connective-tissue cells in culture, Biochem.J. 172, 97 (1978).
24. Miller, S.C., Jee, W.S.S., Kimmel, D.B., and Woodbury, L., Ethane-1-hydroxy-1,1-diphosphonate (EHDP). Effects on incorporation and accumulation of osteoclast nuclei, Calc.Tiss.Res. 22, 243 (1977).
25. Felix, R., Russell, R.G.G., and Fleisch, H., The effect of several diphosphonates on acid phosphohydrolases and other lysosomal enzymes, Biochim.Biophys.Acta 429, 429 (1976).
26. Guilland, D.F., Sallis, J.D., and Fleisch, H., The effect of two diphosphonates on the handling of calcium by rat kidney mitochondria in vitro, Calc.Tiss.Res. 15, 303 (1974).
27. Bisaz, S., Jung, A., and Fleisch, H., Uptake by bone of pyrophosphate, diphosphonates and their technetium derivatives, Clin.Sci. Mol.Med. 54, 265 (1978).
28. Mühlbauer, R.C., Russell, R.G.G., Williams, D.A., and Fleisch,H., The effect of diphosphonates, polyphosphates, and calcitonin on "immoblisation osteoporosis" in rats, Europ.J.Clin.Invest. 1, 336 (1971).
29. Michael, W.R., King, W.R., and Francis, M.D., Effectiveness of diphosphonates in preventing "osteoporosis" of disuse in the rat, Clin.Orthop.Rel.Res. 78, 271 (1971).
30. Black, H.E. and Jee, W.S.S., A histomorphometric and biochemical evaluation of the effects of a diphosphonate in corticosteroid-treated rabbits, Bone Histomorphometry, Armour Montagu, Paris, 157 (1977).
31. Jowsey, J. and Holley, K.E., Influence of diphosphonates on progress of experimentally induced osteoporosis, J.Lab.Clin.Med. 82, 567 (1973).
32. Castronovo, F.P.,Jr. and Callahan, R.J., New bone scanning agent: ^{99m}Tc-labelled 1-hydroxy-ethylidene-1,1-disodium phosphonate, J. Nucl.Med. 13, 823 (1972).
33. Subramanian, G., McAfee, J.G., Blair, R.J., Mehter, A., and Connor, T., ^{99m}Tc-EHDP: A potential radiopharmaceutical for skeletal imaging, J.Nucl.Med. 13, 947 (1972).
34. Tofe, A.J. and Francis, M.D., In vitro optimization and organ distribution studies in animals with the bone scanning agent ^{99m}Tc-Sn-EHDP, J.Nucl.Med. 13, 472 (1972).
35. Sturzenberger, O.P., Swancar, J.R., and Reiter, G., Reduction of dental calculus in humans through the use of a dentifrice containing a crystal-growth inhibitor, J.Periodont. 42, 416 (1971).
36. Bassett, C.A.L., Donath, A., Macagno, F., Preisig, R., Fleisch, H., and Francis, M.D., Diphosphonates in the treatment of myositis ossificans, Lancet 2, 845 (1969).
37. Geho, W.B. and Whiteside, J.A., Experience with disodium etidronate in diseases of ectopic calcification, Clinical Aspects of Metabolic Bone Disease, Excerpta Medica, Amsterdam, 506 (1973).
38. Smith, R., Russell, R.G.G., and Woods, C.G., Myositis ossificans progressiva. Clinical features of eight patients and their response to treatment, J.Bone Jt Surg. 58B, 48 (1976)
39. Stover, S.L., Hahn, H.R., and Miller, J.M., Disodium etidronate in the prevention of heterotopic ossification following spinal

cord injury (preliminary report), Paraplegia 14, 146 (1976).
40. Bijvoet, O.L.M., Nollen, A.J.G., Slooff, T.J.J.H., and Feith, R., Effect of a diphosphonate on para-articular ossification after total hip replacement, Acta Orthop.Scand. 45, 926 (1974).
41. Baumann, J.M., Bisaz, S., Fleisch, H., and Wacker, M., Biochemical and clinical effects of ethane-1-hydroxy-1,1-diphosphonate in calcium nephrolithiasis, Clin.Sci.Mol.Med. 54, 509 (1978).
42. Russell, R.G.G., Smith, R., Preston, C., Walton, R.J., and Woods, C.G., Diphosphonates in Paget's disease, Lancet 1, 894 (1974).
43. Gunčaga, J., Lauffenburger, T., Lentner, C., Dambacher, M.A., Haas, H.G., Fleisch, H., and Olah, A.J., Diphosphonate treatment of Paget's disease of bone. A correlated metabolic, calcium kinetic and morphometric study, Horm.Metab.Res. 6, 62 (1974).
44. De Vries, H.R. and Bijvoet, O.L.M., Results of prolonged treatment of Paget's disease of bone with disodium ethane-1-hydroxy-1,1-diphosphonate (EHDP), Neth.J.Med. 17, 281 (1974).
45. Altman, R.D., Johnston, C.C., Khairi, M.R.A., Wellman, H., Serafini, A.N., and Sankey, R.R., Influence of disodium etidronate on clinical and laboratory manifestations of Paget's disease of bone (osteitis deformans), New Engl.J.Med. 289, 1379 (1973).
46. Canfield, R., Rosner, W., Skinner, J., McWhorter, J., Resnick, L., Feldman, F., Kammerman, S., Ryan, K., Kunigonis, M., and Bohne, W., Diphosphonate therapy of Paget's disease of bone, J.Clin. Endocr.Metab. 44, 96 (1977).
47. Khairi, M.R.A., Johnston, C.C., Altman, R.D., Wellman, H.N., Serafini, A.N., and Sankey, R.R., Treatment of Paget's disease of bone (osteitis deformans). Results of a one-year study with sodium etidronate, JAMA 230, 562 (1974).
48. Alexandre, C. and Meunier, P., Le traitement de la maladie osseuse de Paget par l'éthane-1-hydroxy-1,1-diphosphonate (EHDP), Assoc.Corp.des Etudiants en Médecine de Lyon, Lyon (1977).
49. Bijvoet, O.L.M., Presentation at the III Israel Workshop on Calcified Tissues, Jerusalem (1978).
50. Jowsey, J., Riggs, B.L., Kelly, P.J., Hoffman, D.L., and Bordier, P., The treatment of osteoporosis with disodium ethane-1-hydroxy-1,1-diphosphonate, J.Lab.Clin.Med. 78, 574 (1971).
51. Recker, R.R., Hassing, G.S., Lau, J.R., and Saville, P.D., The hyperphosphatemic effect of disodium ethane-1-hydroxy-1,1-diphosphonate (EHDP™): renal handling of phosphorus and the renal response to parathyroid hormone, J.Lab.Clin.Med. 81, 258 (1973).
52. Walton, R.J., Russell, R.G.G., and Smith, R., Changes in the renal and extrarenal handling of phosphate induced by disodium etidronate (EHDP) in man, Clin.Sci.Mol.Med. 49, 45 (1975).

Summary

Paul L. Munson

Department of Pharmacology, School of Medicine, University of North Carolina, Chapel Hill, N.C. 27514, U.S.A.

We have indeed been fortunate today to hear from six major figures in the discovery and development of new knowledge of the pharmacology of calcium homeostasis and of applications to the therapy of calcium-related diseases. We thank them for the clarifications they have brought us. We also are intrigued by the complexities and remaining mysteries that they have called to our attention.

My assignment was to summarize the Symposium. In a short space I can do so only very inadequately. I will also make some comments on aspects of the subject that I think need emphasis and additional discussion or research.

R.G.G. Russell began the Symposium with a clear, succinct overview of the regulation of calcium metabolism that was notable for its expression of the consensus of informed opinion today. I have only a few reservations.

Russell properly emphasized the fact that the effect of endogenous parathyroid hormone (PTH) on bone to increase resorption and thereby to increase plasma calcium is relatively small in normal man. At least this is true over the short term. I think it does not necessarily follow that the effect of PTH to increase the renal tubular reabsorption of calcium is more important than the effect of PTH on bone in the minute-to-minute and hour-to-hour regulation of plasma calcium. I adhere to the view of Talmage and others that transport of calcium between a bone fluid compartment and extracellular fluid is quickly responsive to PTH (and calcitonin) and thereby plays an important role, equal or more important than that of the kidney, in the regulation of plasma calcium.

Calcitonin (TCT), in adequate doses, is capable of lowering plasma calcium and phosphate in mammals, but I believe, in agreement with Talmage, that it does so more by reducing the net transport of calcium from bone fluid to plasma than by inhibiting bone resorption, which it is also capable of doing, although more slowly. I suggest that it is preferable to think of TCT as an antihypercalcemic rather than as a hypocalcemic hormone. The antihypercalcemic effect would be most important during the postprandial intestinal absorption of calcium.

I agree with Russell if he meant to imply that the effect of PTH to decrease the secretion of H^+ by the kidney is significant only with excessive PTH, as seen in some patients with primary hyperparathyroidism.

Hector De Luca summarized important new discoveries made in the vitamin D field during the past ten years, in which he and his coworkers have been so prominent. The central place of the D metabolite, 1α,25-dihydroxycholecalciferol (1α,25-$(OH)_2D_3$) in calcium metabolism is firmly established and it is now well recognized to be a hormone produced exclusively by the kidney. The production of vitamin D_3 from the precursor in the skin, 7-dehydrocholesterol, by ultraviolet light is not regulated and the next step in metabolism, to 25-hydroxycholecalciferol, mainly in the liver, is rather loosely regulated. PTH is a major but not the only factor in regulation of biosynthesis of the hormone by the kidney. The question may be asked, what is the extent of storage of 1α,25-$(OH)_2D_3$ in the kidney and is its rate of secretion, as well as its biosynthesis, under regulation?

De Luca also discussed the relative biological importance of other metabolites of vitamin D. 24R,25-$(OH)_2D_3$ is of special interest. Does this quantitatively major D metabolite represent merely a protective diversion of D metabolism toward a less active compound or does it have some specialized role to play in calcium metabolism, as discussed in the next paper by Coburn and Brickman?

De Luca thinks of PTH and 1α,25-$(OH)_2D_3$ as acting together to mobilize calcium from bone in hypocalcemia. This raises the old question of the absolute requirement of 1α,25-$(OH)_2D_3$ for the action of PTH on bone, which I regard as not settled yet. Since an increased supply of 1α,25-$(OH)_2D_3$ involves a considerable lag period, whereas an increase in PTH secretion can occur almost instantaneously, I believe that the role of 1α,25-$(OH)_2D_3$ is "permissive" and that the level of PTH is the critical factor.

Jack Coburn (with coauthor Arnold Brickman) presented a comprehensive survey of the clinical pharmacology and therapeutics of the active metabolites of vitamin D_3.

The hormone, 1α,25-$(OH)_2D_3$, is 100 times as active as its immediate precursor, 25-$(OH)D_3$, in increasing calcium absorption in normal man. Although the effect of 1α,25-$(OH)_2D_3$ is much faster than that of D_3, it is by no means immediate, requiring 6 hours or so to become evident. The decrease in urinary cAMP and increase in urinary calcium after 1α,25-$(OH)_2D_3$ suggest that the level of serum PTH was decreased (to explain these effects), but since an increase in serum calcium (to explain the change in PTH secretion) is not observed consistently, the mechanisms involved remain unclear. How does 1α,25-$(OH)_2D_3$ directly and indirectly affect parathyroid function?

The presently available data in man suggest that 1α,25-$(OH)_2D_3$, at the low dose levels adequate to increase intestinal absorption of calcium, does not stimulate bone resorption (or net efflux of calcium out of bone). Hypercalciuria, when seen, is more likely the result of increased absorption of dietary calcium.

Coburn outlined and briefly discussed clinical conditions with reduced actions of vitamin D metabolites. It would hardly be useful to summarize his excellent summary in even fewer words, but I will make a few comments.

The osteomalacia that occurs in patients who take anticonvulsant drugs, yet have normal plasma levels of 1α,25-$(OH)_2D_3$, and which responds to vitamin D or 25-$(OH)D_3$, presents intriguing questions.

Small doses of 1α,25-$(OH)_2D_3$ are adequate to restore to normal the serum calcium of patients with hypoparathyroidism or pseudohypoparathyroidism. Treatment with this hormone, rather than with vitamin D, has the added advantage of a relatively short duration of action (half-time of disappearance of effect, 1 to 3 days).

The use of 1α,25-$(OH)_2D_3$ to counteract the symptoms of secondary hyperparathyroidism in chronic renal failure has produced impressive results. Coburn discussed some of the problems and imperfections that still remain with this therapy. The prospect that 1α,25-$(OH)_2D_3$ may be useful in the treatment or prevention of steroid-induced osteopenia or osteoporosis looks encouraging.

Coburn also discussed the possibility that 24R,25-$(OH)_2D_3$, which circulates at a level 100 times higher than that of 1α,25-$(OH)_2D_3$, may play some important role in normal physiology, such as to synergize with 1α,25-$(OH)_2D_3$ in regulation of parathyroid function.

John Potts, who, with his associates, has made enormously important contributions to the biochemistry of the parathyroid hormone over the past 15 years, summarized the entire subject with concentration on recent exciting advances in the understanding of the biosynthesis and metabolism of the hormone. The detailed information about the amino acid sequence of prepro-, pro-, native circulating hormone, and active core fragment has formed the basis for research on biosynthesis and metabolism. It also has made possible the synthesis of active and inactive fragments, one of which, bPTH-(28-48), is a competitive inhibitor of PTH degradative metabolism. Other fragments synthesized in his laboratory will undoubtedly lead to superior and more specific immuno and receptor binding assays for more accurate delineation of PTH status in health and disease. It is to be hoped that improved methods for quantitative assay of PTH in the plasma of experimental animals will also be developed. We take encouragement in the newer results presented by Potts, which give hope that in the near future the levels of active PTH in plasma can be determined accurately in routine as well as in research laboratories.

Potts's outline indicates that new knowledge of the intracellular PTH biosynthetic events will make it possible to determine the nature of regulation of biosynthesis (and storage), including the relative importance of calcium feedback and other factors. The negative feedback control of PTH secretion by calcium has long been known and is universally accepted. The tightness of this control may be exaggerated. As well demonstrated by Milhaud, Talmage, and coworkers, there are considerable diurnal variations in serum calcium, for which TCT (and indirectly, gastrointestinal hormones) may be largely responsible. Why are these variations not more quickly corrected by changes in PTH secretion? Or is the feedback regulation of PTH rather loose, perhaps beneficially?

G. Milhaud, who, with his coworkers, has been responsible for so many fundamental insights and discoveries about calcitonin, summarized his and coworkers' recent exciting findings about the hormonal action of calcitonin (antagonistic to norepinephrine) on the gill of the salmon, reviewed his use of low doses of porcine calcitonin in the treatment of osteoporosis, still controversial, and briefly sketched new experiments in the suckling newborn rat showing that TCT decreases the elevated plasma triglyceride levels produced naturally by maternal milk or experimentally by oral soybean oil.

The discovery that calcitonin has a major effect on gill function to decrease the influx of water, calcium, and phosphate from seawater may put an end to our mystification at earlier failures to demonstrate an effect of ultimobranchial calcitonin, so highly potent in mammals, in lower species in which it is present in such abundance. How remarkable that calcitonin performs a similar function in mammals and fish, protection against excess calcium, by different mechanisms.

Further developments in the study of the effects of TCT on triglyceride metabolism with their implications for atherosclerosis will be awaited with great interest. Milhaud already presented evidence for the protective action of TCT against experimental atherosclerosis in a rabbit model.

Milhaud's use of low doses of calcitonin in the therapy of osteoporosis has been received with skepticism in many quarters. Reliable evaluation of the efficacy of therapeutic agents in osteoporosis is notoriously difficult. Milhaud has outlined the problems that would impede the conduct of a controlled clinical trial, using the double blind technique. However, a positive outcome of such a trial, even if restricted to analgesia as the therapeutic endpoint, would be greatly influential. The new finding by Milhaud, Moukhtar and coworkers that the mean level of TCT in the plasma of patients with postmenopausal osteoporosis is much lower than that of age-matched postmenopausal women without overt osteoporosis is most impressive and offers a rational basis for the therapy they recommend.

Osteoporosis is a painful, disabling disease with a high incidence, especially in the elderly, that is being attacked in many laboratories and with a variety of therapeutic regimens. The use of TCT at higher doses than those used by Milhaud is being studied, as are low-dose therapy with PTH, $1\alpha,25\text{-}(OH)_2D_3$, and the diphosphonates as well as the earlier proposed estrogens. There may be hope that one or the other of these approaches, with or without adjuvants, will prove to be successful and widely applied.

Our last speaker, Herbert Fleisch, is the leading world expert on diphosphonates and an important contributor to the physiology and pharmacology of bone. The diphosphonates, which he reviewed, are drugs that have profound effects on calcium metabolism. Although, unlike the other substances covered in this Symposium, the diphosphonates are not hormones, they are synthetic compounds that were developed as more stable analogs of a naturally occurring ion, inorganic pyrophosphate. The three diphosphonates of principal current interest are ethanehydroxydiphosphonic acid (EHDP), dichloromethanediphosphonic acid, and aminohydroxypropanediphosphonic acid. These three compounds have similar effects qualitatively, but because of differences in relative potency, depending on the effect measured, they may be used selectively, both as research tools and as potential diagnostic and therapeutic agents. Fleisch outlined an impressive number of newly established and potential clinical applications. The most extensive studies have been on the use of EHDP in the treatment of Paget's disease of bone. Undesirable side effects experienced with the relatively high doses originally employed appear to be absent when a lower dose, still effective in reducing bone turnover, is used. In the U.S.A., the favored therapy of severe, generalized Paget's disease of bone has been salmon calcitonin. Now that EHDP has been approved for medical use in the U.S.A., a larger number of patients will be available to evaluate the relative merits of these two therapeutic agents in Paget's disease.

It should also be mentioned that the diphosphonates have proved to be useful research tools as inhibitors of calcification and in other ways in experimental animals.

Before closing, I would like to call your attention to some research in the area of this Symposium by colleagues at the University of North Carolina.

There are at least three periods in the life of the normal rat in which serum calcitonin is greatly elevated, lactation (Toverud and others), infancy - the suckling period - (Cooper and others), and advancing age (Peng and others). All three of these periods of life have special advantages for the further study of the physiology and pharmacology of TCT. I will give further detail only on the experiments by Toverud _et al_. in the lactating rat.

The lactating rat provides an outstandingly excellent model of calcium metabolism under stress. The calcium concentration of rat milk is 13 times that of human milk. The daily milk output in ml per kg body weight is 5 times that of the human lactating female. The loss of calcium into the milk is 60 times greater per hour per kg body weight in the rat.

The mean calcitonin level in rat serum is substantially (many fold) increased in the lactating rat. At the same time the serum calcium is considerably lower (ca 1 mg/dl) in the lactating rat than in the nonlactating rat, in spite of the great increase in amount of calcium consumed in the food and absorbed from the intestine.

The intake of calcium is increased 3-fold by greater food consumption by the lactating rat. The percent of calcium absorbed is increased 4-fold. The total calcium absorbed is therefore increased more than 10-fold. A logically related observation is that the level of serum $1\alpha,25\text{-}(OH)_2D_3$ is increased 3-fold in the lactating rat.

To outline these undoubtedly interrelated phenomena in the lactating rat:

(1) Food consumption is increased.
(2) Serum calcitonin is increased.
(3) Serum calcium is low.
(4) PTH is increased (inferred).
(5) $1\alpha,25\text{-}(OH)_2D_3$ is increased.
(6) Percent and amount of calcium absorbed are increased.

The sequence and mechanisms involved in the above changes are under further study. We are especially curious to find out how important the increase in calcitonin is for the entire pattern of changes observed. The possible role of prolactin will not be neglected.

In conclusion, I would like to say that as extensive as the coverage of this Symposium has been and as eminent as you recognize our speakers to be, we regret that shortage of time and space prevented us from offering the platform to an additional number of authorities and leading scientific investigators in this field who would have broadened and deepened our appreciation of progress in the understanding of the pharmacology of calcium homeostasis even further.

Invited Lectures

Ca Ion and Muscle Contraction

Setsuro Ebashi

Department of Pharmacology, Faculty of Medicine, University of Tokyo, Bunkyo-ku, Tokyo 113, Japan

Setsuro Ebashi
Department of Pharmacology, Faculty of Medicine,
University of Tokyo, Bunkyo-ku, Tokyo 113, Japan

INTRODUCTION

Ca ion plays crucial roles at various stages of muscle contraction (cf. Ref. 1, 2). The mode of its action is quite different from muscle to muscle. Although it is rather subsidiary, the role of Ca ion in the excitation mechanism of skeletal muscle is not only physiologically important but also noticeable from pathogenetic point of view. It plays more essential role in cardiac muscle; Ca-dependent inward current is undoubtedly connected with the contractile mechanism, directly or indirectly. In vertebrate smooth muscle, Ca ion is nothing but the carrier itself of action current and deeply involved in the activation of contractile processes (Ref. 3, 4). This is also the cases of most invertebrate muscles (Ref. 3).

There is no doubt that contractile processes in all kinds of muscle are regulated by Ca ion, i.e., the contractile state is a rather simple function of the intracellular Ca ion concentration. This principle may be applicable to different kinds of intracellular movement in non-muscle tissues. Furthermore, there is an indication that the intracellular Ca ion regulates the excitability of surface membrane, not only of muscle but also of nerve cells (Ref. 3, 5).*

In view of the situation above, it is impossible to review every aspect of the functions of Ca ion. Therefore, I would like to confine the subjects in this article to a few topics, with which we are concerned for the present time.

---

* It is interesting that Ca spikes and Ca regulation of the excitability were both first established in barnacle muscle (Ref. 3). According to the embryological studies on tunicate (Ref. 6), both Ca and Na channels exist at early stage of development, one of them disappearing during its growth.

REGULATORY SYSTEM IN VERTEBRATE SMOOTH MUSCLE

Brief Historical Sketch

The discovery of the troponin-tropomyosin system in skeletal and cardiac muscle by Ebashi and his colleagues (cf. Ref. 1, 7) had given an impression that the system might be common to all kinds of muscle. However, the subsequent discovery of the myosin-linked system in the scallop striated muscle by Kendrick-Jones et al. (8) has not only dispelled such a belief but also suggested the diversity in the regulatory mechanism in the animal kingdom. Indeed, the regulatory system of vertebrate smooth muscle, though its real mechanism is still the subject of furious controversy (cf. Ref. 9), has been shown to be fundamentally different from those of vertebrate striated muscels and scallop striated muscle.

Soon after the discovery of troponin, Ebashi et al. (10) succeeded in isolating a protein fraction similar to native tropomyosin of skeletal muscle, viz., the troponin-tropomyosin complex of skeletal muscle. This almost convinced them that the regulatory mechanism in smooth muscle in principle should not be different from that in skeletal muscle. This belief was first challenged by Bremel (11) based on the 'competition test' that the vertebrate smooth muscle should be myosin-linked type like scallop muscle. Stimulated by this experiment, we have restarted our investigation on the regulatory mechanism in smooth muscle. In the following, the work carried out in our laboratory by Y. Nonomura, T. Mikawa, T. Toyo-oka, M. Hirata and K. Saida (12-19) in collaboration with Prof. S. Kakiuchi, Osaka University, will be reviewed.

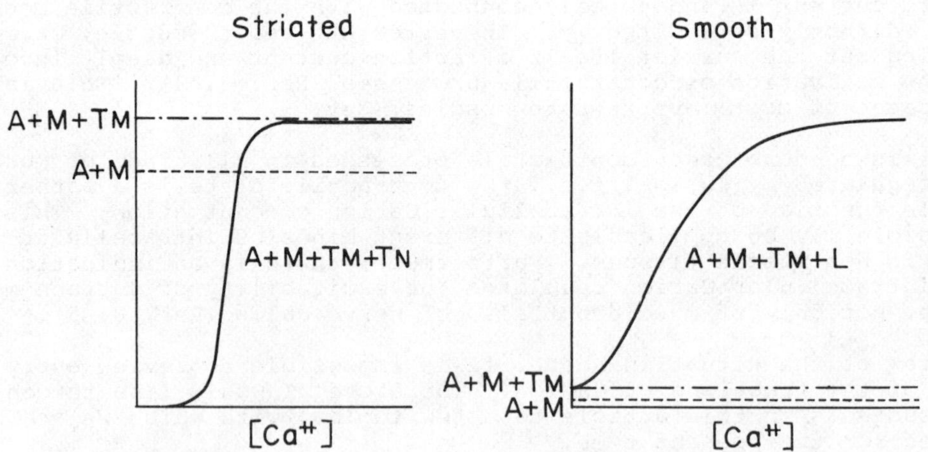

Fig. 1. Schematic Illustration of Myosin-Actin-ATP Interaction of Vertebrate Muscles as a Function of the Free Ca Ion Concentration. A: actin. M: myosin. T_M: tropomyosin. T_N: troponin. L: leiotonin containing acidic protein. Myosin-linked system in scallop muscle shows essentially the same pattern as vertebrate striated muscle (Ref. 17).

Regulatory Proteins of Vertebrate Smooth Muscle

The myosin-actin-ATP interaction in smooth muscle is substantially different from that in striated muscle (Fig. 1). Essential response to MgATP of striated muscle actomyosin with or without tropomyosin at relatively low ionic strength is contraction; the relaxation, i.e., dissociation into actin and myosin, is observable only in the presence of troponin. In sharp contrast with this, the contractile system of smooth muscle without regulatory proteins, or native tropomyosin-like fraction, was not activated by MgATP; tropomyosin did not restore its contractility, so that another factor(s) contained in the native tropomyosin-like fraction must be responsible for its contractility.

From the native tropomyosin-like fraction of gizzard, i.e., an ammonium sulfate fraction of the extract from its muscle mince at a low ionic strength, were isolated two active components, i.e., tropomyosin and a new protein of 80,000 dalton in its molecular weight, named leiotonin.

Leiotonin has distinct properties from troponin. Though it requires collaboration of tropomyosin for its function, it has a fairly strong affinity for actin but scarcely for tropomyosin. Leiotonin is fully effective at a molar ratio to tropomyosin of about one tenth or less. Furthermore, the mode of action of leiotonin is entirely different from that of troponin as suggested above. The myosin-actin-tropomyosin system is activated only in the presence of leiotonin and Ca ion. In other words, the regulatory protein in smooth muscle is a real activator.

At the time when we succeeded in separating leiotonin, we believed that the essential components for contraction of smooth muscle, would be four proteins, viz., myosin, actin, tropomyosin and leiotonin. However, encouraged by the work to be stated in the next subsection, we further analyzed leiotonin preparations and also purified myosin and actin preparations carefully. As a result, we have found that an acidic protein of a lower molecular weight, about 17,000 dalton, is also required for full activation of contractile processes; without this protein, no contractile response could take place (Ref. 20)*.

This acidic protein has a different mobility on the SDS-electrophoresis of Tris-glycine slab gel and a different amino acid composition from that of either troponin C or brain modulator protein. Unlike modulator protein, it is not easily detached from its parent leiotonin molecule. Troponin C cannot replace the regulatory action of this acidic protein, but modulator protein is a good substitute for it. On the other hand, the acidic protein as well as troponin C hardly replace the role of modulator protein in activating phosphodiesterase.

---

* The acidic protein is partially detached from the parent leiotonin molecule during preparation at the step of DEAE-cellulose chromatography (Ref. 15). We now understand reasons why the activity of leiotonin was variable from preparation to preparation and why an acidic fraction separated by DEAE-Sephadex chromatography markedly activated partially desensitized gizzard myosin B.

The position of the acidic protein* resembles troponin C in its relation to leiotonin; it might be plausible to call the latter leiotonin A, whereas the former leiotonin C.

The Role of Phosphorylation of Myosin Light Chain

In 1974 Perry and his colleagues (21) have found that the enzyme which phosphorylates the myosin light chain of skeletal muscle is dependent on Ca ion in a similar manner to the dependence of contractile processes on Ca ion. Bremel, Sobieszek and Small (22) have first introduced the enzyme into smooth muscle studies i.e., the phosphorylation of myosin light chain is the mechanism underlying smooth muscle contraction; the relaxation should be the result of dephosphorylation by a phosphatase specific for the light chain. Hartshorne and his colleagues (24) have subsequently shown that the "native tropomyosin" preparation of Ebashi et al. (10) contains a lot of the Ca-dependent protein kinase and concluded that the physiological function of the fraction is solely explained by the kinase. Chacko and Adelstein and Watanabe and his colleagues have also presented similar ideas (25, 26).

We certainly agree with their opinion in one respect that the gizzard "native tropomyosin" fraction contains a considerable amount of the kinase. However, we cannot accept the opinion to consider the kinase and phosphatase as the device to regulate smooth muscle contraction because of following reasons:

1) In the presence of abundant amount of phosphatase prepared from gizzard, no phosphorylation of gizzard myosin light chain takes place, but ordinary contractile responses, i.e., the superprecipitation of gizzard myosin B in the presence of Ca, can occur (Ref. 16).
2) If the kinase is present in an excess amount, the light chain can be phosphorylated to a great extent even in the absence of Ca ion, but no superprecipitation takes place under this condition (Ref. 17).
3) Leiotonin has no Ca-dependent kinase activity (Fig. 2; Ref. 19).
4) Actomyosin fully phosphorylated by leiotonin-free kinase, does not show any increase in actin-activated ATPase activity of smooth muscle actomyosin (Fig. 2; Ref. 19).
5) Even if myosin in chemically skinned fibers of taenia caecum is replaced by gizzard myosin, the fibers show essentially the same contraction-relaxation cycle depending on Ca ion concentrations; since the procedure to replace myosin should remove most phosphatase (and also a large part of kinase) from the skinned fiber, this result indicates that phosphatase is not involved in relaxation.

---

* A protein very similar to or almost identical with brain modulator protein can be isolated from whole gizzard. This protein cannot be found in myosin B, native-tropomyosin-like fraction, or leiotonin. Thus the acidic protein related to leiotonin can be discriminated from modulator protein in this respect, too. The relation of the acidic protein to troponin C-like protein from gizzard (Ref. 23) is not yet clear.

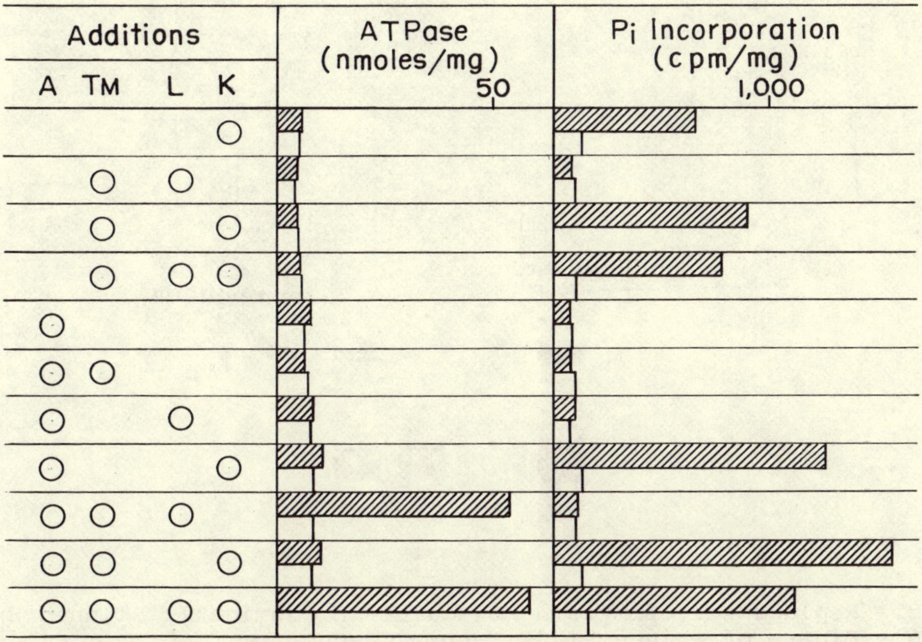

Fig. 2. The ATPase Activity and the Degree of Phosphorylation of Gizzard Myosin in Different Conditions.

▨: 1.3×10^{-5}M Ca^{2+}. □: 1×10^{-7}M Ca^{2+}. K: Ca dependent light chain kinase. For other designations, refer to the legend to Fig. 1.
Reaction mixtures contained: myosin, 310 µg/ml; actin, 250 µg/ml; tropomyosin, 50 µg/ml; leiotonin, 10 µg/ml; kinase, 10 µg/ml; KCl, 50 mM; $MgCl_2$, 8 mM; Tris-maleate, 20 mM (pH 6.8); ATP, 1 mM (in the case of incorporation experiments, [γ-^{32}P]-ATP was added). Temperature 25°C. The reaction was started by adding ATP and after 5 min it was terminated by adding trichloroacetic acid (Originated from Fig. 6 in Ref. 17, but revised for this article).

6) If rabbit skeletal myosin is added to chemically skinned taenia caecum, from which its own myosin had been removed by the same procedure as above, the contraction-relaxation cycle can be induced by changing Ca ion concentration (Fig. 3); this indicates that the regulatory site resides in the thin filament side (the condition used to remove myosin from chemically skinned fibers does not delete tropomyosin from the thin filament (W. Drabikowski and his colleagues, personal communication)).
7) Intact as well as chemically skinned taenia caecum does not show any phosphorylation of light chain during contraction as examined by autoradiograph technique (Fig. 4).
8) The sensitivity to Sr ion of light chain kinase is several fold higher than that of the contractile response; at 1×10^{-4}M, no superprecipitation takes place in spite of full phosphorylation.

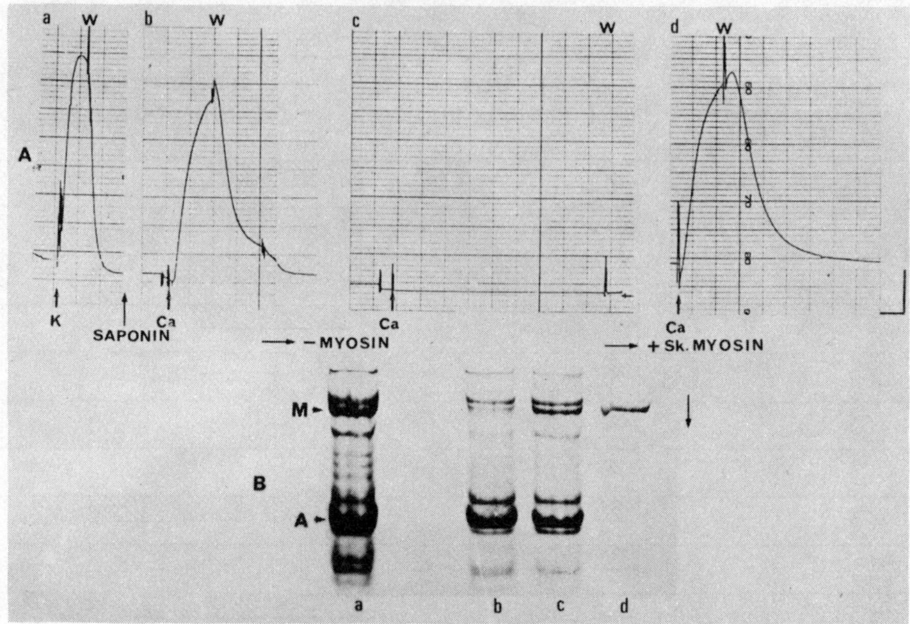

Fig. 3. Replacement of Myosin Molecules in Chemically Skinned Smooth Muscle Fibers of Guinea Pig´s Taenia Caecum.

A. Isometric tension recording. Inserted scales; 10 mg and 1 minute. a: K^+-induced tension of intact fiber. b: Ca^{2+}-induced tension (Ca^{2+} 10^{-5}M) of saponin-treated skinned fibers. c: no tension development with Ca^{2+} (Ca^{2+} 10^{-5}M) after removal of myosin molecules. d: Ca^{2+}-induced tension (Ca^{2+} 10^{-5}M) of irrigated skinned fibers with skeletal muscle myosin.
B. SDS-slab electrophoretic patterns of various states of fibers. a: intact fibers. b: chemichally skinned fibers after removal of myosin. c: irrigated skinned fibers with skeletal muscle myosin. d: skeletal muscle myosin as a marker. (Saida, Mikawa & Nonomura, unpublished data)

9) The profiles of pH-activity curves for the actin-activated ATPase and the phosphorylation of light chain are markedly different from each other.

In the meantime, the essential nature of modulator protein for the light chain kinase activity was noticed first by Yagi (27) and confirmed by several other groups. Hartshorne´s group (28) has shown that the phosphorylation of gizzard myosin and the activation of its actomyosin ATPase by their purified kinase preparation depends on the modulator protein. Hidaka and his colleagues (29) have reported an activating effect of modulator protein on superprecipitation of desensitized vascular actomyosin. As stated in the above, these results can be explained if their kinase or actomyosin system is contaminated by leiotonin from which the acidic protein had been removed.

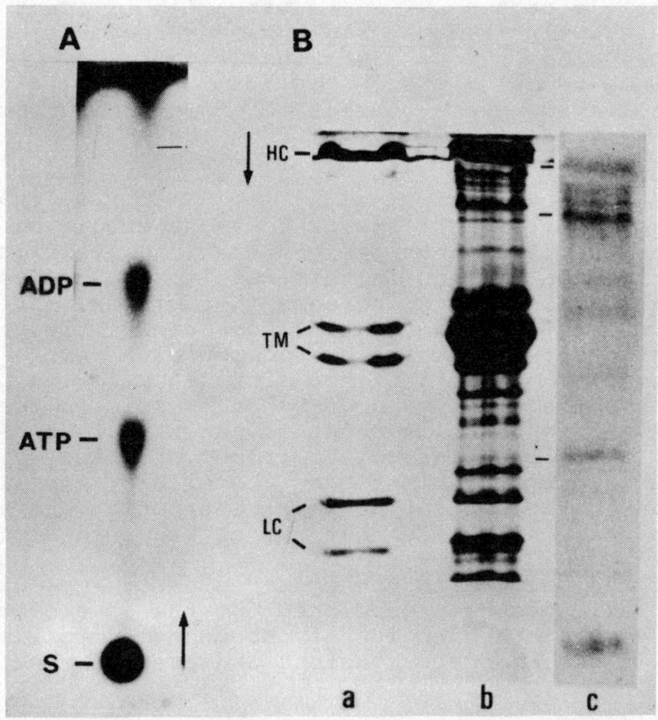

Fig. 4. Autoradiography of Thin Layer Chromatography and SDS-Slab-Gel of Chicken Gizzard.

The tissue incubated with ^{32}P-phosphate was fixed at contracted state, homogenized and then separated into supernatant and precipitate after centrifugation. The incorporation of ^{32}P-phosphate to ATP and ADP was detected by autoradiography of the thin layer chromatograph of the supernatant (A). The transfer of ^{32}P from ^{32}P-ATP to various protein components was shown in SDS-slab electrophoretic patterns of the precipitate (B-b) and their autoradiography (B-c). As it is clearly shown from the pattern of marker proteins (B-a), light chains were not phosphorylated under the conditions. (Mikawa, Nonomura & Ebashi, unpublished data)

Events in Vascular Smooth Muscle

The observations made with gizzard muscle were also confirmed with bovine stomach. The content of Ca dependent kinase in bovine stomach is much less than that in gizzard; we can rather easily remove the kinase from stomach myosin B and the resulting kinase-free myosin B can be activated by ATP in the presence of Ca ion.

The activation of vascular smooth muscle is also dependent on leiotonin, but its relation to actin is somewhat different from the case of gizzard. Aorta leiotonin bound to the actin filament cannot be detached unless actin is denatured by some procedure; pure actin has so far been prepared only by digesting leiotonin by chymotrypsin (Ref. 18). Thus vascular muscle provides pertinent evidence for the thin filament-linked nature of regulation in smooth muscle.

General Comments

It was our tacit belief that skeletal muscle should be the most differentiated muscle so that the penetrating investigation into the mechanism of contraction in this muscle should reveal every secret of contraction. With increase in the knowledge of smooth muscle, however, we now realize that some remarkable properties of skeletal contractile system are not common with smooth muscle contractile system. The most distinct difference is the poor interaction of pure actin and myosin of smooth muscle in the presence of ATP; virtually no contractile response takes place unless provided with the regulatory proteins, viz., tropomyosin and leiotonin including the acidic protein mentioned above (Fig. 1). In other words, the difference in the concept between contraction and regulation is very clear in skeletal muscle, but rather vague in smooth muscle.

We are now of the opinion that the de-repressor type of regulation in skeletal muscle has been developed for the sake of rapidness in contraction; perhaps the precise arrangement of myofilaments in skeletal muscle is designed for constant and precise repetition of contraction.

BRIEF NOTES ON TROPONIN

The Troponin Mechanism

All the arguments on the mechanism how troponin regulates the interaction of actin-filaments with myosin filaments have been based on the model for the thin filament* (Ref. 7; revised by I. Ohtsuki, 30, 31) and the depressive nature of the troponin action (Ref. 1, 7). The ingeneous hypothesis proposed by Huxley (33, 35) and Haselgrove (34), i.e., the dislocation of tropomyosin in the groove of actin strands induced by troponin in the absence of Ca ion would block the myosin-binding site on actin, has still to await further confirmation. Oosawa (36) has proposed an elegant concept that the flexibility of actin filament is a *sine qua non* for contractility and the function of troponin in the absence of Ca ion is to stiffen the actin filament. He and his colleagues have presented several pieces of reasonable evidence for this, but more direct evidence to convince those who would not accept this idea may be required.

---

* Keen attention is now paid to fine structural arrangement of troponin subunits. Ohtsuki´s proposal that troponin T should have wider distribution, about 100 Å, along the axis of the actin filament (Ref. 32) has been confirmed by the immunoelectron microscope using subfragments of troponin T and in good accord with the results of the supersecondary structure prediction by K. Nagano (unpublished).

Final solution of this problem may be obtained by the clarification of much finer structure of the thin filament, viz., the troponin-tropomyosin-actin complex, at the level of single amino acid resolution. Since the amino acid sequences of all these proteins have been clarified, it is not impossible to build up such a fine model as may unify structural and functional aspects.

Modulator Protein and Troponin C

Modulator protein, independently found by Kakiuchi (37) and Cheung (38) in 1970, can replace the role of troponin C to a certain extent. Modulator protein firmly binds to troponin I in the presence of Ca ion but completely detaches from the latter in the absence of Ca ion, so that modulator and troponin I can carry out a full regulatory function (Ref. 39) (the presence of troponin T does not change the situation). However, since troponin C still attaches to troponin I in the absence Ca ion, the complex of troponin C and troponin I cannot be the regulatory system and must require the collaboration of troponin T. Thus the role of troponin T has become somewhat clearer by the behavior of modulator protein.

A question then arises why the troponin system assumes such a complicated system, instead of much simpler form like the modulator-troponin I system. A similar question will be addressed to smooth muscle researchers why the leiotonin system does not utilize modulator protein but is provided with a specific acidic protein. Perhaps, a part of answer resides in the non-specific nature of modulator protein.

Troponin in Invertebrate and Tunicate Muscles

Not only the shellfish muscle but all molluscan muscles are equipped with myosin-linked Ca binding sites for regulation (Ref. 40). This does not mean that molluscan muscles are free of troponin. A considerable amount of troponin has been found in squid mantle muscle (Ref. 41, 42); furthermore, a definite amount of troponin is present even in scallop muscle (Ref. 43). In insect and crustacean muscles, the troponin system is far more important than in molluscan muscles (Ref. 44, 45; in our hands, the myosin-linked system in crustacean muscle is not so clear).

Obinata (46) has shown that the regulation in ascidian smooth muscle is mainly carried out by the troponin system, but this troponin system is not a pure depressor type but partly activating-type like leiotonin. This may present a new phylogenetic problem of the regulatory mechanism.

Phosphorylation of Troponin

Perry and his colleagues have worked out the phosphorylation of troponin subunits (47, 48). Since the extent of phosphorylation of a subunit is greatly influenced by other subunits, phosphorylation is a useful tool for the study of interactions among troponin subunits. Owing to this research, fine structural arrangement of troponin subunits have been clarified to a great extent.

From physiological point of view, however, most phosphorylating reactions appear to have no particular importance. Only exception is the cyclic AMP dependent phosphorylation of serine 20 of cardiac troponin I (Ref. 49). This part is an additional to the skeletal troponin I, so that it is quite plausible that this phosphorylation would influence cardiac function. Indeed, the phosphorylation of this part reduces the sensitivity of actomyosin to Ca ion in *in vitro* system (Ref. 50; also our unpublished result). Its physiological significance, however, is still not clear.

SOME PROBLEMS IN EXCITATION-CONTRACTION COUPLING

The establishment of the role of Ca ion in the contractile processes together with the discovery of ATP-dependent Ca uptake of the sarcoplasmic reticulum (SR) (cf. Ref. 1) has enabled us to depict the whole picture of excitation-contraction coupling (E.C coupling). This has given us an impression that almost everything in this field would have been solved. However, the most important step in E.C. coupling, i.e., the link between the electrical phenomenon in the T-tubule and the Ca release from the SR has not yet been solved at all. Even the fact that the maximum rate of Ca transport into the SR cannot explain the rate of muscle relaxation is not well recognized even by the experts working on the SR (cf. Ref. 51).

Coupling Between T-system and SR

Endo (52) has thoroughly reviewed the present status of the investigation into the physiological mechanism of Ca release from the SR. He excluded the possibility that 'Ca-induced Ca release*, i.e., the mechanism found by himself (53) as well as Podolsky´s group (54), should play crucial role in physiological contraction of skeletal muscle. According to Endo (55), 'depolarization' induced Ca release from the SR is worthy of further consideration.

The 'depolarization' induced Ca release** may assume the presence of a kind of potential difference across the SR membrane. However, the recent observation by Somlyo et al. (56) using X ray micro probe analysis has shown no difference in Na and Cl concentration between the cytoplasm and SR lumen, indicating virtual absence of electrical potential across the SR membrane. This conclusion is rather difficult to reconcile with the observation of Bezanilla (57) with fluorescent dye that the SR exhibit a marked signal indicating the presence of potential change across the SR membrane during contraction and, apparently, does not favor Endo´s concept as to 'depolarization' induced Ca release.

---

* The SR is apt to release its Ca if the Ca ion concentration outside the SR is elevated. This tendency is most marked in cardiac muscle SR, next in slow red muscle SR and much less pronounced in fast white muscle SR.

** Since it is impossible to apply an effective electrical stimulation on the SR membrane because of its minute size, Endo induced the electrical potential difference by changing the composition of electrolytes in the outer medium (cf. Ref. 55). The term 'depolarization' means the change which makes the inside of the SR more negative to the outside.

Although seemingly unlikely, it is not impossible that, in spite of the absence of membrane potential, the SR would be sensitive to the electrical influence which makes the inside of the SR membrane negative to the outside (this was actually the conditions in which Endo made his 'depolarization' experiments). In any case the depolarization in the T-tubule must be mediated to the SR. The discovery of a specific charge movement across the T-tubule membrane (Ref. 58, 59) appeared to give a solution to this problem, but so far it has not yet been directly related to the real Ca release mechanism yet.

Ca Store in Cardiac Muscle

It has been a subject of controversy whether Ca entering cytoplasm from outer medium is the activator itself of contraction in cardiac muscle. If we take the position that most of Ca to induce contraction is derived from the Ca store inside the cardiac cells, the role of entering Ca can be most plausibly explained by 'Ca induced Ca release' mechanism. Fabiato and Fabiato (60) have presented good evidence, though semiquantitative, for this idea by using mechanically skinned cardiac fibers.

The loss of contractility of cardiac muscle incubated in a Ca^{2+} free solution have often been explained to be due to the loss of Ca from its store inside the cell; gradual restoration of contractility after adding Ca ion to the medium was considered to indicate the time course of refilling of the Ca store. However, Endo and Kitazawa (61)* have shown that this is not the case. The cardiac fiber bundle of a diameter about 100 μm, incubated in Ca deficient medium, almost instantaneously restores its previous contractility on addition of Ca ion; results obtained by other workers may be explained by the difficulty in diffusion due to the large diameter of fiber bundle. It has also been shown by Sakai and Kurihara (62) that even after a long incubation in the solution containing Mn and EGTA, nearly maximum contraction can be obtained by rapid cooling. Although the identification of the source of Ca in this case with the physiological Ca store has not yet been made, this is in good accord with the above experiment of Endo and Kitazawa.

Ca Dependent Phosphorylation of Myosin Light Chain

Phosphorylase b kinase, the first enzyme shown to be Ca dependent, may be considered as a model enzyme of excitation-metabolism coupling (Ref. 63, 64). If Ca dependent light chain kinase could show a definite influence on contraction, it should present a new aspect of excitation-contraction coupling. Indeed, phosphorylation takes place during the contraction of intact skeletal fibers (Ref. 65)

---

* This result may also be considered as favourable for the idea that Ca ion from the outer medium is the main fraction of Ca ion to activate the contractile system. Although circumstancial evidence does not favor it, we should not expel the idea until it is finally disproved.

unlike the case of smooth muscle as stated in a previous section*. Unfortunately, however, the phosphorylation of skeletal and cardiac light chains does not show a marked effect on the actin-activated myosin ATPase and superprecipitation (S.V. Perry, personal communication; also our unpublished results).

It is strange that such an enzyme system as related to Ca-dependent light chain kinase cannot find a definite physiological function. One of the reasons for this failure is that we have not yet found appropriate way to examine the physiological significance of the phosphorylation. Another possibility may be that the phosphorylation may not markedly influence the physiological properties of myosin, but is crucial for the metabolic fate of the light chain and related proteins. Further studies based on a broad view of things is urgently required.

ADDITIONAL REMARKS

According to a story, Ca ion had been poisonous to the living organisms in very primitive age, as mercury is still so at present.

TABLE 1 Indispensable Factors for Cell Motility: Mode of Action of Ca Ion in Different Kinds of Cell

<table>
<tr><th></th><th>Pattern of movement</th><th>Ca^{2+}</th><th>ATP</th><th>Metabolic substrate</th><th>Ref.</th></tr>
<tr><td rowspan="2">Muscle (actomyosin)</td><td>molecular</td><td>yes or no[1)]</td><td rowspan="2">yes</td><td rowspan="2">no</td><td rowspan="2"></td></tr>
<tr><td>physiological</td><td>yes</td></tr>
<tr><td rowspan="2">Vorticella (spasmoneme)</td><td>molecular</td><td rowspan="2">yes</td><td rowspan="2">no[2)]</td><td rowspan="2">no</td><td rowspan="2">68,69</td></tr>
<tr><td>physiological</td></tr>
<tr><td rowspan="2">Flagella & cilia (tubulin-dynein)</td><td>forward</td><td>no</td><td rowspan="2">yes</td><td rowspan="2">no</td><td rowspan="2">67</td></tr>
<tr><td>reverse</td><td>yes</td></tr>
<tr><td rowspan="2">Bacterial flagella</td><td>smooth</td><td rowspan="2">no</td><td>no</td><td>yes</td><td rowspan="2">70</td></tr>
<tr><td>tumbling</td><td>yes</td><td>no</td></tr>
</table>

1) 'yes' in vertebrate smooth muscle, but 'no' in striated muscles.
2) required for expelling Ca^{2+} ion from the cell.

---

* This discrepancy can be explained by relatively high content of phosphatase in smooth muscle; the content of the kinase of rabbit skeletal muscle is about 10 times at that of gizzard, but that of phosphatase of skeletal muscle is much the same as that of gizzard (Ref. 66). The content of the kinase in cardiac muscle is very low (Ref. 50).

However, the organism became soon tolerant of it, and finally have learned how to use it.

In Table 1, the way of utilizing Ca ion in muscle and other cells is summarized. Noteworthy is the behavior of vorticella, which makes use of Ca ion in a very explicit manner.

The way how the living organism controls Ca ion as a useful tool might be an important aspect of evolution.

ACKNOWLEDGEMENT

The work in our laboratory quoted in this article was supported in part by the grants from the Muscular Dystrophy Association, Inc., the Ministry of Education, Science and Culture of Japan, the Ministry of Health and Welfare, Japan, and the Iatrochemical Foundation.

REFERENCES

(1) S. Ebashi and M. Endo, Calcium ion and muscle contraction, Progr. Biophys. Mol. Biol. 18, 123-183 (1968).

(2) S. Ebashi, Ca ion as a basis of pharmacological actions, Proc. of IV Internat. Congr. Pharmacol. 1969, vol. I, p 32-55, Schwabe, Basel (1970).

(3) S. Hagiwara, Ca spike, Adv. Biophys. 4, 71-102 (1973).

(4) Y. Nonomura, Y. Hotta and H. Ohashi, Tetrodotoxin and manganese ion: effects on electrical activity and tension in taenia coli of guinea pig, Science 152, 97-99 (1969).

(5) R.W. Meech, The sensitivity of helix aspersa neurones to injected calcium ions, J. Physiol. 237, 259-277 (1974).

(6) H. Okamoto, K. Takahashi and M. Yoshii, Membrane currents of the tunicate egg under the voltage clamp condition, J. Physiol. 254, 607-638 (1976).

(7) S. Ebashi, M. Endo and I. Ohtsuki, Control of muscle contraction, Quart. Rev. Biophys. 2, 351-384 (1969).

(8) J. Kendrick-Jones, W. Lehman and A.G. Szent-Györgyi, Regulation in molluscan muscles, J. Mol. Biol. 54, 313-326 (1970).

(9) R. Casteels, T. Godfraind and J.D. Rüegg (1977), Excitation Contraction Coupling in Smooth Muscle, Elsevier/North-Holland, Amsterdam.

(10) S. Ebashi, H. Iwakura, H. Nakajima, R. Nakamura and Y. Ooi, New structural proteins from dog heart and chicken gizzard, Biochem. Z. 345, 201-211 (1966).

(11) R.D. Bremel, Myosin-linked calcium regulation in vertebrate smooth muscle, Nature 252, 405-407 (1974).

(12) S. Ebashi, T. Toyo-oka and Y. Nonomura, Gizzard troponin, J. Biochem. 78, 859-861 (1975).

(13) S. Ebashi, A simple method of preparing actin-free myosin from smooth muscle, J. Biochem. 79, 229-231 (1976).

(14) S. Ebashi, Y. Nonomura, T. Toyo-oka and E. Katayama, Regulation of muscle contraction by the calcium-troponin-tropomyosin system, Calcium in Biological Systems, Symposium XXX p. 349-360, Cambridge Univ. Press, London (1976).

(15) T. Mikawa, T. Toyo-oka, Y. Nonomura and S. Ebashi, Essential factor of gizzard 'Troponin' fraction, J. Biochem. 81, 273-275 (1977).

(16) T. Mikawa, Y. Nonomura and S. Ebashi, Does phosphorylation of myosin light chain have direct relation to regulation in smooth muscle? J. Biochem. 82, 1789-1791 (1977).

(17) S. Ebashi, T. Mikawa, M. Hirata and Y. Nonomura, The regulatory role of calcium in muscle, In 'Calcium Transport and Cell Function', Ann. N.Y. Acad. Sci. vol. 307, p. 307-321 (1978).

(18) M. Hirata, T. Mikawa, Y. Nonomura and S. Ebashi, Ca^{2+} Regulation in vascular smooth muscle, J. Biochem. 82, 1793-1796 (1977)

(19) S. Ebashi, T. Mikawa, M. Hirata, T. Toyo-oka and Y. Nonomura, Regulatory proteins in smooth muscle, In Ref. 9, p 325-334 (1977).

(20) T. Mikawa, Y. Nonomura, M. Hirata, s. Ebashi and S. Kakiuchi, Involvement of an acidic protein in the regulation of smooth muscle contraction by the leiotonin-tropomyosin system, J. Biochem. submitted.

(21) E. Pires, S.V. Perry and M.A.W. Thomas, Myosin light chain kinase, a new enzyme from striated muscle, FEBS lett. 41, 292-296 (1974).

(22) R.D. Bremel, A. Sobieszek and J.V. Small, Regulation of actin-myosin interaction in vertebrate smooth muscle, In The Biochemistry of Smooth Muscle, ed. N.L. Stephens, p. 533-549, Univ. Park Press, Baltimore (1977). This publication is a report of the meeting held in 1975.

(23) R.J.A. Grand, S.V. Perry and R.A. Weeks, The troponin-like components of smooth muscle, In Ref. 9, p. 335-341 (1977).

(24) M.O. Aksoy, D. Williams, E.M. Sharkey and D.J. Hartshorne, Relationship between Ca^{2+} sensitivity and phosphorylation of gizzard actomyosin, Biochem. Biophys. Res. Commun. 69, 35-41 (1976).

(25) S. Chacko, M.A. Conti and R.S. Adelstein, Effect of phosphorylation of smooth muscle myosin on actin activation and Ca^{2+} regulation, Proc. Natl. Acad. Sci. U.S.A. 74, 129-133 (1977).

(26) M. Ikebe, T. Aiba, H. Onishi and S. Watanabe, Calcium sensitivity of contractile proteins from chicken gizzard muscle, J. Biochem. 83, 1643-1656 (1978).

(27) Y. Yagi and M. Yazawa, Identification of an activator protein for myosin light chain kinase as the Ca^{2+}-dependent modulator protein, J. Biol. Chem. 1338-1340 (1978).

(28) R. Dabrowska, J.M.F. Sherry, D.K. Aromatorio and D.J. Hartshorne, Modulator protein as a comopnent of the myosin light chain kinase from chicken gizzard, Biochemistry 17, 253-257 (1978).

(29) H. Hidaka, T. Yamaki, H. Inoue,M. Asano and T. Totsuka, Effect of vascular relaxing agents on superprecipitation of vascular actomyosin, Jap. J. Pharmacol. 28, Suppl. (1978) in press.

(30) S. Ebashi, Calcium ions and muscle contraction, Nature 240, 2, 7-218 (1972).

(31) S. Ebashi, Regulatory mechanism of muscle contraction with special reference to the Ca-troponin-tropomyosin system, Essays in Biochemistry, 1-30 (1974).

(32) S. Ebashi, I. Ohtsuki, R. Tsukui and T. Fujii, The role of regulatory proteins in contractile mechanism, In Exploratory Concepts in Muscular Dystrophy II p. 419-429, Excer. Med., Amsterdam (1973).

(33) H.E. Huxley, Structural changes in the actin and myosin containing filaments during contraction, Cold Spring Harbor Symp. Quant. Biol. 37, 361-376 (1972).

(34) H.C. Haselgrove, X-ray evidence for a conformational change in the actin containing filaments of vertebrate striated msucle, Cold Spring Harbor Symp. Quant. Biol. 75, 33-35 (1972).

(35) T. Wakabayashi, H.E. Huxley, L.A. Amos and A. Klug, Three-dimensional image reconstruction of actin-tropomyosin complex and actin-tropomyosin-troponin T-troponin I complex, J. Mol. Biol. 93, 477-497 (1975).

(36) F. Oosawa and S. Asakura (1975), Thermodynamics of the polymerization of protein, Academic Press.

(37) S. Kakiuchi, R. Yamazaki and H. Nakajima, Properties of a heat-stable phosphodiesterase activating factor isolated from brain extract, Proc. Japan Acad. 46, 587-592 (1970).

(38) W.Y. Cheung, Cyclic 3´,5´-Nucreotide phosphodiesterase: Demonstration of an activator, Biochem. Biophys. Res. Commun. 38, 533-538 (1970).

(39) G.W. Amphlett, T.C. Vanaman and S.V. Perry, Effect of the troponin C-like protein from bovine brain (Brain Modulator Protein) on the Mg^{2+}-stimulated ATPase of skeletal muscle actomyosin, FEBS lett. 72, 163-168 (1977).

(40) W. Lehman, J. Kendrick-Jones and A.G. Szent-Györgyi, Myosin-linked regulatory systems: Comparative studies, Cold Spring Harbor Symp. 37, 319-330 (1972).

(41) T. Tsuchiya, T. Kaneko and J.J. Matsumoto, Calcium sensitivity of mantle muscle of squid, J. Biochem. 83, 1191-1193 (1978)

(42) K. Konno, Two regulatory system in squid mantle muscle, J. Biochem. in press.

(43) A. Goldberg and W. Lehman, Troponin-like proteins from muscles of the scallop, Aequipecten irradians, Biochem. J. 171, 413-418 (1978).

(44) W. Lehman, B. Bullard and K. Hammond, Calcium-dependent myosin from insect flight muscles, J. Gen. Physiol. 63, 553-563 (1974).

(45) W. Lehman, Calcium ion-dependent myosin from decapod-crustacean muscles, Biochem. J. 163, 291-296 (1977).

(46) T. Obinata, Two components from ascidian smooth muscle, J. Biochem. submitted.

(47) S.V. Perry and H.A. Cole, Phosphorylation of troponin and the effects of interactions between the components of the complex, Biochem. J. 141, 733-743 (1974).

(48) H.A. Cole and S.V. Perry, The phosphorylation of troponin I from cardiac muscle, Biochem. J. 149, 525-533 (1975).

(49) R.J.A. Grand, J.M. Wilkinson and L.E. Mole, The amino acid sequence of rabbit cardiac troponin I, Biochem. J. 159, 633-641 (1976).

(50) H.A. Cole, N. Freason, A.J.G. Moir, S.V. Perry and R.J. Solaro, Phosphorylation of cardiac myofibrillar proteins, in 'Recent Adv. Stud. Card. Struct. Metab.' Eds., T. Kobayashi, T. Sano and N.S. Dhalla, vol. 11, p. 111-120, Univ. Park Press, Baltimore (1978).

(51) S. Ebashi, Excitation-contraction on coupling, Ann. Rev. Physiol. 38, 293-313 (1976)

(52) M. Endo, Calcium release from the sarcoplasmic reticulum, Physiol. Rev. 57, 71-108 (1977).

(53) M. Endo, M. Tanaka and Y. Ogawa, Calcium induced release of calcium from the sarcoplasmic reticulum of skinned skeletal muscle fibers, Nature 228, 34-36 (1970).

(54) L.E. Ford and R.J. Podolsky, Regenerative calcium release within muscle cells, Science 167, 58-59 (1970).

(55) M. Endo and Y. Nakajima, Release of calcium induced by "Depolarisation" of the sarcoplasmic reticulum membrane, Nature New Biol. 246, 216-218 (1973).

(56) A.V. Somlyo, H. Shuman and A.P. Somlyo, Composition of sarcoplasmic reticulum in situ by electron probe X-ray microanalysis, Nature 268, 556-558 (1977).

(57) F. Bezanilla and P. Horowicz, Fluorescence intensity changes associated with contractile activation in frog muscle stained with nile blue A, J. Physiol. 246, 709-735 (1975).

(58) M.F. Schneider and W.K. Chandler, Voltage dependent charge movement in skeletal muscle; a possible step in excitation-contraction coupling, Nature 242, 244-246 (1973).

(59) W.K. Chandler, F.R. Rakowski and M.F. Schneider, Effects of glycerol treatment and maintained depolarization on charge movement in skeletal muscle, J. Physiol. 254, 285-316 (1976).

(60) A. Fabiato and F. Fabiato, Contractions induced by a calcium-triggered release of calcium from the sarcoplasmic reticulum of single skinned cardiac cells, J. Physiol. 249, 469-495 (1975).

(61) M. Endo and T. Kitazawa, E-C coupling studies on skinned cardiac fibers, Biophysical Aspects of Cardiac Muscle, eds. M. Morod and M. Tabatabai, 1977, in press.

(62) T. Sakai and S. Kurihara, The rapid cooling contraction of toad cardiac muscle, Jap. J. Physiol. 24, 649-666 (1974).

(63) E. Ozawa, K. Hosoi and S. Ebashi, Reversible stimulation of muscle phosphorylase b kinase by low concentrations of calcium ions, J. Biochem. 61, 531-533 (1967).

(64) E. Ozawa, Activation of Muscular Phosphorylase b kinase by a minute amount of Ca ion, J. Biochem. 71, 321-331 (1972).

(65) K. Bárány and M. Bárány, Phosphorylation of the 18,000-dalton light chain of myosin during a single tetanus of frog muscle, J. Biol. Chem. 252, 4752-4754 (1977).

(66) N. Frearson, B.W.W. Focant and S.V. Perry, Phosphorylation of the light chain components of myosin from smooth muscle, FEBS Lett. 63, 27-32 (1976).

(67) Y. Naitoh, Ionic control of the reversal response of cilia in paramecium caudatum: A calcium hypothesis, J. Gen. Physiol. 51, 85-103 (1968).

(68) H. Hoffmann-Berling, Der Mechanismus eines neuen, von der Muskelkontraktion verschiedenen Kontraktionszyklus, Biochim. Biophys. Acta 27, 247-255 (1958).

(69) W.B. Amos, A reversible mechanochemical cycle in the contraction of Vorticella, Nature 229, 127-128 (1971).

(70) S.H. Larsen, J. Adler, J.J. Gargus and R.W. Hogg, Chemomechanical coupling without ATP: The source of energy for motility and chemotaxis in bacteria, Proc. Nat. Acad. Sci. U.S.A. 71, 1239-1243 (1974).

The Sodium Channels in Excitable Membranes

R.D. Keynes

Physiological Laboratory, Cambridge CB2 3EG, England

In 1952, Hodgkin & Huxley (1) put forward a mathematical description of the time and voltage dependence of the ionic permeability changes that underlie conduction of the nervous impulse. This served admirably for their primary objective of reconstructing the conducted action potential from voltage-clamp data, and the accuracy and general applicability of the Hodgkin-Huxley equations have stood the test of time in a most impressive manner. But as they themselves stressed, these equations do not lead directly to unique models for the molecular structure of the ionic gates, because they are compatible with more than one type of physical process. In order to arrive at a better understanding of the internal organization of the ionic channels in electrically excitable membranes, we need to know more about the mode of operation of the gating mechanism than we can expect to learn just from measuring ionic currents.

The voltage-sensitivity of the system implies that it must be a charged structure, and therefore that its opening and closing must be accompanied by a small movement of charge within the membrane, the so-called *gating current*. Hodgkin & Huxley (1) were unable to detect the gating current, because with the techniques available to them it was always swamped by the much larger ionic current. However, the advent of tetrodotoxin (TTX) to block the sodium channels, and tetraethylammonium (TEA) or caesium ions to block the potassium channels, enabled Armstrong & Bezanilla (2) and Keynes & Rojas (3,4) to make successful measurements of the sodium gating current in the squid giant axon. Because the potassium channels open more slowly, and there are probably fewer of them, it has not yet proved possible to measure the potassium gating current. The object of this article is briefly to review the present situation with regard to gating current measurements, and to consider the progress that has been made towards the evolution of a molecular model for the sodium channel by carrying out a parallel investigation of the properties of the ionic and gating currents.

The phenomenon with which we are concerned is basically a displacement current, that is to say a current that flows across a capacitance when the potential is changed stepwise by ΔV. But whereas for a simple capacitance the total transfer of charge varies linearly with ΔV, the integral of the gating current may be expected to display decidedly nonlinear characteristics. Assuming that with the aid of TTX applied externally, and TEA^+ or Cs^+ ions introduced by perfusion, the passage of ionic current across the membrane can be blocked completely or reduced to very small proportions, the technical problem is to separate the asymmetrical component of the displacement current from the appreciably larger symmetrical component. In the earliest studies (3), this was done by adding together the current records for a positive test pulse, consisting of the gating current plus the capacity transient, and for a negative control pulse of identical amplitude, consisting of the capacity transient alone. However, it has since become evident that this 'equal and opposite' procedure is open to the objection that there is liable to be a substantial backward transfer of gating charge during the control pulse, which will distort the leading edge of the records that it yields. Most practitioners now superimpose the control pulse on a negative prepulse that takes the starting potential to -120 or -150 mV, though according to Bullock & Schauf (5) a holding potential of -100 mV suffices to avoid the difficulty in *Myxicola* axons. In order not to damage the membrane by pulsing it too far in the negative direction, the 'divided pulse' procedure introduced by Armstrong & Bezanilla (2) is used, in which each record for a positive test pulse starting from the holding potential is added to the records for, say, five negative control pulses of exactly one fifth the size starting from -150 mV. Recorded in this fashion, the sodium gating current is revealed as rising almost instantaneously to its peak, and then decaying exponentially. Sometimes, as in Fig. 4 of Keynes & Rojas (3), a single exponential fits the traces very closely, but on other occasions there is a palpable second component ten or twenty times slower than the main one, and also a pedestal of ionic leakage current. The origin of the slow component is still a matter for speculation, and all that can be said about it at present is that it generally seems to be least conspicuous when the temperature is low and the holding potential is large.

Keynes & Rojas (3) provided some evidence that, most fortunately for the purpose of recording the gating current, the act of blocking the selectivity filter of the sodium channel with TTX does not interfere with the operation of the gating mechanism. Dr Kimura and I have recently confirmed this (6) by measuring the gating current in an axon whose potassium channels were blocked as usual with caesium, but whose sodium channels were patent, the flow of sodium current being eliminated by taking the membrane potential exactly to E_{Na}, the Nernst potential for Na^+ ions, at which the driving force acting on the sodium is zero. Keeping all other conditions constant, TTX was then added externally. Within the accuracy of the measurements, there was no change in the amount of charge transferred, and the only consistent effect of the TTX was slightly to slow down the gating current relaxation.

The two parameters of the gating current that need to be correlated quantitatively with the properties of the ionic current are the total charge transfer for a given voltage-clamp pulse, which provides information about the steady-state distribution of the gating charge, and the relaxation time constant, which tells us about the kinetics of the system. Plots of the charge distribution against potential (6) show that the foot of the S-shaped curve lies at around -150 mV, while its steepest point is in the neighbourhood of -25 mV. On the assumption that the individual gating particles obey Boltzmann's Law for a partition between two discrete energy levels separated by a single energy barrier, the value of kT/z'e turns out to be close to 19 mV (4), whence each particle has an effective valency z' of 1.3. It should be appreciated that this figure only represents the sum of all the charge carried by side groups, multiplied by the fraction of the electric field through which each one moves, and therefore provides no guidance as to precisely how the mobile charge is disposed within the molecule. Its chief importance lies in the proof that it affords that the opening of the sodium channels involves the cooperation of more than one gating particle. Since the much steeper voltage dependence of the rise in the sodium permeability of the membrane corresponds to kT/z'e = 6.5 mV (4), it must be concluded that each channel incorporates at least 3 voltage-controlled subunits. It will be remembered that Hodgkin & Huxley (1) arrived at the same conclusion on purely kinetic grounds.

Although for small pulses the forward transfer of gating charge at 'on' is roughly equal to the backward transfer at 'off', when the potential returns to the holding level, the ratio Q_{off}/Q_{on} drops well below unity when the system is driven harder. But whereas Q_{off} usually reaches a flat plateau for values of V_p beyond +20 mV, Q_{on} often fails to achieve saturation even at +50 mV. It has become clear from the recent work of Armstrong & Bezanilla (7) and Meves & Vogel (8) that this behaviour is caused by a partial immobilization of the gating particles when inactivation of the sodium channels sets in. Because no one has observed a component of the gating current that displays anything resembling Hodgkin & Huxley's h kinetics, the idea that inactivation is brought about by the independent movement of charged blocking particles must now be abandoned. Instead, it may be suggested that following the interaction of the three gating subunits that opens the channel, there is a further conformational change that closes it again, which is not accompanied by an appreciable charge movement. The subunits in the final inactivated state would only be capable of reverting rather slowly to the resting condition on restoration of the original electric field. But while this explanation is qualitatively plausible, much experimentation remains to be done in order to sort out current disagreements over the kinetic and steady-state relationships between inactivation and immobilization of the gating subunits. The difficult question of the homogeneity of the charge movement that makes up Q_{on} also has to be resolved, even if it can no longer be doubted that the bulk of it is in some way intimately involved in the operation of the sodium gating system. For if the opening of each channel requires the transfer of 4 electronic charges across the membrane, then it follows immediately that a Q_{max} of 34 nCoulombs/cm^2 (4) corresponds to a channel density of just over 500/μm^2. This fits well with Levinson & Meves's (9) determination of the number of TTX binding sites in the squid giant axon, and with other estimates of the number of sodium channels, and leaves little room for contamination of the sodium gating current by other sources of asymmetrical displacement current.

The facts that I have considered so far are best described by picturing the sodium channel as a macromolecule which undergoes a series of transitions between five different conformational states, thus:

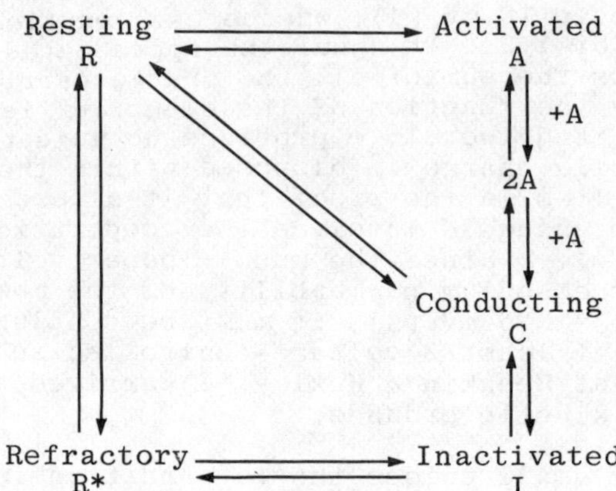

In this diagram, the horizontal steps are postulated to be the ones whose rate constants are strongly voltage-dependent, and which therefore generate most of the gating current. The rate constants of the vertical steps appear to be appreciably less voltage-dependent, rendering them relatively silent in electrical terms.

The idea that an activated state is interposed between the resting and conducting states was originally put forward (4,10) in order to account for the variation in magnitude of the gating current tail time constant with pulse size and duration, for if $\tau_{C \to R}$ is 50% larger than $\tau_{A \to R}$, the observed behaviour of $\tau_{gc,off}$ is readily understood. A critical test of this proposition is to compare τ_{gc} with the sodium conductance time constant τ_m, as defined by Hodgkin & Huxley (1). The values reported by Keynes & Rojas (4) suggested that the two time constants were roughly the same over the whole range of pulse potentials, but an investigation of the kinetics of activation of the sodium channels with improved techniques has now led me to revise this view (6,11). For this purpose, a computerized voltage-clamp system has been employed both to correct the records of the early time course of the rise in the sodium current for the capacity and gating current transients, and to fit the data objectively to the full expansion of the Hodgkin-Huxley equations, and so disentangle the m and h processes. We have also set out to measure τ_m both for opening and for closing the channels, by approaching certain values of V_p from either direction. The results show, in good agreement with both Hodgkin & Huxley (1) and Keynes & Rojas (4), that at 6.3°C τ_m attains a well defined maximum of about 500 μsec in the vicinity of -30 mV, both for positive and negative-going pulses. But although as before τ_{gc} was found to be not very different from τ_m for positive values of V_p, the new measurements indicate that it varies a good deal less steeply with potential, and at the same temperature has a maximum somewhat under 200 μsec at around -10 mV. This reduction in the estimated size of τ_{gc} is largely the consequence of using the computer to fit the sum of two exponentials to all records where a slow component was present, and regarding the fast component as the relevant one, rather than lumping the two together and measuring by eye. The fact that τ_m is two or three times larger than τ_{gc} for small depolarizing pulses seems at first sight to be consistent with the model, if the interaction of the three activated subunits to open the channel requires a finite time for its completion. However, it has yet to be seen whether a system obeying linear kinetics will correctly reproduce all the features of the kinetics and steady-state properties of the sodium channels that are observed experimentally.

REFERENCES

(1) A.L.Hodgkin and A.F.Huxley. A quantitative description of membrane current and its application to conduction and excitation in nerve. *J. Physiol.* 117, 500-554 (1952).

(2) C.M.Armstrong and F.Bezanilla. Charge movement associated with the opening and closing of the activation gates of the Na channels. *J. gen. Physiol.* 63, 533-552 (1974).

(3) R.D.Keynes and E.Rojas. Kinetics and steady-state properties of the charged system controlling sodium conductance in the squid giant axon. *J. Physiol.* 239, 393-434 (1974).

(4) R.D.Keynes and E.Rojas. The temporal and steady-state relationships between activation of the sodium conductance and movement of the gating particles in the squid giant axon. *J. Physiol.* 255, 157-189 (1976).

(5) J.O.Bullock and C.L.Schauf. Combined voltage-clamp and dialysis of *Myxicola* axons: behaviour of membrane asymmetry currents. *J. Physiol.* 278, 309-324 (1978).

(6) R.D.Keynes and J.E.Kimura. (in preparation)

(7) C.M.Armstrong and F.Bezanilla. Inactivation of the sodium channel. II. Gating current experiments. *J. gen. Physiol.* 70, 567-590 (1977).

(8) H.Meves and W.Vogel. Inactivation of the asymmetrical displacement current in giant axons of *Loligo forbesi*. *J. Physiol.* 267, 377-393 (1977).

(9) S.R.Levinson and H.Meves. The binding of tritiated tetrodotoxin to squid giant axons. *Phil. Trans. R. Soc.* B 270, 349-352 (1975).

(10) R.D.Keynes. Organization of the ionic channels in nerve membranes. In *The Nervous System*, vol. 1, *The Basic Neurosciences*, ed. D.B.Tower. New York: Raven Press. (1975).

(11) R.D.Keynes and J.E.Kimura. Activation of the sodium channels in the squid giant axon. *J. Physiol.* (in the press) (1978).

Cyclic Nucleotides

Cyclic Nucleotides and Muscle Function: Introduction

Joel G. Hardman

Department of Pharmacology, Vanderbilt University, School of Medicine, Nashville, Tennessee 37232, U.S.A.

More than two decades have passed since the discovery of adenosine 3',5'-cyclic phosphate (cAMP) by Sutherland and Rall (1). During this period, an abundance of evidence has accumulated to indicate a key role for cAMP as a regulator of numerous biological processes, especially those under control by a variety of hormones and neurotransmitters (2). cAMP is formed from ATP by a reaction catalyzed by the enzyme, adenylate cyclase, which is principally located on plasma membranes in mammalian cells (2,3). A number of hormones and neurotransmitters can increase the rate of formation of cAMP through interaction with specific membrane receptors. The mechanism or mechanisms whereby an agonist-receptor interaction leads to increased activity of adenylate cyclase is still unknown, but new advances give cause for optimism that in the not-too-distant future such mechanisms will become much more clearly understood (4-7). cAMP is degraded to 5'-AMP by one or more phosphodiesterases, which also may be regulated, probably indirectly, by certain hormones and neurotransmitters and which can be inhibited by a number of pharmacological agents (8-10). All known effects of cAMP in mammalian cells are initiated through the activation by the nucleotide of a protein kinase (11-13). This enzyme, or family of enzymes, catalyzes the transfer of the terminal phosphate from ATP to certain proteins, which undergo, as a result of the phosphorylation, alterations in activity or function.

Guanosine 3',5'-cyclic phosphate (cGMP) was identified in biological material several years after the discovery of cAMP and has been extensively investigated as another potential regulatory nucleotide (2,14). cGMP is formed from GTP by a reaction catalyzed by the enzyme, guanylate cyclase, which exists in both soluble and membrane-associated forms in most mammalian cells (14-16). The degradation of cGMP is analogous to that of cAMP, being catalyzed by cyclic nucleotide phosphodiesterase(s) (8,9,14). A cGMP-dependent protein kinase has been found in many mammalian tissues (17,18), and specific protein substrates for this enzyme have been demonstrated, but their functions have not yet been identified (19-22). It has been suggested that cGMP also may regulate cellular process through mechanisms that do not involve a protein kinase (14). Although a variety of hormones and neurotransmitters can raise cGMP levels in intact cells through an apparent calcium-dependent increase in guanylate cyclase activity, these agents do not appear to activate the enzyme directly (14,15), and it is not at all clear that cGMP mediates all, or any, of their actions. A number of pharmacological agents can increase cGMP levels in intact cells in a calcium-independent manner and can activate guanylate cyclase in cell-free systems. This group of agents, which includes several nitroso compounds, will be discussed in two of the papers in this symposium.

Both cAMP and cGMP, the enzymes involved in their biosynthesis and degradation and cAMP- or cGMP-dependent protein kinases (or both) occur in skeletal, cardiac and smooth muscle. The most completely understood effect of cAMP in muscle (or any other tissue) is on glycogenolysis in skeletal muscle (11,12). This involves the now-familiar "phosphorylation cascade", wherein cAMP activates protein kinase, which catalyzes the phosphorylation and activation of phosphorylase-B kinase, which catalyzes the phosphorylation and activation of glycogen phosphorylase, while at the same time the protein kinase catalyzes the phosphorylation and inactivation of glycogen synthase. The same events appear to take place in heart muscle and may occur in smooth muscle, although the latter has not been studied very thoroughly. There is presently little if any evidence that indicates a role for cGMP in the regulation of muscle carbohydrate metabolism. cGMP-dependent protein kinase can catalyze the phosphorylation of muscle phosphorylase-B kinase and glycogen synthase *in vitro*, but it seems unlikely that these are substrates for the enzyme under physiological conditions (23).

During the last several years, there has been considerable interest in the possible regulation of muscle contraction by cAMP and cGMP. Most of the interest in this subject has focused on myocardium and smooth muscle. A mass of evidence now supports the concept that cAMP is involved in the positive inotropic effects of *beta*-adrenergic agonists and some other agents on the heart (2,24,25). Precisely how cAMP alters myocardial contractility, however, is still unsettled, and there is still debate about whether or not cAMP is the only means by which *beta*-adrenergic agents act on the heart. cAMP-dependent protein kinase has been shown to catalyze the phosphorylation of components of cardiac sarcolemma and sarcoplasmic reticulum, with resultant alterations in the binding or transport of calcium (24-27). cAMP-dependent protein kinase also can catalyze the phosphorylation of a component of the cardiac contractile apparatus, although the functional significance of this phosphorylation is unclear (28-30). Acetylcholine can raise cGMP in the myocardium, and it has been proposed that cGMP is a mediator of the negative inotropic effects of acetylcholine (31,32). Some recent evidence, however, is not consistent with this view (33-35).

There is now considerable evidence to indicate the involvement of cAMP in smooth muscle relaxation produced by *beta*-adrenergic agonists and other agents (2,36-38), although some data are not compatible with such a role being general and necessary (39,40). cGMP has been proposed to be involved in promoting smooth muscle contraction (41,42), on the one hand, or relaxation, on the other (43). These conflicting points of view reflect data that show associations of either contraction or relaxation with elevations in cGMP in response to a variety of neurotransmitters and pharmacological agents. Other data, however, have shown dissociations between cGMP and either type of contractile response (39,44-46). Suffice it to say that while there are many studies that show interesting associations between contractile responses to various agonists and elevations in cAMP or cGMP levels in smooth muscle, cause-effect relationships have not been firmly established.

The papers to be presented in this symposium will not cover all facets of the possible roles of cyclic nucleotides in muscle function. Rather, selected topics will be covered which represent several areas where there is more than common interest, brisk investigation and some controversy. These topics include the possible role of cGMP in smooth muscle function, the regulation of the biosynthesis of both cGMP and cAMP, the involvement of cations in neurotransmitter action on the heart, the regulation of membrane permeability to cations by cAMP and the phosphorylation and regulation of contractile proteins. The authors of these papers are and have been pioneers and leaders in their respective fields. It is hoped that from this symposium we will gain a better appreciation of what is actually known about these various topics, perhaps some perspective of what we may think we know but really don't, and an awareness of where future research should be focused.

REFERENCES

1. E.W. Sutherland and T.W. Rall, The relation of adenosine-3',5'-phosphate and phosphorylase to the actions of catecholamines and other hormones, Pharm. Rev. 12, 265 (1960).

2. G.A. Robison, R.W. Butcher and E.W. Sutherland, (1971) Cyclic AMP, Academic Press, New York.

3. J.P. Perkins, Adenyl cyclase, Adv. Cyclic Nucleotide Res., 3, 1 (1973).

4. M.E. Maguire, E.M. Ross and A.G. Gilman, β-Adrenergic receptor: Ligand binding properties and the interaction with adenylyl cyclase, Adv. Cyclic Nucleotide Res., 8, 1 (1977).

5. B.B. Wolfe, T.K. Harden and P.B. Molinoff, In vitro study of β-adrenergic receptors, Ann. Rev. Pharm. Tox., 17, 575 (1977).

6. R.J. Lefkowitz and L.T. Williams, Molecular mechanisms of activation and desensitization of adenylate cyclase coupled beta-adrenergic receptors, Adv. Cyclic Nucleotide Res., 9, 1 (1978).

7. E.M. Ross, T. Haga, A.C. Howlett, J. Schwarzmeier, L.S. Schleifer and A.G. Gilman, Hormone-sensitive adenylate cyclase: Resolution and reconstitution of some components necessary for regulation of the enzyme, Adv. Cyclic Nucleotide Res., 9, 53 (1978).

8. M.M. Appleman, W.J. Thompson and T.R. Russell, Cyclic nucleotide phosphodiesterases, Adv. Cyclic Nucleotide Res., 3, 65 (1973).

9. J.N. Wells and J.G. Hardman, Cyclic nucleotide phosphodiesterases, Adv. Cyclic Nucleotide Res., 8, 119 (1977).

10. M. Chasin and D.N. Harris, Inhibitors and activators of cyclic nucleotide phosphodiesterase, Adv. Cyclic Nucleotide Res., 7, 225 (1976).

11. E.G. Krebs, Protein kinases, Curr. Top. Cell. Regul., 5, 99 (1972).

12. H.G. Nimmo, and P. Cohen, Hormonal control of protein phosphorylation, Adv. Cyclic Nucleotide Res., 8, 145 (1977).

13. P. Greengard, Phosphorylated proteins as physiological effectors, Science, 199, 146 (1978).

14. N.D. Goldberg and M.K. Haddox, Cyclic GMP-metabolism and involvement in biological regulation, Ann. Rev. Biochem., 46, 823 (1977).

15. D.L. Garbers, T.D. Chrisman and J.G. Hardman, Guanylate cyclase, in: Eukaryotic Cell Function and Growth: Regulation by Intracellular Cyclic Nucleotides. (J.E. Dumont, B.L. Brown and N.J. Marshall, eds) Plenum Press, New York and London, page 155, (1976).

16. H. Kimura and F. Murad, Two forms of guanylate cyclase in mammalian tissues and possible mechanisms for their regulation, Metabolism, 24, 439 (1975).

17. J.F. Kuo, M. Shoji, and W.-N. Kuo, Mammalian cyclic GMP-dependent protein kinase and its stimulatory modulator, Adv. Cyclic Nucleotide Res., 9, 199 (1978).

18. J.D. Corbin and T.M. Lincoln, Comparison of cAMP- and cGMP-dependent protein kinases, Adv. Cyclic Nucleotide Res., 9, 159 (1978).

19. J.E. Casnellie, and P. Greengard, Guanosine 3':5'-cyclic monophosphate-dependent phosphorylation of endogenous substrate proteins in membranes of mammalian smooth muscle, Proc. Nat. Acad. Sci. U.S.A., 71, 1891 (1974).

20. H.R. DeJonge, Cyclic nucleotide-dependent phosphorylation of intestinal epithelium proteins, Nature, 262, 590 (1976).

21. H.E. Ives, J.E. Casnellie, P. Greengard and J.D. Jamieson, Cyclic GMP-dependent protein phosphorylation in isolated aortic smooth muscle cells, J. Cell Biol., 75, 325a (1977).

22. D.J. Schlichter, J.E. Casnellie and P. Greengard, An endogenous substrate for cGMP-dependent protein kinase in mammalian cerebellum, Nature, 273, 61 (1978).

23. T.M. Lincoln, and J.D. Corbin, On the role of cAMP and cGMP-dependent protein kinases in cell function, J. Cyclic Nucleotide Res., 4, 3 (1978).

24. M.L. Entman, The role of cyclic AMP in the modulation of cardiac contractility, Adv. Cyclic Nucleotide Res., 4, 163 (1974).

25. R.W. Tsien, Cyclic AMP and contractile activity in heart, Adv. Cyclic Nucleotide Res., 8, 363 (1977).

26. E.-G. Krause, H. Will, B. Schirpke, and A. Wollenberger, Cyclic AMP-enhanced protein phosphorylation and calcium binding in a cell-membrane-enriched fraction from myocardium, Adv. Cyclic Nucleotide Res. 5, 473 (1975).

27. A.M. Katz, M. Tada and M.A. Kirchberger, Control of calcium transport in the myocardium by the cyclic AMP-protein kinase system, Adv. Cyclic Nucleotide Res. 5, 453 (1975).

28. H.A. Cole and S.V. Perry, The phosphorylation of troponin I from cardiac muscle, Biochem. J. 149, 525 (1975).

29. P. England, Correlation between contraction and phosphorylation of the inhibitory subunit of troponin in perfused rat heart, FEBS Lett. 50, 57 (1975).

30. J.T. Stull and J.E. Buss, Phosphorylation of cardiac troponin by cyclic adenosine 3':5'-monophosphate-dependent protein kinase, J. Biol. Chem. 252, 851 (1977).

31. W.J. George, R.D. Wilkerson and P.J. Kadowitz, Influence of acetylcholine on contractile force and cyclic nucleotide levels in the isolated perfused rat heart, J. Pharmacol. Exp. Therap. 184, 228 (1973).

32. W.J. George, L.J. Ignarro, R.J. Paddock, L. White and P.J. Kadowitz, Oppositional effects of acetylcholine and isoproterenol on isometric tension and cyclic nucleotide concentrations in rabbit atria, J. Cyclic Nucleotide Res. 1, 339 (1975).

33. J. Diamond, R.E. Ten Eick and A.J. Trapani, Are increases in cyclic GMP levels responsible for the negative inotropic effects of acetylcholine in the heart, Biochem. Biophys. Res. Comm. 79, 912 (1977).

34. G. Brooker, Dissociation of cyclic GMP from the negative inotropic action of carbachol in guinea pig atria, J. Cyclic Nucleotide Res. 3, 407 (1977).

35. M.J. Mirro, A.M. Watanabe and J.C. Bailey, Dissociation between changes in cyclic GMP levels and electrophysiological effects of acetylcholine in guinea pig atria, Clin. Res. 26, 253A (1978).

36. H.-P. Bär, Cyclic nucleotides and smooth muscle, Adv. Cyclic Nucleotide Res. 4, 195 (1974).

37. R. Andersson, K. Nilsson, J. Wikberg, S. Johansson, E. Mohme-Lundholm and L. Lundholm, Cyclic nucleotide and the contraction of smooth muscle, Adv. Cyclic Nucleotide Res. 5, 491 (1975).

38. D.H. Namm and J.P. Leader, Occurrence and function of cyclic nucleotides in blood vessels, Blood Vessels 13, 24 (1976).

39. J. Diamond, Role of cyclic nucleotides in control of smooth muscle contraction, Adv. Cyclic Nucleotide Res. 9, 327 (1978).

40. M.-F. Vesin and S. Harbon, The effects of epinephrine, prostaglandins, and their antagonists on adenosine cyclic 3',5'-monophosphate concentrations and motility of the rat uterus, Mol. Pharm. 10, 457 (1974).

41. T.-P. Lee, J.F. Kuo and P. Greengard, Role of muscarinic cholinergic receptors in regulation of guanosine 3',5'-cyclic monophosphate content in mammalian brain, heart muscle and intestinal smooth muscle, Proc. Nat. Acad. Sci. USA 69, 3287 (1972).

42. E.W. Dunham, M.K. Haddox, and N.D. Goldberg, Alteration of vein cyclic 3':5' nucleotide concentrations during changes in contractility, Proc. Nat. Acad. Sci. USA 71, 815 (1974).

43. K.-D. Schultz, K. Schultz, and G. Schultz, Sodium nitroprusside and other smooth muscle-relaxants increase cyclic GMP levels in rat ductus deferens, Nature 265, 750 (1977).

44. J. Diamond and R.A. Janis, Increases in cyclic GMP levels may not mediate relaxant effects of sodium nitroprusside, verapamil and hydralazine in rat vas deferens, Nature 271, 472 (1978).

45. G. Angles d'Auriac and M. Worcel, Cellular levels of cAMP and cGMP in rat uterus smooth muscle. Effects of angiotensin, carbachol and various metabolic conditions, in: Smooth Muscle Pharmacology and Physiology (M. Worcel and G. Vassort, eds), Les Colloques de l'Institut National de la Santé et de la Recherche Médicale. 50, 101 (1975).

46. D. Leiber, M.-F. Vesin and S. Harbon, Regulation of guanosine 3',5'-cyclic monophosphate levels and contractility in rat myometrium, FEBS Lett. 80, 183 (1978).

The Possible Role of Cyclic GMP in the Actions of Hormones and Drugs on Smooth Muscle Tone: Effects of Exogenous Cyclic GMP Derivatives

G. Schultz, K.-D. Schultz, E. Böhme and V.A.W. Kreye

Pharmakologisches Institut and II. Physiologisches Institut, Universität Heidelberg, D-6900 Heidelberg, Germany

The cytoplasmic calcium concentration appears to be the primary intracellular signal in the control of smooth muscle tone (1,2). Various extracellular factors including neurotransmitters and hormones can increase the tissue tone by promoting the influx of calcium from inactive extra- and intracellular pools into the cytoplasm. Cyclic nucleotides, which have been shown to be important intracellular signals in the control of various cellular functions, appear to modulate the control of smooth muscle tone by calcium. The findings, on which the assumption of such modulating roles for cyclic AMP and especially for cyclic GMP are based, are summarized in the following.

CYCLIC AMP AND SMOOTH MUSCLE FUNCTION

Cyclic AMP is believed to be the intracellular mediator of the relaxing effects of ß-adrenergic agonists and some other hormones and neurotransmitters (for review see Refs. 3-6). This assumption is mainly based on studies in which time- and dose-relationships of the effects of various hormonal relaxants on tissue tone and cyclic AMP levels were compared and on other ones in which relaxant effects of exogenous cyclic AMP derivatives, especially dibutyryl cyclic AMP, were demonstrated (see Table I). It is not clear how cyclic AMP causes relaxation. A hyperpolarizing effect of cyclic AMP has been shown in vascular smooth muscle (7,8). Cyclic AMP can stimulate the phosphorylation of some membrane proteins (9) and of filamin (10) in the ductus deferens. Furthermore, this nucleotide has been shown to stimulate calcium uptake by microsomal preparations from vascular (11-13), intestinal (14) and uterine (15) smooth muscle. The latter findings favor the assumption that cyclic AMP causes tissue relaxation by accelerated removal of cytoplasmic calcium.

CYCLIC GMP AND SMOOTH MUSCLE FUNCTION

Effects of contraction-producing agents on cyclic GMP levels

Various neurotransmitters and hormones that causes smooth muscle contraction increase the intracellular cyclic GMP concentration (for review see Refs. 16,17). It was originally suggested that the increase in the cyclic GMP level may be causally involved in the contractile responses of the tissue to the stimuli (14,18,19). However, the evidence for this assumption becomes weaker and weaker. Although both responses to hormonal stimulation, increase in the tissue tone and in the cyclic GMP level,

Table I: Effects of various cyclic nucleotide derivatives on noradrenaline-induced contractions of the rat ductus deferens.

	change of noradrenaline-induced peak tension		change of noradrenaline-induced peak tension
control	+ 20 %	cGMP	+ 16 %
dibutyryl cAMP	- 86 %	dibutyryl cGMP	+ 6 %
8-Br-cAMP	+ 11 %	8-Br-cGMP	- 55 %
		8-Br-2':3'-cGMP	+ 4 %

Contractile responses to 1 μM noradrenaline were recorded by repeated stimulation (generally 4 times) with changes of medium after peak tension had been reached. After 20 min of incubation with the added nucleotide (10 μM), responses to repeated noradrenaline stimulation were again recorded. Data indicate changes of the mean response to noradrenaline after nucleotide addition compared to the mean response before addition. Sodium bromide (1 to 1,000 μM) did not significantly affect the response to noradrenaline as compared to controls. The number of tissues was 8 to 13 per group.

appear to be secondary to elevated cytoplasmic calcium, quantitative and temporal correlation of these two responses are poor. As an example, the effects of cholinergic and α-adrenergic stimulants on the rat ductus deferens are shown in Fig. 1. Whereas acetylcholine had only a small contracting effect in this tissue, cyclic GMP levels were rapidly increased up to 4-fold (20). Different observations were made with α-adrenergic agonists. Phenylephrine, added at the maximally contracting concentration of 0.1 mM, had no effect on the cyclic GMP level within the first minute (21), whereas the effect of this agonist on tissue tone was several-fold higher than that of acetylcholine. Similar discrepancies were observed in several other tissues under a variety of conditions (for review see Refs. 6,22).

Effects of smooth muscle-relaxants on cyclic GMP levels

A variety of agents capable of preventing or reducing smooth muscle contraction increased the cyclic GMP concentration (23-25). Besides various antihypertensives and vasodilators, which are in part assumed to act as "calcium-antagonists" (26), nitroglycerin and sodium nitrite increased cyclic GMP levels. Sodium nitroprusside and hydroxylamine were the most potent stimulators of cyclic GMP formation in vascular and other smooth muscular tissues.

EFFECTS OF EXOGENOUS CYCLIC GMP DERIVATIVES ON TISSUE TONE

Since cyclic GMP levels are increased by agents that produce contraction and other ones that cause relaxation of smooth muscle, we have attempted to elucidate the role of cyclic GMP in the actions of these compounds by studying the effect of exogenous cyclic GMP on tissue tone.

Rat ductus deferens

When cyclic GMP or more lipophilic derivatives were applied to the isolated rat ductus deferens, the basal tissue tone was unaffected. In contrast, contractile

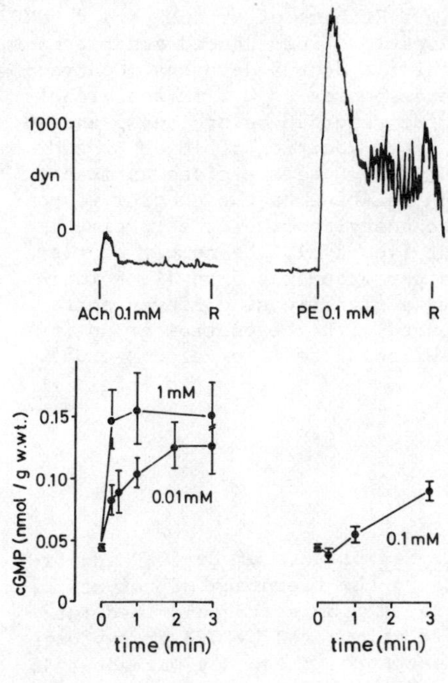

Fig. 1: Effects of acetylcholine and phenylephrine on the tone and the cyclic GMP level of the rat ductus deferens. Contractile responses to acetylcholine (ACh) and phenylephrine (PE), both added at the maximally contracting concentration of 0.1 mM, were recorded under isometric conditions in Krebs-Ringer-bicarbonate buffer (upper panels). Cyclic GMP levels, given as mean ± SEM of 6 to 10 determinations, were measured in separate tissue samples incubated under isotonic conditions in Tris-HCl-buffered medium, with acetylcholine and phenylephrine added at the concentrations indicated (lower panels). Cyclic AMP levels were not changed under these conditions. Data taken from Refs. 20 and 21.

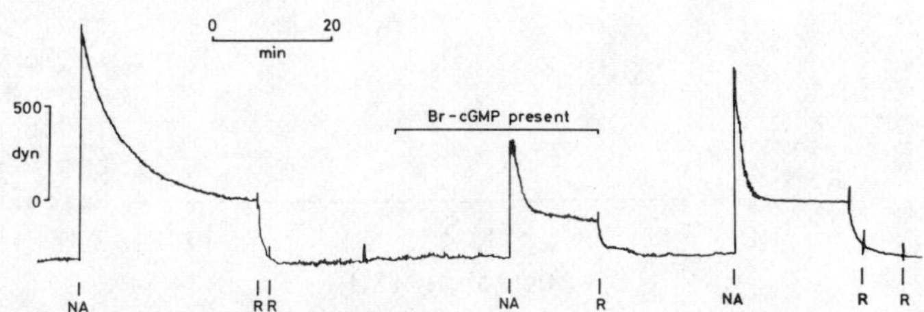

Fig. 2: Effect of Br-cGMP on the contractile response of the rat ductus deferens to noradrenaline. Isometric responses to 10 μM noradrenaline were recorded in Krebs-Ringer-bicarbonate buffer without or with 10 μM Br-cGMP added as indicated. R indicates a change to fresh medium.

responses of the tissue were lowered by Br-cGMP[#]. The phasic and tonic responses to 10 μM noradrenaline were reduced by about 50% when the ductus deferens was preincubated for about 20 min with 10 μM Br-cGMP (Fig. 2).
This relaxant effect was specific for the 8-bromo derivative of cyclic GMP (see Table I). Cyclic GMP itself, its dibutyryl derivative, and the 8-bromo derivatives of 2':3'-cGMP and of cyclic AMP as well as Br-GMP and 8-Br-guanosine were ineffective.
Br-cGMP (10 μM) added for 10 and 20 min to ductus deferentes had no effect on cyclic AMP levels (27). This finding makes it unlikely that Br-cGMP, which is a poor

[#] Abbreviations. 8-Bromo guanosine 3':5'-monophosphate, Br-cGMP; 8-bromo guanosine 2':3'-monophosphate, Br-2':3'-cGMP; 8-bromo guanosine 5'-monophosphate, Br-GMP.

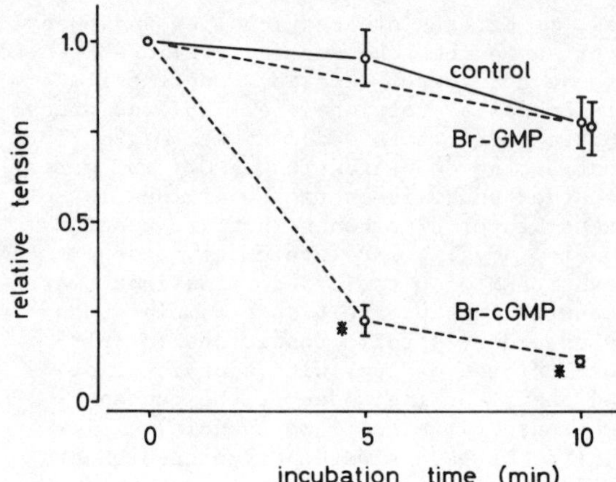

Fig. 3: Effects of Br-cGMP and Br-GMP on acetylcholine-induced contractions of the rat ductus deferens. Contractile responses to 0.1 mM acetylcholine were recorded before and 5 and 10 min after addition of 10 μM Br-cGMP or Br-GMP. Data are given as mean ± SEM relative to the control response to acetylcholine at zero time point (4oo dyn). The number of tissues per group was 5 to 11. * indicates a significant difference as compared with the control group in the Wilcoxon test (p value $\leqq$ 0.05).

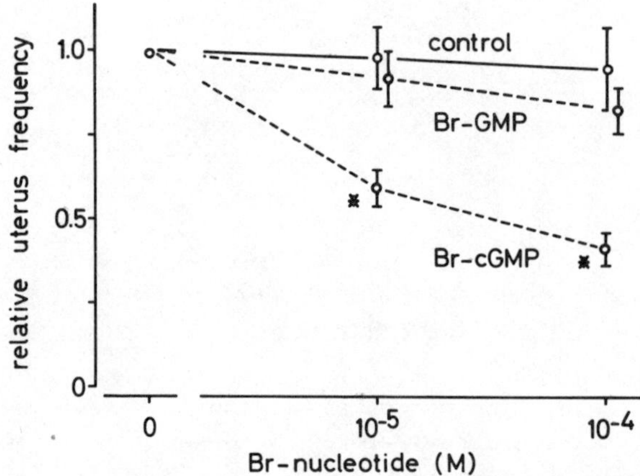

Fig. 4: Effects of Br-cGMP and Br-GMP on the frequency of rat uteri. Tissues from estrogen-primed rats were stimulated by 0.2 μM oxytocin. The control frequency was about 15 per 30 min. Nucleotides were added at 10 μM concentration for 30 min, followed by another 30 min period at 100 μM. Data are mean ± SEM of 7 to 10 observations per group. * indicates a significant difference in the Wilcoxon test (p value $\leqq$ 0.05) when compared with the control value.

substrate for cyclic nucleotide phosphodiesterases (28), has an indirect relaxant effect by inhibiting the degradation of endogenous cyclic AMP.
The response of the ductus deferens to acetylcholine was more potently depressed by Br-cGMP than that to noradrenaline (Fig. 3). Br-cGMP (10 μM) reduced acetylcholine-induced contraction by 75% within 5 min, and Br-GMP was inactive.
The effect of Br-cGMP to reduce the tonic response of the ductus deferens to K^+ (35 mM K_2SO_4) was relatively weak (27). Br-cGMP added to 10 and 100 μM concentrations reduced the tissue tone by 20 and 30%, respectively.
The effect of Br-cGMP on the calcium-induced increase in tissue tone was studied in calcium-deprived, K^+-depolarized tissues. When the calcium concentration was increased stepwise from 0.1 to 10 mM, Br-cGMP (10 μM) reduced the increase in tissue tone at all calcium concentrations tested, whereas Br-GMP and Br-guanosine had slightly stimulatory effects.

Rat uterus

The effect of Br-cGMP on uterine frequency was studied in oxytocin-stimulated uterus horns from estrogen-primed rats (Fig. 4). Br-cGMP at 10 and 100 μM concentrations reduced the uterine frequency by 40 and 60%, respectively; the tension developed during the rhythmic contractions was not significantly changed. Br-GMP had no effect on the uterus frequency.

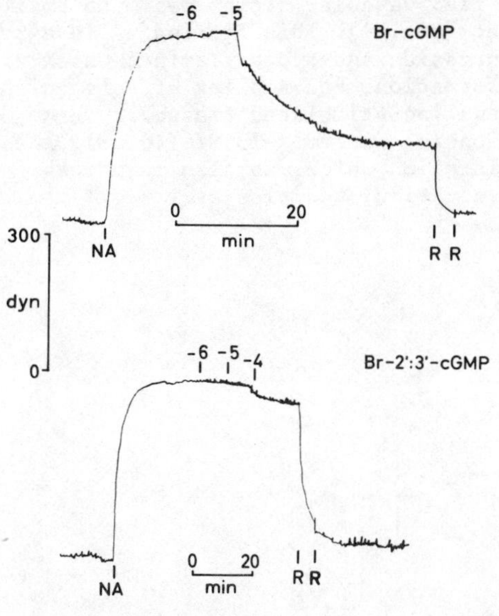

Fig. 5: Effects of Br-cGMP and Br-2':3'-cGMP on the noradrenaline-stimulated rat aorta. Helical strips of rat aortae were stimulated by 10 nM noradrenaline, and nucleotides were added after the tissue tone had reached the plateau of the tonic response. The molar concentration of Br-cGMP (upper panel) and Br-2':3'-cGMP (lower panel) added to the medium is indicated in logarithmic form. R indicates a change to fresh medium.

Vascular tissues

Br-cGMP was a potent relaxant of noradrenaline-stimulated rat aortic strips (Fig. 5). When the nucleotide was added during the tonic phase of the response to 10 nM noradrenaline, there was a rapid relaxation by about 10, 45 and 90% with 1, 10 and 100 µM Br-cGMP, respectively. Br-2':3'-cGMP and Br-GMP were ineffective. Noradrenaline-stimulated rabbit aortic strips were similarly sensitive to Br-cGMP (Fig. 6). The relaxation of tissue stimulated by 10 nM noradrenaline was about 10, 35 and 90% with 1, 10 and 100 µM Br-cGMP, respectively. The relaxant effect of Br-cGMP was less pronounced in noradrenaline (0.1 µM)-stimulated hog spleen arteries. This tissue was relaxed by about 10 and 75% with 10 and 100 µM Br-cGMP, respectively.

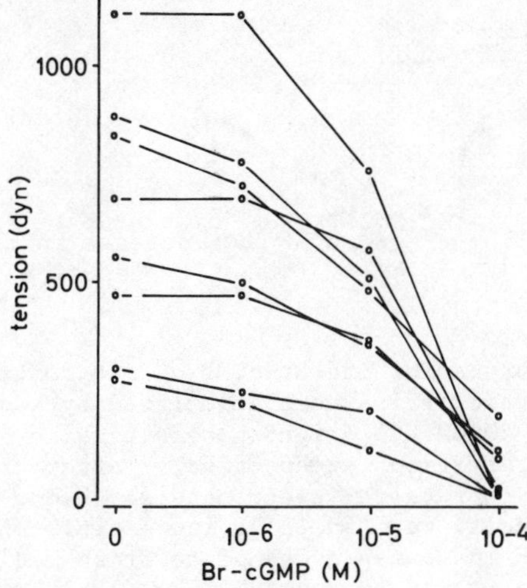

Fig. 6: Effect of Br-cGMP on noradrenaline-stimulated strips of rabbit aorta. Aortic strips were stimulated by 10 nM noradrenaline, and Br-cGMP was added on the plateau of the tonic response at increasing concentrations as shown in Fig. 5. Tensions of individual tissues are given at the Br-cGMP concentrations indicated on the abscissa.

The relaxant effect of Br-cGMP on K^+-depolarized vascular strips was much smaller than that on noradrenaline-stimulated tissues (Fig. 7). This finding is in agreement with the observation that sodium nitroprusside and hydroxylamine, which are potent stimulants of endogenous cyclic GMP formation, had smaller effects on the K^+-depolarized rat aorta than on the noradrenaline-stimulated tissue.
In rat aorta contracted by the divalent cation-ionophore, A-23187 (10 µM), the addition of 10 and 100 µM Br-cGMP caused relaxation which was also much less pronounced than that observed in noradrenaline-stimulated tissues.

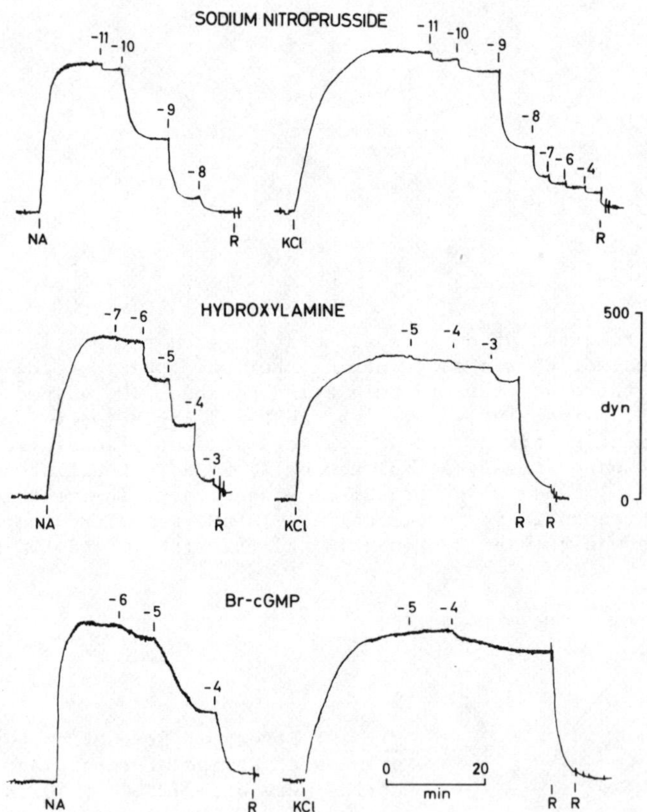

Fig. 7: Effects of sodium nitroprusside, hydroxylamine and Br-cGMP on rat aortic strips stimulated by noradrenaline and K^+. Aortic strips were stimulated by addition of 10 nM noradrenaline (left panels) and 40 mM KCl (right panels), which are about 55 and 70% maximally effective concentrations, respectively. After the tonic response had fully developed, sodium nitroprusside (upper panels), hydroxylamine (middle panels) or Br-cGMP (lower panels) were added at increasing concentrations (indicated in logarithmic form). R indicates a change to fresh medium.

CONCLUDING REMARKS

The various findings of hormone- and drug-induced alterations in cyclic GMP levels and of relaxing effects of Br-cGMP strongly suggest that cyclic GMP is actively involved in the control of smooth muscle function. Whereas cyclic GMP-dependent protein kinases, which are potently stimulated by Br-cGMP, have been studied in a variety of mammalian tissues (29), the occurrence of such an enzyme and its physiological substrates have not been demonstrated in smooth muscle. In spite of this ignorance, it is likely that protein phosphorylation is involved in cyclic GMP-controlled processes in smooth muscle since a stimulatory effect of cyclic GMP on phosphorylation of membrane proteins has been shown in the ductus deferens (9).

Data are accumulating that there exists a negative correlation between cyclic GMP levels and smooth muscle tone under the influence of contracting und relaxing agents, and the present results with Br-cGMP correspond with these findings. The results obtained by different approaches in studying the possible role of cyclic GMP for smooth muscle function support the hypothesis (20,23) that cyclic GMP can reduce tissue excitation and excitability by neurotransmitters and hormones (Fig. 8). Our former suggestion that cyclic GMP may lower hormone-induced excitation by acting as a negative feedback inhibitor on the influx of calcium (23) is supported by the findings that Br-cGMP reduced the calcium-induced increase in the tone of K^+-depolarized ductus deferentes and that the intracellular cyclic GMP level was increased by various relaxants that are assumed to act as "calcium-antagonists".

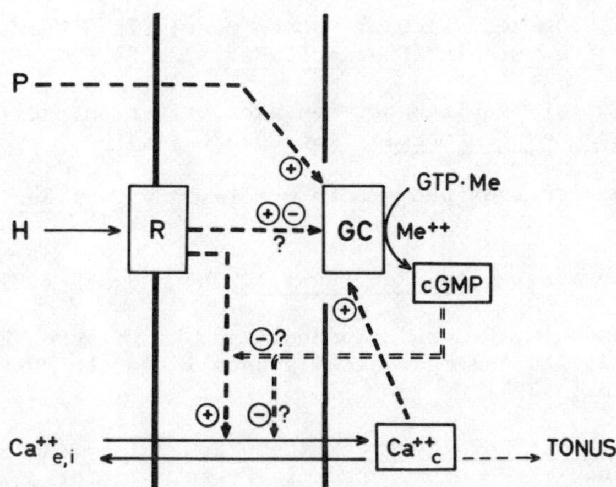

Fig. 8: Possible interactions of cyclic GMP with tissue excitation and with the control of smooth muscle tone by calcium. Neurotransmitters and other hormonal stimuli (H) interact with membrane-bound receptors (R) and cause increased calcium influx from inactive extra- or intracellular pools ($Ca^{++}_{e,i}$) into the cytoplasmic pool (Ca^{++}_{c}). Hormone-induced increases in cyclic GMP formation appear to be secondary to increased Ca^{++}_{c} and probably other intermediate mechanisms. It is not clear whether the soluble, the particulate or both forms of the guanylate cyclase (GC) are under hormonal control. Many smooth muscle-relaxing drugs (P) apparently stimulate cyclic GMP formation more directly, and their effects on guanylate cyclase do not involve calcium. Intracellular cyclic GMP appears to lower smooth muscle excitation and excitability when its concentration is increased by stimulating hormones and relaxing drugs, respectively. It is not yet clear whether cyclic GMP reduces calcium influx into cytoplasm or interferes with another membrane function, which may be the permeability to another ion contributing to the excitation process. Modified from Ref. 22.

However, Br-cGMP and the potent stimulants of cyclic GMP formation, sodium nitroprusside and hydroxylamine, were poor relaxants of K^+-depolarized tissues the contraction of which is generally assumed to depend on calcium ions entering the cell from extracellular environment. Moreover, sodium nitroprusside although increasing cyclic GMP levels in the ductus deferens (23) did not relax the noradrenaline-stimulated tissue (30). The notion that cyclic GMP may act as an inhibitor of calcium influx is additionally weakened by findings indicating that sodium nitroprusside does not affect membrane calcium influx (31-33). On the other hand, sodium nitroprusside and nitroglycerin have been shown to reduce the noradrenaline-stimulated ^{36}chloride efflux from the rat aorta (34). Therefore, it is possible that cyclic GMP affects tissue excitation or excitability by a change of membrane function other than and possibly primary to calcium permeability. The cyclic GMP-controlled membrane function, which still needs to be defined, may contribute to different degrees to the excitation process in various smooth muscle tissues when stimulated by hormonal and non-hormonal factors.

ACKNOWLEDGEMENTS

The studies performed in the authors' laboratories were supported by grants of the Deutsche Forschungsgemeinschaft. The expert technical assistance of Miss Isolde Villhauer is gratefully acknowledged.

REFERENCES

(1) A.P. Somlyo and A.V. Somlyo, Vascular smooth muscle.II. Pharmacology of normal and hypertensive vessels, Pharmacol.Rev. 22, 249-353 (1970).

(2) L. Hurwitz and A. Suria, The link between agonist action and response in smooth muscle, Ann.Rev.Pharmacol. 11, 303-326 (1971).

(3) H.-P. Bär, Cyclic nucleotides and smooth muscle, Adv.Cycl.Nucl.Res. 4, 195-237 (1974).

(4) C.L. Prosser, Smooth muscle, Ann.Rev.Physiol. 36, 503-535 (1974).

(5) R. Andersson, L. Lundholm, E. Mohme-Lundholm and K. Nilsson, Role of cyclic AMP and Ca^{++} in metabolic and mechanical events in smooth muscle, Adv.Cycl. Nucl.Res. 1, 213-229 (1972).

(6) G. Schultz in: Asthma II: Physiology, immunopharmacology and treatment, ed. by L.M. Lichtenstein and K.F. Austen, Academic Press, New York, 77-89 (1977).

(7) A.P. Somlyo, A.V. Somlyo and V. Smiesko, Cyclic AMP and vascular smooth muscle, Adv.Cycl.Nucl.Res. 1, 175-194 (1972).

(8) A.V. Somlyo, G. Haeusler and A.P. Somlyo, Cyclic adenosine monophosphate: Potassium-dependent action on vascular smooth muscle membrane potential, Science 169, 490-491 (1972).

(9) J.E. Casnellie and P. Greengard, Guanosine3':5'-cyclic monophosphate-dependent phosphorylation of endogenous substrate proteins in menbranes of mammalian smooth muscle, Proc.Natl.Acad.Sci.U.S.71, 1891-1895 (1974).

(10) D. Wallach, P. Davies, P. Bechtel, M. Willingham and I. Pastan, Cyclic AMP-dependent phosphorylation of the actin-binding protein filamin, Adv.Cycl.Nucl.Res. 9, 371-379 (1978).

(11) M. Baudoin-Legros and P. Meyer, Effects of angiotensin, catecholamines and cyclic AMP on calcium storage in aortic microsomes, Brit.J.Pharmacol. 47, 377-385 (1973).

(12) D.F. Fitzpatrick and A. Szentivanyi, Stimulation of calcium uptake into aortic microsomes by cyclic AMP and cyclic AMP-dependent protein kinase, Naunyn-Schmiedeberg's Arch.Pharmacol. 298, 255-257 (1977).

(13) R.C. Bhalla, R.C. Webb, D. Singh and T. Brock, Role of cyclic AMP in aortic microsomal phosphorylation and calcium uptake, Amer.J.Physiol. 234, H 508-514 (1978).

(14) R. Andersson, K. Nilsson, J. Wikberg, S. Johansson. E. Mohme-Lundholm and L. Lundholm, Cyclic nucleotides and the contraction of smooth muscle, Adv.Cycl.Nucl.Res. 5, 491-518 (1975).

(15) K. Nishikori, T. Takenaka and H. Maeno, Stimulation of microsomal calcium uptake and protein phosphorylation by adenosine cyclic 3':5'-monophosphate in rat uterus, Molec.Pharmacol. 13, 671-678 (1977).

(16) G. Schultz and J.G. Hardman in: Eukaryotic cell function and growth, Regulation by intracellular cyclic nucleotides, ed. by J.E. Dumont, B.L. Brown and N.J.Marshall, Plenum Press, New York and London, 667-683 (1976).

(17) N.D. Goldberg and M.K. Haddox, Cyclic GMP metabolism and involvement in biological regulation, Ann.Rev.Biochem. 46, 823-896 (1977).

(18) E.W. Dunham, M.K. Haddox and N.D. Goldberg, Alteration of vein cyclic 3':5' nucleotide concentrations during change of contractility, Proc.Natl.Acad.Sci. U.S. 71, 815-819 (1974).

(19) T.-P. Lee, J.F. Kuo and P. Greengard, Role of muscarinic cholinergic receptors in the regulation of guanosine 3':5'-cyclic monophosphate content in mammalian brain, heart muscle and intestinal smooth muscle, Proc.Natl.Acad-Sci. U.S. 69, 3287-3291 (1972).

(20) G. Schultz, J.G. Hardman, K. Schultz, C.E. Baird and E.W. Sutherland, The importance of calcium ions for the regulation of guanosine-3':5'-monophosphate levels, Proc.Natl.Acad.Sci. U.S. 70, 3889-3893 (1973).

(21) G. Schultz, K. Schultz and J.G. Hardman, Effects of norepinephrine on cyclic nucleotide levels in the ductus deferens of the rat, Metabolism 24, 429-437 (1975).

(22) G. Schultz in: Hormones and cell regulation, Vol 2, ed. by J. Dumont and J. Nunez, Elsevier-North Holland, Biochemical Press, 107-117 (1978).

(23) K.-D. Schultz, K. Schultz and G. Schultz, Sodium nitroprusside and other smooth muscle relaxants increase cyclic GMP levels in rat ductus deferens, Nature 265, 750-751 (1977).

(24) S. Katsuki and F. Murad, Regulation of adenosine cyclic 3':5'-monophosphate and guanosine cyclic 3':5'-monophosphate levels and contractility in bovine tracheal smooth muscle, Molec.Pharmacol. 13, 330-341 (1977).

(25) J. Diamond and T.G. Holmes, Effects of potassium chloride and smooth muscle relaxants on tension and cyclic nucleotide levels in rat myometrium, Can.J.Physiol.Pharmacol. 53, 1099-1107 (1975).

(26) A. Fleckenstein, G. Grün, H. Tritthart, K. Byon and P. Harding, Uterus-Relaxation durch hochaktive Ca^{++}-antagonistische Hemmstoffe der elektro-mechanischen Koppelung wie Isoptin (Verapamil, Iproveratril) Substanz D 600 and Segontin (Prenylamin). Klin.Wschr. 49, 32-41, (1971).

(27) K.-D. Schultz, Effects of 8-Br-cGMP on the tonus of rat ductus deferens and aorta, Naunyn-Schmiedeberg's Arch.Pharmacol. 297 Suppl.II, R 12 (1977).

(28) G. Michal, K. Mühlegger, N. Nelboeck, C. Thiessen and G. Weimann, Cyclophosphates VI. Cyclophosphates as substrates and effectors of phosphodiesterase, Pharmacol.Res.Comm. 6, 203-252 (1974).

(29) J.F. Kuo, M. Shoji and W.N. Kuo, Molecular and physiopathologic aspects of mammalian cyclic GMP-dependent protein kinase, Ann.Rev.Pharmacol. 18, 341-355 (1978).

(30) V.A.W. Kreye, G.D. Baron, J.B. Lüth and H. Schmidt-Gayk, Mode of action of sodium nitroprusside on vascular smooth muscle, Naunyn-Schmiedeberg's Arch. Pharmacol. 288, 381-402 (1975).

(31) G. Häusler, S. Thorens in: Vascular neuroeffector mechanisms, ed. by J.A. Bevan, G. Burnstock, B. Johansson, R.A. Maxwell, O.A. Nedergard, S. Karger, Basel, 232-241 (1976).

(32) V.A.W. Kreye, J.B. Lüth in: Ionic actions on vascular smooth muscle, ed. by E. Betz, Springer, Berlin-Heidelberg-New York, 145-149 (1976).

(33) R. Fermum, U. Klinner, P. Meisel, Versuche zum Wirkungsmechanismus von Gefäßspasmolytika. I. Wirkung von Nitroprussid-Natrium, Nitroglycerin, Prenylamin und Verapamil an der arretierten Kalium-induzierten Kontraktur isolierter Koronararterien, Acta biol.med.germ. 35, 1347-1358 (1976).

(34) V.A.W. Kreye, R. Kern and I. Schleich in: Excitation-contraction coupling in smooth muscle, ed. by R. Casteels et al., Elsevier-North Holland Biomedical Press, 145-150 (1977).

Effect of Nitro-Compound Smooth Muscle Relaxants and Other Materials on Cyclic GMP Metabolism

Ferid Murad, Chandra K. Mittal, William P. Arnold and J. Mark Braughler

Division of Clinical Pharmacology, University of Virginia
Charlottesville, Virginia 22908, U.S.A.

INTRODUCTION

Although numerous agents such as choline esters, histamine, alpha adrenergic agents, etc can increase cyclic GMP levels in various tissues (1-4), the precise mechanism of these effects and the roles of cyclic GMP are not known. Many analogies exist between the cyclic GMP and cyclic AMP systems. These include the cyclases that catalyze similar reactions, and the cyclic nucleotide specific phosphodiesterases and protein kinases. The similarities in the two systems have undoubtedly led many investigators to design experiments and models based on past experience and information with cyclic AMP. At present it appears that the use of analogies of the two systems has hindered the originality of approaches to cyclic GMP formation and functions. New approaches in cyclic GMP research are needed. In this communication a model is proposed for the regulation of cyclic GMP formation in tissues and possible functions of the nucleotide. As summarized below the regulation of guanylate cyclase by redox events and free radicals is rather complex and at present is apparently unique for the cyclic GMP system. Perhaps with future research, these novel regulatory mechanisms will have broader applications to other enzyme systems. Hopefully, some of the information gleaned from cyclic GMP research will also result in new approaches to the area of cyclic AMP. Intuitively one would still like to believe that with the evolutionary development of these two systems there should be numerous areas of similarity and analogy.

This laboratory and others have attempted to correlate cyclic GMP accumulation in tissues with various hormones and drugs with some biochemical or physiological event in an effort to define some functions of cyclic GMP (see reviews 5,6). Unfortunately this has not been a very fruitful approach to the area of cyclic GMP. Nevertheless, such studies are necessary and have certainly provided important clues regarding functions of cyclic AMP.

About 5 to 6 years ago we began to work with guanylate cyclase and thought that an understanding of its precise regulation could provide important clues and ideas regarding possible functions of cyclic GMP. We would like to review these studies and propose a general model for the regulation of guanylate cyclase activity in tissues. This model has provided us and others with a new framework for the design of experiments in this area. We would also like to propose that an important function of cyclic GMP in tissues is to act in a feedback fashion to influence the redox state of tissues and free radicals since these appear to be important regulators of cyclic GMP synthesis.

Multiple Forms of Guanylate Cyclase and their Properties

Most tissues contain at least two forms of guanylate cyclase with different properties. These forms can be separated with gel filtration or centrifugation of homogenates (7-9). The properties of the soluble and particulate forms of guanylate cyclase and subcellular distribution of the particulate enzyme have been reviewed previously (8,10,11). Some of the properties of soluble and particulate guanylate cyclase are summarized in Table 1. Whether the soluble and particulate enzymes are different or similar proteins can not be resolved until the physicochemical and kinetic properties of highly purified preparations are examined. Recent studies suggest that soluble and particulate forms of the enzyme are different antigenically (13)

Table 1

Some of the properties of soluble and particulate guanylate cyclase from rat heart, lung and liver
(Adapted from references 6,10-12).

	Soluble	Particulate
Apparent size	ca. 150,000	ca. 300,000
Km for GTP	12-65 μM	50-100 μM
Metal-nucleotide sites	1	2 or more
Ka for free Mn^{2+}	ca. 0.2 mM	ca. 0.2 mM
Activity with Mg^{2+}	ca. 10% of Mn^{2+}	ca. 10% of Mn^{2+}
Effect of Ca^{2+}	generally stimulates	generally inhibits
Effect of detergents	stimulates 30-100%	stimulates > 300%
50% inhibition by ATP	ca. 0.4 mM	> 1 mM
Cyclic AMP formation	not detectable	not detectable
50% inhibition by p-chloromercuriphenyl sulfonic acid	10-100 nM	1-10 μM

Activation of Guanylate Cyclase with Azide and Nitro Compounds

In the process of examining the kinetic properties of crude preparations, we found that azide, hydroxylamine and a variety of other agents activated guanylate cyclase (14-16). Activation with azide was observed with soluble and particulate preparations from rat liver and kidney and particulate preparations from cerebral cortex and cerebellum but not with a variety of other preparations. The tissue specificity could be attributed to the requirement of a macromolecular "azide activator factor" for azide effects and/or the absence of protein inhibitors in preparations (14-16). While the azide activator factor in liver supernatant fractions is probably catalase, other enzymes can also fulfill this requirement (17,18). Peroxidase and some cystochromes can also replace the requirement for catalase or the activator protein (17,19). The inhibitors present in some preparations have also been characterized and appear to be hemoglobin and myoglobin (20).

A variety of other compounds can also activate guanylate cyclase. These include nitrite ion, nitroglycerin, nitroprusside (21) nitrosamines (22) and nitrosoureas (23). Activation by this later group of agents is not tissue specific and has no requirement for protein factors such as catalase. As summarized in Figure 1 all of these materials can form the reactive nitric oxide radical in incubations. While some materials such as azide or hydroxylamine require enzymatic conversion by catalase or other proteins, many of these compounds can either spontaneously or under the appropriate redox conditions also form nitric oxide. Nitric oxide can activate all preparations of guanylate cyclase that we have examined (24). The only exceptions to date have included guanylate cyclase preparations from cell cultures (Balb 3T3, lymphoma S49 and human skin fibroblasts) (25). Most of the guanylate cyclase activity in these cell culture preparations is particulate and presumably the preparations possess some unknown inhibitor of the activation.

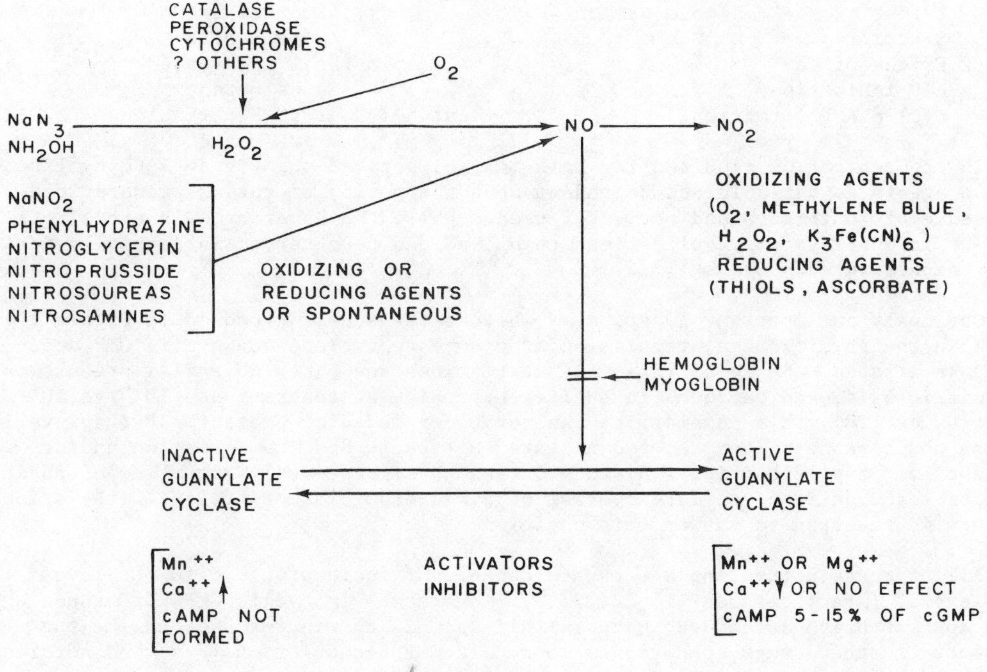

Fig. 1 Activation of guanylate cyclase by nitric oxide-forming compounds (From ref. 11)

Activation of guanylate cyclase by these agents is readily reversible. This can be accomplished with the removal of the activator or by the addition of hemoglobin or myoglobin which reverses the activation (20,24). The properties of guanylate cyclase are markedly altered with activation as summarized in Table 2. While V max is increased markedly, the Km for GTP may or may not be changed and is dependent upon the enzyme preparation examined. Native guanylate cyclase prefers Mn^{2+} as a cation cofactor and Mg^{2+} may be 5 to 20% as effective as Mn^{2+}. Guanylate cyclase also requires excess free cation for maximal activity. With activation the cation requirements are changed. Mg^{2+} is as effective as Mn^{2+} and the apparent requirement for free Mn^{2+} is lost (26). The ability to increase guanylate cyclase activity with high concentrations of Ca^{2+} is also lost after activation. While ATP inhibition of inactive and active enzyme is similar, activated guanylate cyclase can catalyze the formation of cyclic AMP (27). The significance of this new pathway for cyclic AMP formation is not known. However, investigators should be alerted to the possibility of cyclic AMP formation by guanylate cyclase in crude preparations of adenylate cyclase.

Table 2

Alteration in properties of guanylate cyclase with activation.
(From ref. 12)

Property	Remarks
Apparent size	unaltered
Km for GTP	unaltered or increased
Requirement for free Mn^{2+}	decreased
Effect of Mg^{2+} as sole cation	as effective as Mn^{2+}
Stability	more labile
Effect of Ca^{2+}	markedly decreased
ATP inhibition	unaltered
Cyclic AMP formation	ca. 5-15% of cGMP formation

The concentration required for half-maximal activation of guanylate cyclase by these agents is variable and dependent upon the purity of enzyme preparations, presence of inhibitors and activator used. While the apparent Ka's are 1 to 100 μM in most systems, values less than 1 μM can be observed in partially purified and reconstituted systems (17,27).

Obviously an important question is whether the scheme proposed in Figure 1 can explain the physiological regulation of guanylate cyclase and cyclic GMP accumulation in tissues. Since nitrite, nitrate, amines and other potential precursors of nitric oxide can be found in sufficiently high concentrations (10^{-6} to 10^{-3}M) in tissues (28), this possibility can not be excluded at present. Perhaps various drugs and hormones alter the redox state in tissues and the formation and/or metabolism of nitric oxide. A study of the physicochemical properties of purified inactive and active guanylate cyclase obtained from tissues treated with various agents is required to answer this question.

All of these activating agents are capable of increasing cyclic GMP levels in a variety of intact tissues (15,22,29,30). Increases in cyclic GMP with these agents are concentration dependent, occur within seconds to minutes and, unlike the effects of other drugs or hormones on cyclic GMP accumulation (2,4), do not require Ca^{2+} in the incubation medium. Thus, Ca^{2+} dependent and independent mechanisms exist in tissues for cyclic GMP formation. As would be predicted, the increases in cyclic GMP levels with these nitric oxide forming agents are enhanced by cyclic nucleotide phosphodiesterase inhibitors (29). These agents could, therefore, be useful to increase cyclic GMP levels in a tissue and correlate these with the ensuing physiological events. Such studies could permit investigators to examine processes regulated by cyclic GMP or calcium and their interactions. Increases in cyclic GMP levels in tracheal smooth muscle segments with nitroprusside, nitroglycerin and other nitric oxide forming agents are associated with relaxation of preparations (29,30). Since 8-bromocyclic GMP also produces relaxation of these preparations, the effects of nitro compounds on smooth muscle motility may be due to guanylate cyclase activation and cyclic GMP accumulation (29,30). Obviously additional experiments are required to test this hypothesis.

These nitro compounds also increase cyclic GMP levels in preparations of guinea pig atria without altering the degree or rate of contraction (30). These studies do not support the view that increases in cyclic GMP in heart preparations with choline esters mediate the effects of acetylcholine (1). The effects of acetylcholine on contractility and cyclic GMP accumulation in smooth muscle (29) and heart (31) can also be dissociated with dose-response curves and time courses. With tracheal smooth muscle acetylcholine causes increased contractility before increases in cyclic GMP and lower concentrations are required for contractile effects (29). Similar dose-response and temporal dissociations have been observed with acetylcholine effects on heart preparations (31).

Activation of Guanylate Cyclase with Hydroxyl Radical

Since the conversion of azide to nitric oxide also requires H_2O_2, the formation of H_2O_2 from superoxide anion ($O_2^{\cdot -}$) by superoxide dismutase in incubations and effects on guanylate cyclase activity were also examined (32). Preparations of partially purified liver soluble guanylate cyclase free of catalase activity could be activated with superoxide dismutase (32). This effect was prevented with catalase to scavenge the H_2O_2 formed or inhibitors of superoxide dismutase such as cyanide or thiols (32). However, the activation by superoxide dismutase could be enhanced with nitrate reductase to increase superoxide anion formation. Since these effects were prevented with antioxidants and hydroxyl free radical scavengers, it appears that hydroxyl radical as well as nitric oxide radical can activate guanylate cyclase (32). The formation of hydroxyl radicals in incubations probably occurs via the Haber-Weiss (33) or Fenton (34) reaction as summarized in Figure 2.

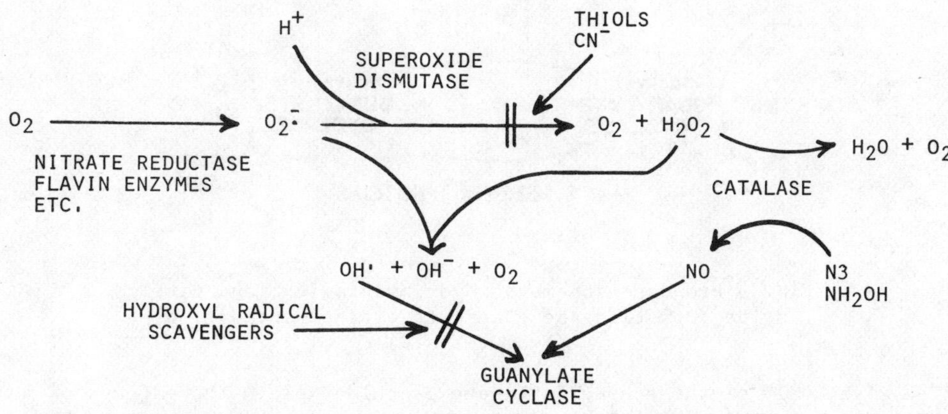

Fig. 2 Activation of guanylate cyclase by superoxide dismutase and hydroxyl radical. (From ref. 11)

Since superoxide anion, hydroxyl radical and other reactive oxygen species are quite ubiquitous in tissues and their formation occurs under many conditions associated with increases in cyclic GMP, we have proposed that these reactive oxygen species are most likely the physiological regulators of guanylate cyclase activity (11,12,32). For example the formation of superoxide anion and/or hydroxyl radicals are associated with phagocytosis (35), fatty acid oxidation (36), prostaglandin formation (37) and platelet aggregation (38), processes that have been associated with cyclic GMP accumulation.

Proposed Scheme of Free Radical Formation and Guanylate Cyclase Activation

The scheme summarized in Figure 3 could explain the effects of numerous agents and processes on guanylate cyclase activity and cyclic GMP accumulation.

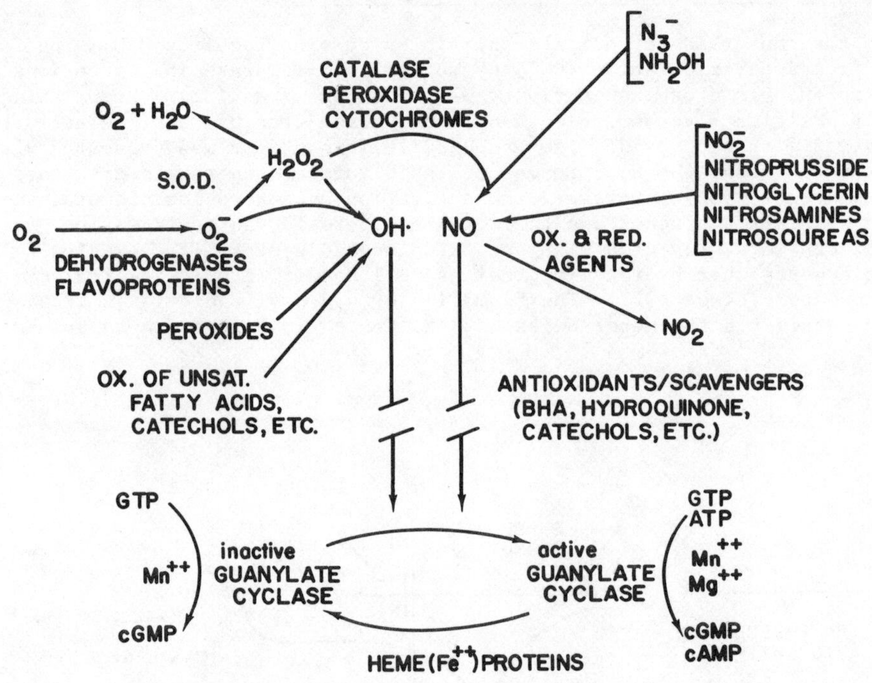

Fig. 3 Proposed scheme of hydroxyl radical and nitric oxide formation and guanylate cyclase activation. (From ref. 12)

The right half of the scheme in Figure 3 could explain the effects of many pharmacologic agents including the nitro compound smooth muscle relaxants. As discussed above these agents increase cyclic GMP accumulation in a calcium independent manner. The left half of the scheme could represent the various physiological and hormonal mechanisms to increase cyclic GMP. While the precise coupling of hormone-receptor interactions and other membrane events to free radical formation and guanylate cyclase activation are not presently known, we would like to offer some speculations and suggestions. The effects of hormones on cyclic GMP levels are calcium dependent. This could result from the effects of calcium on phospholipase (39) and/or lipoxygenase (40). The release of fatty acids from membrane phospholipids and/or the formation of endoperoxides and prostaglandins could influence guanylate cyclase activity through alterations in redox and/or hydroxyl radical formation. Unsaturated fatty acids may also activate guanylate cyclase directly as has been suggested (41,42).

Effects of Nitric Oxide, Hydroxyl Radical and Fatty Acids on Purified Guanylate Cyclase

As summarized in Tables 3 and 4 nitric oxide, superoxide dismutase and unsaturated fatty acids can activate highly purified liver soluble guanylate cyclase (purified approximately 1000-fold) (43). Thus, activation by free radicals and unsaturated fatty acids appears to have no other requirements The purified enzyme can catalyze the formation of cyclic AMP from ATP and this activity is also increased with free radicals (43). Activation of guanylate cyclase with free radicals is a unique system for enzyme regulation. Generally oxidative free radicals decrease enzyme activity. The precise mechanism of free radical activation must await physicochemical studies of pure enzyme, but presumably an oxidizing event occurs that is reversible. While activation can be reversed, the mechanism and requirements for reduction of the enzyme are also unknown.

Table 3

Activation of Purified Guanylate Cyclase

Cyclic GMP formation by purified guanylate cyclase was determined with the agents listed at the concentrations indicated using Mn^{2+} or Mg^{2+} as cofactor. (From ref. 43).

Addition		nmol cyclic GMP/mg/min	
		Mn^{2+}	Mg^{2+}
None		34.0	4.0
Na nitroprusside	(0.1mM)	29.2	6.1
Na nitroprusside	(0.01mM)	37.3	5.6
Nitric oxide	(167μl)	40.3	30.2
Na arachidonate	(50μM)	29.0	6.5
Na arachidonate	(25μM)	31.5	9.0
Na arachidonate	(10μM)	33.5	4.1

Clearly manipulation of the redox state of tissues and cell-free preparations can lead to guanylate cyclase activation and cyclic GMP formation. Such alterations in redox and free radicals could explain the physiological mechanisms involved in regulating cyclic GMP synthesis. Free radical regulation of an enzyme's activity could well have broader application to other enzyme systems and the model proposed could have even greater application. While this remains to be seen, the fundamental mechanisms underlying this sequence of events should prove to be quite intriguing and certainly provide us and others with useful probes to examine cyclic GMP formation and function.

A Proposed Function for Cyclic GMP as a Feedback Regulator of Redox

If guanylate cyclase is, indeed, a monitor of the redox state of cells and free radical formation, one might predict that the product of this process, namely cyclic GMP, should function to feedback and regulate the redox state of cells. Although difficult to study, this hypothesis is testable and has offered a new approach to the study of a function for cyclic GMP. We wish to propose that cyclic GMP is not a tightly coupled second messenger as we have come to understand

Table 4

Activation of Purified Guanylate Cyclase with Superoxide Dismutase

Enzyme prepared with preparative electrophoresis (Expt. 1) or GTP-affinity chromatography (Expt. 2) was incubated with 1 mg/ml of superoxide dismutase and/or nitrate reductase using 4mM $MgCl_2$ and 1mM GTP. (From ref. 43).

	nmol cyclic GMP/ mg/min
Experiment 1	
None	4.0
Superoxide dismutase	4.3
Nitrate reductase	4.2
Superoxide dismutase + nitrate reductase	5.0
Experiment 2	
None	3.8
Superoxide dismutase	4.5
Nitrate reductase	4.0
Superoxide dismutase + nitrate reductase	5.8

the role of cyclic AMP. Rather cyclic GMP formation is a more distal event in hormonal interaction with tissues. The increase in cyclic GMP probably occurs through a very complicated set of events that involve alterations in redox state and free radical formation. The increase in cyclic GMP may not be necessarily involved in transmitting the hormonally-induced function to some process. Temporally, cyclic GMP accumulation may be a later event whose function is to feedback and readjust the redox state of a cell that occurs with its interaction with a hormone or drug. This proposed realteration of the redox state and free radical formation by cyclic GMP could be an important function since these reactive free radicals if left unchecked would be quite destructive to the function and integrity of most proteins and membranes. While heretical to some, a new approach to the study of cyclic GMP function apart from the second messenger concept is presently needed and suggested by the proposed model.

REFERENCES

(1) W.J. George, J.B. Polson, A.G. O'Toole and N.D. Goldberg, Elevation 3',5'-cyclic phosphate in rat heart after perfusion with acetylcholine, Proc. Nat. Acad. Sci. USA 66, 398 (1970).

(2) R.I. Clyman, A.S. Blacksin, J.A. Sandler, V.C. Manganiello and M. Vaughan, The role of calcium in regulation of cyclic nucleotide content in human umbilical artery, J. Biol. Chem. 250, 4718 (1975).

(3) F. Murad and H. Kimura, Cyclic nucleotide levels in incubations of guinea pig trachea, Biochim. Biophys. Acta 343, 275 (1974).

(4) G. Schultz, J.G. Hardman, K. Schultz, C.E. Baird and E.W. Sutherland, The importance of calcium ions for the regulation of guanosine 3',5'-cyclic monophosphate levels, Proc. Nat. Acad. Sci. USA 70, 3889 (1973).

(5) N.D. Goldberg and M.K. Haddox, Cyclic GMP metabolism and involvement in biological regulation, Ann. Rev. Biochem. 46, 823 (1977).

(6) F. Murad and G.D. Aurbach, Cyclic GMP in metabolism: Interrelationship of biogenic amines, hormones and other agents, in The Year in Metabolism 1977, ed. N. Freinkel, Plenum, New York 1-32 (1978).

(7) H. Kimura and F. Murad, Evidence for two different forms of guanylate cyclase in rat heart, J. Biol. Chem. 249, 6910 (1974).

(8) H. Kimura and F. Murad, Two forms of guanylate cyclase in mammalian tissues and possible mechanisms for their regulation, Metabolism 24, 439 (1975).

(9) T.D. Chrisman, D.L. Garbers, M.A. Parks, and J.G. Hardman, Characterization of particulate and soluble guanylate cyclases from rat lung, J. Biol. Chem. 250, 374 (1975).

(10) F. Murad, C.K. Mittal, W.P. Arnold, K. Ichihara, J.M. Braughler and M. El-Zayat, Properties and regulation of guanylate cyclase: Activation by azide nitro compounds and hydroxyl radical and effects of heme containing proteins, in Molecular Biology and Pharmacology of Cyclic Nucleotides, ed. G. Folco and R. Paoletti, Elsevier, North Holland, 33-42 (1978).

(11) F. Murad, C.K. Mittal, W.P. Arnold, S. Katsuki and H. Kimura, Guanylate cyclase: Activation by azide, nitrocompounds, nitric oxide and hydroxyl radical and inhibition by hemoglobin and myoglobin, Adv. Cyclic Nucleotide Res. 9, 145 (1978).

(12) C.K. Mittal and F. Murad, Properties and oxidative regulation of guanylate cyclase, J. Cyclic Nucleotide Res. 3, 381 (1977).

(13) D.L. Garbers, Sea urchin sperm guanylate cyclase antibody: Cross reactivity with various rat tissue guanylate cyclases, J. Biol. Chem. 253, 1898 (1978).

(14) H. Kimura, C.K. Mittal and F. Murad, Activation of guanylate cyclase from rat liver and other tissues by sodium azide, J. Biol. Chem. 250, 8016 (1975).

(15) H. Kimura, C.K. Mittal and F. Murad, Increases in cyclic GMP levels in brain and liver with sodium azide, an activator of guanylate cyclase, Nature (London) 257, 700 (1975).

(16) C.K. Mittal, H. Kimura and F. Murad, Requirement for a macromolecular factor for sodium azide activation of guanylate cyclase, J. Cyclic Nucleotide Res. 1, 261 (1975).

(17) C.K. Mittal, H. Kimura and F. Murad, Purification and properties of a protein required for sodium azide activation of guanylate cyclase, J. Biol. Chem. 252, 4384 (1977).

(18) N. Miki, M Nagano and K. Kuriyama, Catalase activates cerebral guanylate cyclase in the presence of sodium azide, Biochem. Biophys. Res. Commun. 72, 952 (1976).

(19) F. Murad, C.K. Mittal and W.P. Arnold, Guanylate cyclase: Properties and regulation, Proc. of the V Internat. Congress of Endocrinology, Excerpta Medica (1976).

(20) C.K. Mittal, W.P. Arnold and F. Murad, Characterization of protein inhibitors of guanylate cyclase activation from rat heart and bovine lung J. Biol. Chem. 253, 1266 (1978).

(21) S. Katsuki, W.P. Arnold, C.K. Mittal and F. Murad, Stimulation of guanylate cyclase by sodium nitroprusside, nitroglycerin and nitric oxide in various tissues and comparison to the effects of sodium azide and hydroxylamine, J. Cyclic Nucleotide Res. 3, 23 (1977).

(22) F. DeRubertis and P.A. Craven, Calcium-independent modulation of cyclic GMP and activation of guanylate cyclase by nitrosamines, Science 193, 897 (1976).

(23) D.L. Vesely, R.E. Rovere and G.S. Levey, Activation of guanylate cyclase by streptozotocin and 1-methyl-1-nitrosourea, Cancer Res. 37, 28 (1977).

(24) W.A. Arnold, C.K. Mittal, S. Katsuki and F. Murad, Nitric oxide activates guanylate cyclase and increases guanosine 3',5'-cyclic monophosphate levels in various tissue preparations, Proc. Nat. Acad. Sci. USA 74, 3203 (1977).

(25) K. Ichihara, M. El-Zayat, C.K. Mittal and F. Murad (in preparation).

(26) H. Kimura, C.K. Mittal and F. Murad, Appearance of magnesium guanylate cyclase activity in rat liver with sodium azide activation, J. Biol. Chem. 251, 7769 (1976).

(27) C.K. Mittal and F. Murad, Formation of adenosine 3',5'-monophosphate by preparations of guanylate cyclase from rat liver and other tissues, J. Bio. Chem. 252, 3136 (1977).

(28) S.R. Tannenbaum, D. Feit, V.R. Young, P.D. Land and W.R. Bruce, Nitrite and nitrate are formed by endogenous synthesis in the human intestine, Science 200, 1487 (1978).
(29) S. Katsuki and F. Murad, Regulation of adenosine cyclic 3',5'-monophosphate and guanosine cyclic 3',5'-monophosphate levels and contractility in bovine tracheal smooth muscle, Molecular Pharmacology 13, 330 (1977).
(30) S. Katsuki, W.P. Arnold, and F. Murad, Effects of sodium nitroprusside, nitroglycerin and sodium azide on levels of cyclic nucleotides and mechanical activity of various tissues, J. Cyclic Nucleotide Res. 3, 239 (1977).
(31) G. Brooker, Dissociation of cyclic GMP from the negative inotropic action of carbachol in guinea pig atria, J. Cyclic Nucleotide Res. 3, 407 (1977).
(32) C.K. Mittal and F. Murad, Activation of guanylate cyclase by superoxide dismutase and hydroxyl radical: A physiological regulator of guanosine 3',5'-monophosphate formation, Proc. Nat. Acad. Sci. USA 74, 4360 (1977).
(33) F. Haber and J. Weiss, The catalase decomposition of hydrogen peroxide by ion salts, Proc. Roy. Soc. Edinburgh Sect. A. 147, 332 (1934).
(34) J.H. Boxendale, M.G. Evans, and G.S. Park, Trans. Faraday Soc. 42, 155 (1946).
(35) A.T. Tauber and B.M. Babior, Evidence for hydroxyl radical production by human neutrophils, J. Clin. Invest. 60, 374 (1977).
(36) J.L. Bolland, Proc. Roy. Soc. London A186, 218 (1946).
(37) R.W. Egan, J. Paxton and F.A. Kuehl, Mechanism of irreversible self-deactivation of prostaglandin synthetase, J. Biol. Chem. 251, 7329 (1976).
(38) J.G. While, G.H.R. Rao and J. M. Gerrard, Effects of nitroblue tetrazolium and vitamin E on platelet utrastructure, aggregation and secretion, Am. J. Pathal. 88, 387 (1977).
(39) A. Derksen and P. Cohen, Patterns of fatty acid release from endogenous substrates by human platelet homogenates and membranes, J. Biol. Chem. 250, 9342 (1975).
(40) R.B. Koch, Calcium ion activation of lipoxidase, Arch. Biochem. Biophys. 125, 303 (1968).
(41) D. Wallach and I. Pastan, Stimulation of guanylate cyclase of fibroblasts by free fatty acids, J. Biol. Chem. 251, 5802 (1976).
(42) D.V. Glass, W. Frey, D.W. Carr and N.D. Goldberg, Stimulation of human platelet guanylate cyclase by fatty acids, J. Biol. Chem. 252 1279 (1977).
(43) J.M. Braughler, C.K. Mittal and F. Murad, Purification of soluble guanylate cyclase from rat liver (Submitted).

Adenylate Cyclase in Heart and Skeletal Muscle

G.I. Drummond, Jean Dunham, Ponnal Nambi and M. Sano

Biochemistry Group, Department of Chemistry, University of Calgary, Calgary T2N 1N4, Canada

ABSTRACT

Adenylate cyclase was present in microsomal fractions of ventricular muscle. The enzyme in this fraction could be distinguished from activity in the easily sedimentable (plasma membrane fraction) on the basis of response to nonionic detergents, phospholipase A incubation and rate of activation by guanine nucleotide and epinephrine. The data suggest that a small percentage of adenylate cyclase in the heart resides within membranes of the sarcoplasmic reticulum. Adenylate cyclase in plasma membranes prepared from fast skeletal muscle was stimulated by guanine nucleotide (Gpp(NH)p) and catecholamines. On the basis of agonist and antagonist potency with respect to enzyme activation and catecholamine binding it is concluded that in skeletal muscle, adenylate cyclase is coupled to adrenergic receptors of the β_2 type.

INTRODUCTION

Adenylate cyclase [ATP pyrophosphate lyase (cyclizing), EC 4.6.1.1] is associated with cellular membranes in mammalian tissues. Many studies beginning with those of Sutherland and his associates (1,2) have supported the idea that the enzyme is located exclusively or predominantly in the plasma membrane. Such a location is in accord with its role in forming cyclic AMP intracellularly in response to hormones which interact with their receptors at the external cell surface. Indeed, localization in the plasma membrane was an important criterion in the "second messenger" hypothesis of cyclic AMP action (3,4). Evidence exists however, that in some tissues the enzyme is not confined to the plasma membrane; it may be located at intracellular sites. Determining the exact location of the enzyme in many tissues is difficult because of limitations in the methodology for membrane isolation. This is particularly true of contractile tissue; isolation of membranes from these tissues in a high degree of purity and yield is extremely difficult. With regard to the heart, several studies that have involved fractionation of homogenates by differential centrifugation and preparation of at least partially purified plasma membranes (5-9) have provided evidence that the enzyme is predominantly in the surface membrane. The procedure we have developed for isolation of plasma membranes from guinea pig ventricle (8) involves homogenization to cut the fibers at right angles to the long axes, followed by removal of soluble proteins, mitochondria and microsomes by repeated washing with centrifugation at gravitational forces decreasing from 520 to 120 × g. Contractile protein is then extracted into 1.25 M KCl. The KCl-extracted residue is fractionated on a discontinuous sucrose gradient; adenylate cyclase is confined to a region of the gradient which contains ouabain-sensitive Na^{+}+K^{+}-ATPase. The yield of adenylate cyclase in this preparation is approximately 70% (F^{-}-stimulated activity) representing a 10- to 15-fold increase in specific activity over the starting homogenate. In support of a plasma membrane locale, Wollenberger and Schulz (10), using histochemical techniques based on deposition of pyrophosphate (a product of the adenylate cyclase reaction) as the lead salt, found that the enzyme occurred along the plasma membrane and the transverse tubules and at the tight junction regions of the intercalated discs. Some investigators have suggested that this enzyme is also present in the sarcoplasmic reticulum of the heart. For example, Entman et al. (11) reported several years ago that a microsomal fraction from canine cardiac muscle that actively accumulated Ca^{2+} contained adenylate cyclase; similar results were reported by Katz et al. (12) and by Sulakhe and Dhalla (13). In the latter study, the percentage of total activity residing in the microsomal fraction of dog and rabbit hearts was calculated to be 9 and 15% respectively. Microsomal preparations from heart are considered to represent predominantly membranes of the sarcoplasmic reticulum. The possibility has existed however that microsomal preparations are contaminated with fragments of the plasma membrane which could account for the relatively small amounts of adenylate cyclase found in these fractions by the above investigators. Such contamination could result from excessive disruption of the surface membrane during homogenization.

In skeletal muscle, the exact distribution of adenylate cyclase is even more uncertain. In an early study the enzyme was reported to be present in microsomal fractions of rabbit skeletal muscle (14) and in another study activity was reported in mitochondrial and microsomal fractions (15). Using histochemical techniques Wollenberger and his associates (10,16) have concluded that in fast skeletal muscle (psoas), the enzyme is located chiefly at the lateral sacs of the sarcoplasmic reticulum at their points of contact with the transverse tubules. In red skeletal muscle (soleus) the enzyme seemed to occur primarily along the sarcolemma (plasma membrane) (10,16). Some years ago (17) we developed a procedure for isolation of plasma membranes from rabbit leg (fast) muscle that contained a good yield of adenylate cyclase. Various parameters such as phase contract microscopy, "marker" enzyme activities and chemical composition, suggested that the preparation constituted plasma membrane in a high degree of purity. Such preparations represented an increase in adenylate cyclase specific activity of 10- to 20-fold over the starting homogenate with a yield of activity of 30 to 40%. On the basis of these studies we have felt that a significant fraction of the enzyme in fast skeletal muscle resides within the plasma membrane. Recently Caswell et al. (18) have fractionated skeletal muscle (sacrospinalis) microsomal preparations on density gradients and have found that adenylate cyclase as well as catecholamine receptors in this fraction are located within the transverse tubules; no activity was found in the longitudinal reticulum or terminal cisternae. They conclude that the action of epinephrine in skeletal muscle is mediated through receptors and adenylate cyclase in the external membrane.

In this report we present evidence for a discrete adenylate cyclase in microsomal membranes of rabbit heart. Studies on the nature of catecholamine stimulation of the enzyme in plasma membrane preparations from rabbit skeletal muscle are included.

METHODS

Preparation of cardiac fractions. Ventricular muscle from female rabbits was homogenized in 5 volumes of 0.25 M sucrose, 20 mM Tris-HCl, 1 mM dithiothreitol, 5 mM $MgCl_2$, 1 mM ethylenediamine tetraacetic acid, pH 7.5 in a Polytron PT 10 homogenizer for 30 seconds at a rheostat setting of 4 and then for 2 seconds at maximal velocity. The suspension was diluted with an equal volume of buffer and centrifuged at 25,000 × g for 10 minutes. The supernatant fluid was removed, the pellet was suspended in 10 volumes of buffer and homogenized for two 15 second intervals at maximum velocity. The suspension was centrifuged as before; the resulting pellet was suspended in 10 volumes of buffer (based on initial tissue weight) and this constituted the washed particle fraction. The two supernatants were combined and centrifuged at 100,000 × g for 1 hour. The pellet was suspended in a volume of buffer equivalent to 1/25 that used for the initial homogenization; this constituted the 100,000 × g particle or microsomal fraction. Assays were performed immediately.

Preparation of skeletal muscle plasma membranes. Membranes were prepared essentially by the method of Severson et al. (17) with minor modifications.

Adenylate cyclase was assayed by the method of Salomon et al. (19) with modifications we have previously described (20). Catecholamine binding was quantitated using [^{3}H]dihydroalprenolol (21).

RESULTS

Adenylate cyclase in cardiac fractions. Specific activity of adenylate cyclase in the washed particle fraction was 25 and 255 pmol cyclic AMP formed min^{-1} mg^{-1} for basal and F^--stimulated activity respectively; corresponding values for the microsomal fraction were 40 and 200 pmol min^{-1} mg^{-1}. Although the percentage of

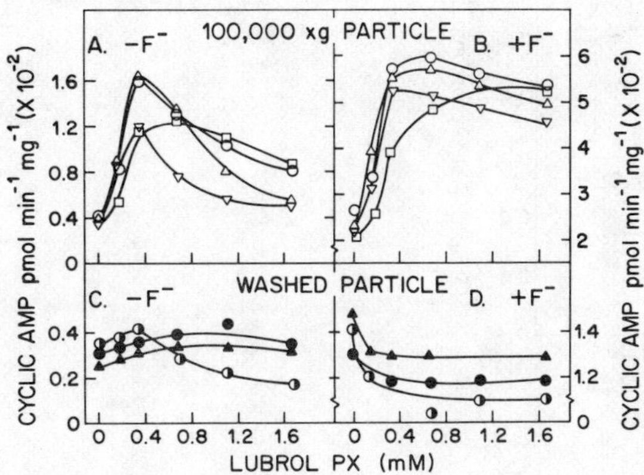

Fig. 1. Effect of Lubrol PX on adenylate cyclase in rabbit heart fractions.

Lubrol PX was present in the assay at the concentration indicated; a concentration of 0.33 mM is equivalent to 0.02% (w/v). Plates A and C, basal; B and D, F^--stimulated activity. Each symbol represents a different membrane preparation. Assay temperature 20°C. (From J. Mol. Cell. Cardiol. 10, 317 (1978).)

total activity recovered in the microsomal fraction was low, 5%, as compared with 70% in the washed particle fraction, the microsomal enzyme differed significantly in several respects from that in the washed particle (plasma membrane) fraction. For example, both basal and F^--stimulated activity in microsomal fractions was significantly increased by low concentrations of Lubrol PX, a nonionic detergent, added to the assay (Fig. 1). In contrast basal activity in the washed particle fraction was only slightly increased by the detergent and F^--stimulated activity was diminished. The effect of Lubrol was not confined to rabbit heart particles; when guinea pig ventricles were fractionated in an analogous manner the response of the enzyme in the two fractions to the detergent was virtually identical to that in rabbit heart. The different effect of the detergent on enzyme activity in the two fractions did not simply reflect a difference in membrane particle size. Thus, when a washed particle preparation was exhaustively homogenized in buffer containing 1.25 M KCl which disrupted all membrane structures and dissolved much of the contractile protein, the effect of the detergent was unchanged. The effect of Triton X-100, another nonionic detergent was examined. Again a striking stimulation of enzyme activity, both basal and F^--activated, occurred in the 100,000 × g pellet fraction (Fig. 2) while basal activity in the washed particle fraction was only slightly increased and F^--stimulated activity was decreased.

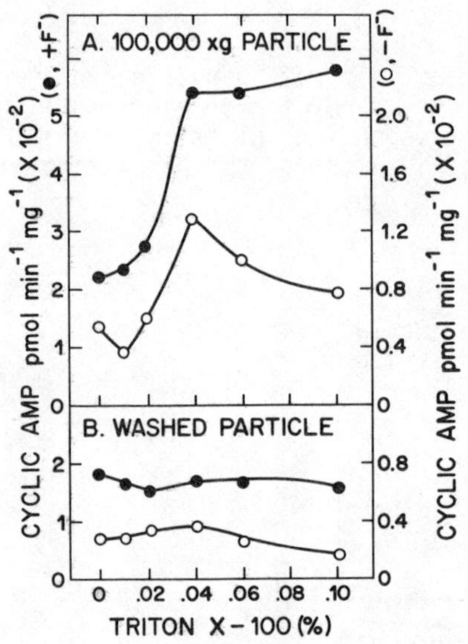

Fig. 2. Effect of Triton X-100 on rabbit heart cyclase.

Conditions are as in Fig. 1, except Triton X-100 was present in the assay at the concentrations indicated; (○) basal; (●) F^--stimulated activity. (From J. Mol. Cell. Cardiol. 10, 317, (1978).)

Enzyme activity in the two fractions was affected differently when membranes were incubated with phospholipase A before assay. Both basal and F^--stimulated activity in the 100,000 × g fraction was increased following incubation with small amounts of phospholipase A; then declined when larger amounts were used (Fig. 3). In contrast basal activity in the washed particle fraction was depressed by phospholipase A while F^--stimulated activity was unaffected. Activation of the enzyme in both fractions by epinephrine was also examined. This was done in the presence of the GTP analog, 5'guanylimidodiphosphate (Gpp(NH)p) (22). Again enzyme activity in the two fractions differed. In both preparations epinephrine alone produced slight activation. When membranes were incubated with Gpp(NH)p prior to assay, the washed particle enzyme was slowly activated; in the presence of the nucleotide, epinephrine produced a dramatic and rapid increase in activity (Fig. 4A). In contrast, the microsomal enzyme was more rapidly activated by Gpp(NH)p alone and

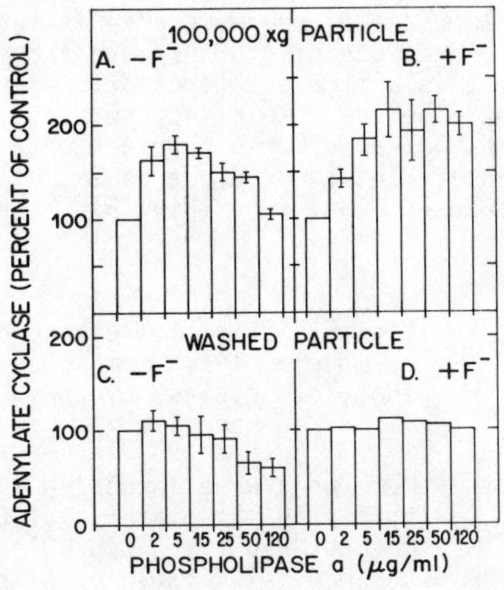

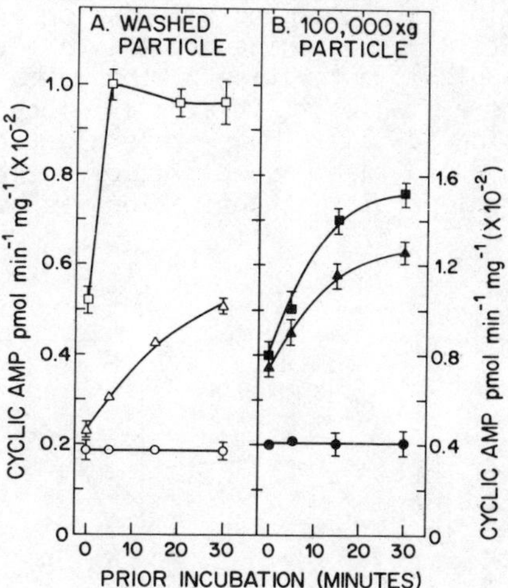

Fig. 3. (left) Effect of phospholipase A on adenylate cyclase activity in cardiac fractions.

Aliquots of membrane suspensions containing 0.2 mM $CaCl_2$ and 100 mM KCl, were incubated with phospholipase A (Vipera russelli) in the concentrations indicated, for 30 minutes at 30°C. The suspensions were then centrifuged at 100,000 × g for 1 hour, the pellets were suspended and assayed at 20°C. (From J. Mol. Cell. Cardiol. 10, 317 (1978).)

Fig. 4. (right) Activation of cardiac adenylate cyclase by Gpp(NH)p in the presence and absence of epinephrine.

Membranes were incubated with 0.1 mM Gpp(NH)p (△,▲) 0.1 mM Gpp(NH)p plus 50 μM epinephrine (□,■), or without addition (○,●), at 20°C for the times indicated, in the adenylate cyclase assay mix without ATP. ATP was then added to begin the assay which was conducted at 20°C for 5 minutes. (From J. Mol. Cell. Cardiol. 10, 317 (1978).)

epinephrine produced only a small acceleration in rate of activation (Fig. 4B).

These studies confirm that adenylate cyclase is present in microsomal fractions prepared from heart. Moreover, the activity is distinctly different from that in the more easily sedimentable (plasma membrane) fraction on the basis of the properties described here. The data should not be interpreted to indicate two kinetically distinct enzyme species; the differences noted may more properly reflect differences in structure and composition of the membrane in which the enzyme is situated. The data allow the suggestion that a small fraction of adenylate cyclase in the heart is located in microsomal membranes, possibly membranes of the sarcoplasmic reticulum.

Adenylate cyclase in skeletal muscle plasma membranes. We have examined the nature of catecholamine stimulation in skeletal muscle plasma membranes in the presence of the guanine nucleotide, Gpp(NH)p. A variety of studies have shown that Gpp(NH)p activation of the mammalian enzyme is a time and temperature dependent process which is irreversible; hormones increase the rate of active state formation (23,24). When skeletal muscle plasma membranes were incubated with Gpp(NH)p and then washed free of nucleotide by sedimentation before assay, a time dependent persistent activation was evident at both 30 and 37°C (Fig. 5). When

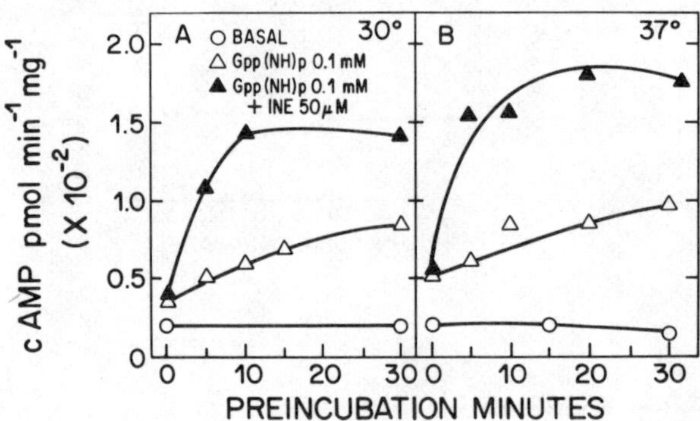

Fig. 5. Activation of skeletal muscle plasma membrane adenylate cyclase by Gpp(NH)p and INE

Membranes were incubated in buffer in the presence of 0.1 mM Gpp(NH)p (△), Gpp(NH)p plus 50 μM (-)-INE (▲), or without any addition (○) at 30 and 37°C for the times indicated, then diluted 20-fold, sedimented at 100,000 × g for 30 min, washed once more, resuspended and assayed at 30°C.

isoproterenol ((-)-INE) was present in addition to Gpp(NH)p, activation was greatly accelerated and reached completion in 10 minutes at 30°. Using these

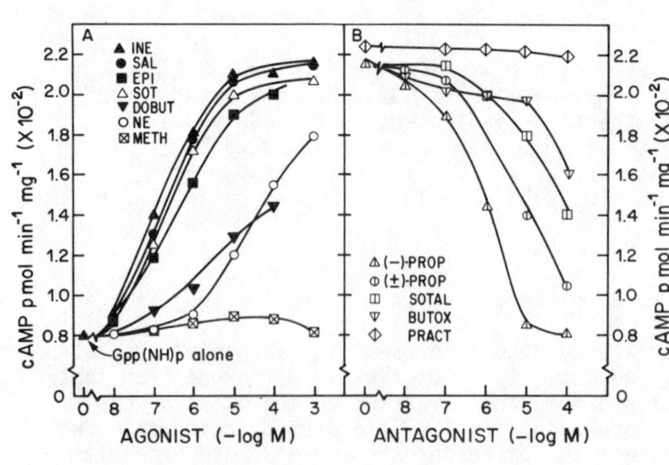

Fig. 6. Effect of agonists and antagonists on activation of skeletal muscle adenylate cyclase in the presence of Gpp(NH)p.

Conditions were as for Fig. 5 except that agonists and antagonists were present as indicated. In plate A, the agonist was 50 μM (-)-INE. Abbreviations: INE, (-)-isoproterenol; sal, salbutamol; epi, epinephrine; sot, soterenol; dobut, dobutamine; NE, norepinephrine; meth, methoxamine; prop, propranolol; sotal, sotalol; butox, butoxamine; pract, practolol.

conditions we examined the potency of a variety of adrenergic agonists to accelerate Gpp(NH)p activation. Fig. 6A shows that INE was the most potent (K_a 0.4 μM), the β_2 agonists salbutamol and soterenol were almost as potent as INE (K_a 0.75 and 0.80 μM respectively), epinephrine was less potent (K_a 2 μM), and norepinephrine was much less potent (K_a 100 μM). The cardioselective agent dobutamine was much less potent than INE and seemed to act as a partial agonist. The α adrenergic agonist methoxamine was inactive. The stimulatory action of INE was strongly inhibited by propranolol (Fig. 5B); the selective β_2 antagonist butoxamine was also inhibitory while the β_1 antagonist practolol was ineffective as was the α adrenergic antagonist phentolamine (data not shown). Binding of catecholamine to the membranes was investigated using the antagonist [^{3}H]dihydroalprenolol ([^{3}H]DHA) (21). "Specific" binding amounted to 60% of the total binding and occurred rapidly reaching completion in 3 minutes at 37°C. Binding of [^{3}H]DHA to the membranes was antagonized by a variety of adrenergic agonists (Fig. 7A). Again INE, soterenol and salbutamol were most potent, K_i values of 1, 0.5 and 2 μM respectively; epinephrine was less potent, K_i 8 μM while norepinephrine was much less potent; K_i 100 μM. These values are in good agreement with the K_a values for activation of adenylate cyclase. The α agonist methoxamine did not inhibit binding. Several adrenergic antagonists also competed for binding sites (Fig. 7B); (-)-alprenolol was most potent, followed by (-)-propranolol and the β_2 antagonist butoxamine. The β_1 antagonist practolol was inactive (data not shown) and the α adrenergic antagonist phentolamine did not compete for binding sites. These results indicate that sympathomimetic amine stimulation of skeletal muscle adenylate cyclase involves activation of adrenergic receptors of the β_2 type.

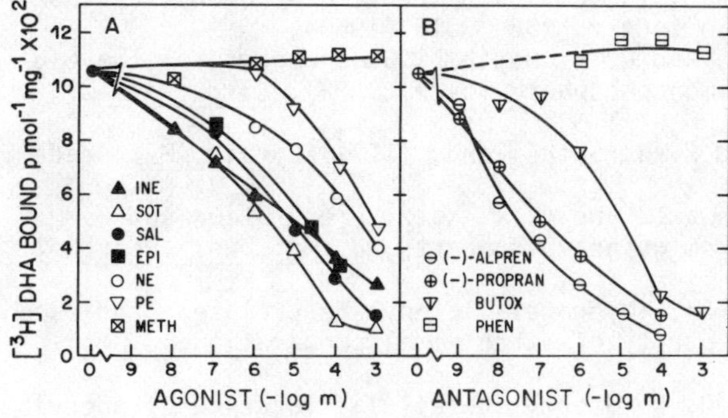

Fig. 7. Effect of adrenergic agonists and antagonists on [^{3}H]dihydroalprenolol binding.

"Specific" [^{3}H]DHA binding was measured as described in Materials and Methods and agonists and antagonists were present in the concentrations indicated. Abbreviations as in Fig. 6 and in addition PE, phenylephrine; phen, phentolamine.

DISCUSSION

Our studies with cardiac preparations suggest that adenylate cyclase in the heart is primarily located in the plasma membrane but a small portion (about 5%) of the activity appears in the microsomal fraction; this activity can be distinguished from that in the easily sedimentable fraction by the action of detergents, by phospholipase A treatment, and by rate of activation by Gpp(NH)p and epinephrine. Since microsomal fractions consist primarily of membranes of the sarcoplasmic reticulum, it might be reasonable to suggest that the activity we have measured may be present in these membranes. A functional role for adenylate cyclase at such an internal site is suggested by the proposed role of cyclic AMP in Ca^{2+}

transport in the heart. It has been established (25,26) that cardiac microsomes are phosphorylated by cyclic AMP-dependent protein kinase; phosphorylated membranes accumulate increased amounts of Ca^{2+}. Tada et al. (25,27) have suggested that catecholamine activation of adenylate cyclase in the sarcoplasmic reticulum results in cyclic AMP mediated phosphorylation of membrane sites leading to increased Ca^{2+} accumulation which could explain catecholamine induced abbreviation of systole and enhanced contractility. Relevant to these findings Entman et al. (28,29) have recently reported evidence for an adenylate cyclase-glycogenolytic complex in cardiac sarcoplasmic reticulum.

Our work with fast skeletal muscle leads us to conclude that a significant portion of adenylate cyclase is located within the plasma membrane in this tissue. We have not investigated microsomal fractions. The studies of Caswell et al. (18) are of interest in that they indicate that β-adrenergic receptors and adenylate cyclase in this fraction are actually associated with transverse tubules (invaginations of the plasma membrane) rather than with sarcoplasmic reticulum. In the plasma membrane adenylate cyclase seems coupled to adrenergic receptors which, on the basis of the present studies, may be classified as of the β_2 type. Lefkowitz (30) has reached similar conclusions in his studies on canine diaphragm muscle. It is significant that glycogenolysis in skeletal muscle is also mediated via β_2 adrenergic receptors.

REFERENCES

1. E. W. Sutherland, T. W. Rall and T. Menon, Adenyl cyclase. I. Distribution, preparation and properties, J. Biol. Chem. 237, 1220 (1962).
2. P. R. Davoren and E. W. Sutherland, The cellular location of adenyl cyclase in the pigeon erythrocyte, J. Biol. Chem. 238, 3016 (1963).
3. E. W. Sutherland, G. A. Robison, and R. W. Butcher, Some aspects of the biological role of adenosine 3',5'-monophosphate (cyclic AMP), Circ. Res. 37, 279 (1968).
4. G. A. Robison, R. W. Butcher and E. W. Sutherland, (1971) Cyclic AMP, Academic Press, New York.
5. M. Tada, J. O. Finney, M. H. Schwartz and A. M. Katz, Preparation and properties of plasma membrane from guinea pig hearts, J. Mol. Cell. Cardiol. 4, 417 (1972).
6. A. M. Watanabe and H. R. Besch, Jr., Myocardial adenylate cyclase: Studies on the relationship of activity to purity of sarcolemmal preparation, J. Mol. Cell. Cardiol. 7, 563 (1975).
7. V. H. Engelhard, D. A. Plut and D. R. Storm, Subcellular location of adenylate cyclase in cardiac muscle, Biochim. Biophys. Acta 451, 48 (1976).
8. E. C. Hui, M. Drummond and G. I. Drummond, Calcium accumulation and cyclic AMP-stimulated phosphorylation in plasma membrane enriched preparations of myocardium, Arch. Biochem. Biophys. 173, 415 (1976).
9. P. J. St. Louis and P. V. Sulakhe, Isolation of sarcolemmal membranes from cardiac muscle, Int. J. Biochem. 7, 547 (1976).
10. A. Wollenberger and W. Schulze, (1976) Recent Advances in Cardiac Structure and Function Vol. 9, p. 101, Eds. P. E. Roy and N. S. Dhalla, University Park Press, Baltimore.
11. M. L. Entman, G. S. Levey and S. E. Epstein, Demonstration of adenyl cyclase in canine sarcoplasmic reticulum, Biochem. Biophys. Res. Commun., 35, 728 (1969).
12. A. M. Katz, M. Tada, D. I. Repke, J. Iorio and M. A. Kirchberger, Adenylate cyclase: its probable location in the sarcoplasmic reticulum as well as sarcolemma of the canine heart, J. Mol. Cell. Cardiol. 6, 73 (1974).
13. P. V. Sulakhe and N. S. Dhalla, Adenylate cyclase of heart sarcotubular membranes, Biochim. Biophys. Acta 293, 379 (1973).

14. M. Rabinowitz, L. Desalles, J. Meisler and L. Lorand, Distribution of adenyl cyclase activity in skeletal muscle fractions, Biochim. Biophys. Acta 97, 29 (1965).
15. K. Seraydarian and W. F. H. M. Mommaerts, Density gradient separation of sarcotubular vesicles and other particulate constituents of rabbit muscle, J. Cell. Biol. 26, 641 (1965).
16. W. Schulze, E. G. Krause and A. Wollenberger, Cytochemical demonstration and localization of adenylate cyclase in skeletal and cardiac muscle. Advan. Cyclic Nucleotide Res. 1, 249 (1972).
17. D. L. Severson, G. I. Drummond and P. V. Sulakhe, Adenylate cyclase in skeletal muscles: Kinetic properties and hormonal stimulation, J. Biol. Chem. 247, 2949 (1972).
18. A. H. Caswell, S. P. Baker, H. Boyd, L. T. Potter and M. Garcia, β-Adrenergic receptor and adenylate cyclase in transverse tubules of skeletal muscle, J. Biol. Chem. 253, 3049 (1978).
19. Y. Salomon, C. Londos and M. Rodbell, A highly sensitive adenylate cyclase assay, Anal. Biochem. 58, 541 (1974).
20. G. I. Drummond and J. Dunham, Adenylate cyclase in cardiac microsomal fractions, J. Mol. Cell. Cardiol. 10, 317 (1978).
21. C. Mukherjee, M. G. Caron, M. Coverstone and R. J. Lefkowitz, Identification of adenylate cyclase-coupled β-adrenergic receptors in frog erythrocytes with (-)-[^{3}H]alprenolol, J. Biol. Chem. 250, 4869 (1975).
22. C. Londos, Y. Salomon, M. C. Lin, J. P. Harwood, M. Schramm, J. Wolff and M. Rodbell, 5'-guanylimidodiphosphate, a potent activator of adenylate cyclase systems in eukaryotic cells, Proc. Natl. Acad. Sci. USA 71, 3087 (1974).
23. M. Rendell, Y. Salomon, M. C. Lin, M. Rodbell and M. Berman, The hepatic adenylate cyclase system. III. A mathematical model for the steady state kinetics of catalysis and nucleotide regulation, J. Biol. Chem. 250, 4253 (1975).
24. S. Jacobs, V. Bennett and P. Cuatrecasas, Kinetics of irreversible activation of adenylate cyclase of fat cell membranes by phosphonium and phosphoramidate analogs of GTP, J. Cyclic Nucleotide Res. 2, 205 (1976).
25. M. A. Kirchberger, M. Tada, D. I. Repke and A. M. Katz, Cyclic adenosine 3', 5'-monophosphate-dependent protein kinase stimulation of calcium uptake by canine cardiac microsomes, J. Mol. Cell. Cardiol. 4, 673 (1972).
26. P. J. Laraia and E. Morkin, Adenosine 3',5'-monophosphate-dependent membrane phosphorylation, Circ. Res. 35, 298 (1974).
27. M. Tada, M. A. Kirchberger, D. I. Repke and A. M. Katz, The stimulation of calcium transport in cardiac sarcoplasmic reticulum by adenosine 3',5'-monophosphate dependent protein kinase, J. Biol. Chem. 249, 6174 (1974).
28. M. L. Entman, K. Kaniike, M. A. Goldstein, T. E. Nelson, E. P. Bornet, W. Futch and A. Schwarz, Association of glycogenolysis with cardiac sarcoplasmic reticulum, J. Biol. Chem. 251, 3140 (1976).
29. M. L. Entman, E. P. Bornet, A. J. Garber, A. Schwarz, G. S. Levey, D. C. Lehotay and L. A. Bricker, The cardiac sarcoplasmic reticulum-glycogenolytic complex. A possible effector site for cyclic AMP, Biochim. Biophys. Acta 449, 228 (1977).
30. R. J. Lefkowitz, Heterogeneity of adenylate cyclase-coupled β-adrenergic receptors, Biochem. Pharmacol. 24, 583 (1975).

Cyclic Nucleotide Accumulation and Hormonal Modulation of Cardiac Metabolism and Contraction

Joan Heller Brown, Laurence L. Brunton, J. Scott Hayes, James B. Reese and Steven E. Mayer

Division of Pharmacology, School of Medicine, University of California at San Diego, La Jolla, CA 92093, U.S.A.

INTRODUCTION

In recent years this laboratory has directed its attention toward defining in biochemical terms the effects of β-adrenergic agents on the inotropic state and on the metabolism of cardiac muscle. Figure 1 summarizes the well-known effects of catecholamines and suggests some points of experimental attack. A prime question about which much of our work revolves is, "Is the intracellular accumulation of cyclic AMP necessary or sufficient for these metabolic and inotropic responses?" The data bearing on this question are as ponderous as they are controversial and have recently been well reviewed (1). We will confine this communication to three separate lines of investigation in our laboratory that pertain to the regulation of cardiac metabolism and contractility by agents that influence cyclic nucleotide metabolism.

The first of these stems from observations that the activation of glycogenolysis by catecholamines in perfused guinea pig and rat hearts is inhibited by high K^+ (56mM) medium (2). Mayer et al. proposed that the primary effect of K^+ depolarization was on catecholamine-stimulable cyclic AMP synthesis, since this early step was blocked in hearts perfused with high K^+ medium (2, 3). In more recent experiments on guinea pig papillary muscles, a lower K^+ concentration (22mM) shifted the concentration-dependence of catecholamine-stimulated cyclic AMP formation to the right, resulting in a partial dissociation of cyclic AMP accumulation from enhanced contractility (4). These data demonstrate an apparent "uncoupling" of β-adrenergic receptors from cyclic AMP synthesis but do not suggest the mechanism of this effect of K^+. Indeed, this effect of K^+ may be unique to cardiac tissue since in other tissues elevated $[K^+]_0$ increases cyclic AMP formation and does not block the effect of catecholamines on cyclic AMP accumulation (5, 6). Our current investigations with isolated mouse atria focus on the mechanisms by which elevated $[K^+]_0$ alters cyclic AMP accumulation in cardiac tissue.

Our second line of investigation extends the observations of Keely on the actions of epinephrine and prostaglandin E_1 (PGE_1) on glycogen metabolizing enzymes in the isolated perfused rat heart (7). Keely found that whereas both hormones caused accumulation of cyclic AMP and activation of protein kinase, only epinephrine caused phosphorylase *b* to *a* conversion. We have performed similar experiments on perfused rat hearts, measuring cyclic AMP accumulation, the activation state of protein kinase, the activation states of two substrates of protein kinase (phosphorylase *b* kinase and glycogen synthase), phosphorylase activity, and, as a measure of contractile state, dP/dt. A comparison of the effects of PGE_1 and isoproterenol (INE) on these parameters offers some insights into the specificity of hormonal regulation of cardiac contractility and metabolism.

The third topic evolves from an observation made during studies of β-adrenergic effects on the enzymes regulating glycogen metabolism in perfused guinea pig and rat hearts. In the course of these studies we noted that in the guinea pig myocardium, isoproterenol led to phosphorylase activation without the anticipated increase in the pH 6.8:8.2 activity ratio of phosphorylase b kinase. We present data explaining this observation and demonstrating several differences between guinea pig cardiac phosphorylase b kinase and the skeletal muscle enzyme.

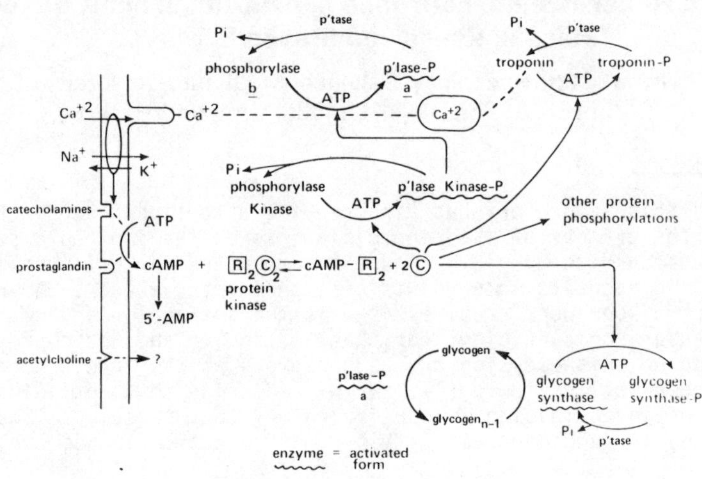

Fig. 1 Regulation of glycogen metabolism in muscle

METHODS

For assay of cyclic AMP, 30 to 50 mg of frozen tissue powder or individual atria were homogenized in 10% TCA containing 7000 cpm (0.25 pmol) of $[^3H]$ cyclic AMP to monitor for recovery. The supernatant fluid was purified over Dowex 50 AG Wx4 (200-400 mesh). Cyclic AMP was quantified by the protein binding assay of Gilman (8). Cyclic AMP-dependent protein kinase was assayed as described by Keely et al. (9) by measuring the transfer of ^{32}P from $[\gamma\text{-}^{32}P]$ ATP to histone using a 10-fold dilution of heart cytosol as the source of the enzyme. Extent of activation is expressed as the ratio of activities in the absence of cyclic AMP and the presence of 2 μM cyclic AMP. Phosphorylase kinase activity was determined by measuring the incorporation of ^{32}P into phosphorylase b using the filter paper method described by Reimann et al. (10). The standard reaction mixture contained 20 μl of 0.125 M Tris, 0.125 M B-glycerophosphate buffer, pH 6.8 or 8.6 (final pH 8.2), 0.6 mg of phosphorylase b (95μM final), 30 nmols of ATP, 1 μmol of magnesium acetate, and 10μl of a 10-fold (pH 6.8) or 20-fold (pH 8.2) dilution of heart cytosol as the source of the enzyme in a total volume of 63 μl. Glycogen phosphorylase was assayed fluorometrically as described previously (11). The results are expressed as the ratio of activity without AMP to activity assayed with AMP. Glycogen synthase was assayed in a soluble extract of frozen heart powder by the method of Thomas et al. (12). For determination of the $A_{0.5}$ for glucose-6-P, the concentration of UDP-glucose was 0.5mM. For determination of activity ratios (-G6P:+10mM G6P) the concentration of UDP-glucose was 5mM. For assay of phosphatase activity, heart powder was homogenized (1:5) in Tris-Cl (50mM, pH 7.6), mercaptoethanol (15mM), phenylmethylsulfonylfluoride (1mM), EDTA (2mM), and EGTA (0.2mM), with charcoal (10mg/ml) and centrifuged for 20 minutes at 20,000 x g. The supernatant was combined with phosphorylase a (1mg/ml) and $MnCl_2$(7mM) and incubated at 30°. Aliquots were removed at appropriate times and assayed for phosphorylase activity as described above. β-glycerol-PO_4 (50mM) and NaF (20mM) were used as phosphatase inhibitors in several experiments.

POTASSIUM DEPOLARIZATION AND CYCLIC AMP ACCUMULATION IN MOUSE ATRIA

Isolated right and left atria from adult male Swiss-Webster mice were incubated in oxygenated Krebs-Henseleit buffer (13) at 37° for 30 minutes, and then in the appropriate experimental media with a phosphodiesterase inhibitor (100 μM methylisobutylxanthine) for 20 minutes. Isoproterenol (INE) or vehicle (ascorbate) was then added for 1 minute, after which the tissue was frozen in freon cooled in liquid nitrogen.

Although increasing $[K^+]_0$ from 6mM to 23mM did not alter cyclic AMP accumulation in response to INE, elevation of $[K^+]_0$ to 57mM markedly decreased the response to all concentrations of INE (Fig. 2). This effect of elevated K^+ was rapid; exposure of atria to 57mM K^+ for as little as 2 minutes resulted in inhibition of the INE response. The K^+ effect was also reversible; normal responsiveness to INE was restored when atria exposed to elevated K^+ were returned to normal Krebs-Henseleit buffer before being challenged with INE.

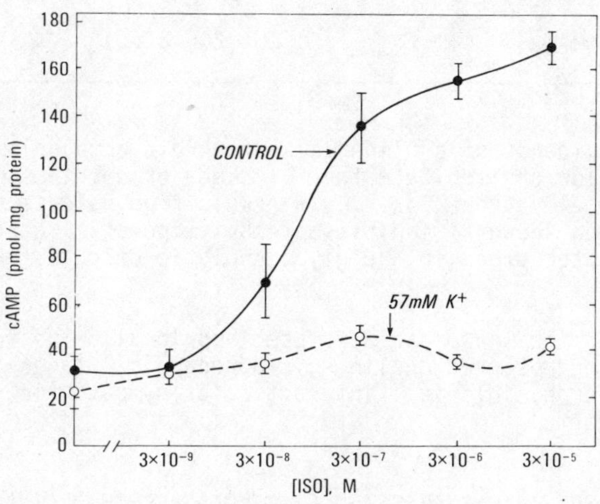

Fig. 2 Effect of K^+ on β-adrenergic response of isolated mouse atria

Since a decrease in Na^+ accompanied the increase in K^+ (to maintain isotonicity), we examined the hormonal responsiveness of atria placed in media in which the sodium concentration was reduced (Table 1). Cyclic AMP accumulation in response to INE was unaffected by lowered $[Na^+]_0$ when sucrose replaced Na^+. However, replacement of NaCl by choline chloride greatly diminished cyclic AMP accumulation due to INE. This diminution of responsiveness appeared to be due to muscarinic effects of choline since it was abolished by atropine.

An unexpected observation was that atropine reversed the blockade of catecholamine responsiveness produced by elevated K^+ (Table 1). The simplest interpretation of this finding is that the effect of high K^+ is indirect and is mediated through muscarinic receptors. This interpretation is supported by evidence that elevated K^+ causes acetylcholine release from cardiac nerve endings (14). Our experiments employing the cholinergic agonist carbamylcholine (carbachol) provide further evidence for a muscarinic cholinergic effect on adrenergic responses of atria. Addition of carbachol (30 μM) prior to or with INE decreased the normal six-fold increase in cyclic AMP accumulation by 80%. The effect of carbachol, like those of elevated K^+ and choline, was blocked by atropine.

TABLE 1 Ionic Composition and Substitution on Isoproterenol-Induced Cyclic AMP Accumulation in Mouse Atria

Modification of Krebs-Henseleit	Control	Isoproterenol (30μM)	Isoproterenol (30μM) + Atropine (30μM)
None (144mM Na^+, 6mM K^+)	34.2±2.1(18)	252±36(17)	195±33(3)
High K^+ (93mM Na^+, 57mM K^+)	31.4±4.5(9)	83±4(18)	209±24(6)
Low Na^+ (93mM Na^+, 6mM K^+ 51mM choline)	34.2±3.0(8)	69±5(17)	232±29(10)
Low Na^+ (93mM Na^+, 6mM K^+ 102mM sucrose)	___	266±32(6)	___

Values are means ± S.E.M. (n).

These data provide evidence of cholinergic-adrenergic antagonism at the level of cyclic AMP accumulation and indicate that blockade of cardiac metabolic responses to catecholamines by elevated $[K^+]_0$ could result from release of acetylcholine. Acetylcholine has been shown to inhibit catecholamine-stimulated cyclic AMP formation, as well as later steps in the glycogenolytic cascade, in perfused hearts (15, 16).

Further studies using the mouse atrial system should allow us to determine the molecular basis of the observed cholinergic-adrenergic antagonism and to assess its importance in the physiological interaction of sympathetic and parasympathetic influences on the heart.

HORMONE-SPECIFIC EXPRESSION OF CARDIAC PROTEIN KINASE ACTIVITY

The effects of isoproterenol (INE) and PGE_1 were compared in Langendorff perfused hearts (from male Sprague-Dawley rats, 225-300g) paced at 240 beats per minute. Hearts were perfused with Krebs-Henseleit buffer according to Neely (13) at 32° for 30 minutes then with control buffer or with buffer containing drug for 2 minutes; hearts were then freeze-clamped. Contractility (dP/dt) was monitored before and during drug exposure using an intraventricular pressure transducer. Cyclic

AMP and the enzymes governing glycogen metabolism were assessed in appropriate homogenates of the powdered frozen heart tissue as described in Methods. Table 2 summarizes the results of these studies.

INE caused all the anticipated changes: enhancement of dP/dt, increased intracellular cyclic AMP, increases in the activation states of protein kinase, phosphorylase b kinase, and phosphorylase, and conversion of glycogen synthase to a less active form. We assessed the activation state of glycogen synthase by measurement of its apparent affinity for glucose-6-phosphate (G6P; Table 2). The increase in $A_{0.5}$ for G6P in the face of an unchanged activity ratio (-G6P:+G6P) reinforces the earlier conclusion of this laboratory that the apparent affinity of glycogen synthase for its allosteric modifier is a more sensitive index of its activation state and extent of phosphorylation than is the -G6P:+G6P activity ratio (17).

In contrast to these effects of INE, PGE_1 increased cyclic AMP accumulation and the protein kinase activity ratio but caused no detectable changes in the activities of two protein kinase substrates (phosphorylase b kinase and glycogen synthase), nor activation of phosphorylase, nor any change in dP/dt (Table 2). Thus when cyclic AMP formation was stimulated by PGE_1 there was a failure of the resultant protein kinase activation to be expressed.

To determine whether PGE_1 could hinder the activation of glycogenolysis by INE, we exposed hearts (n=2) to INE plus PGE_1. All of the values from these hearts were similar to those from hearts treated with INE alone, suggesting that PGE_1 had no insurmountable inhibitory effect on the glycogenolytic pathway. Note, also, that the effects of INE plus PGE_1 were not additive. Provided that the drug concentrations were maximal and pending collection of data from additional hearts, non-additivity would argue against the actions of INE and PGE_1 being on separate cell populations.

TABLE 2 Effects of Isoproterenol and PGE_1 on the Glycogenolytic Cascade in Perfused Rat Heart

	Control	INE (80nM)	PGE (2μM)	INE (80nM) PGE_1(2μM)
dP/dt‡	1	1.7	1	1.4
Cyclic AMP (pmol/mg protein)	4.6 ± 0.2	9.5 ± 0.7*	6.1 ± 0.2*	9.2
Protein Kinase (-cAMP:+cAMP)	0.18 ± 0.01	0.35 ± 0.04*	0.24 ± 0.03*	0.37
Phos b Kinase (pH 6.8:8.2)	0.15 ± 0.01	0.19 ± 0.01**	0.14 ± 0.01	0.19
Phosphorylase (-AMP:+AMP)	0.05 ± 0.01	0.31 ± 0.03*	0.05 ± 0.01	0.32
Glycogen Synthase				
($A_{0.5}^{G6P}$, mM)	0.45 ± 0.05	0.82 ± 0.09*	0.40 ± 0.04	0.99
(-G6P:+G6P)	0.08 ± 0.01	0.11 ± 0.02	0.11 ± 0.02	0.03

Values are means ± S.E.M. of determinations from 8 - 11 hearts
* = $p < 0.005$, ** = $p < 0.01$ compared to control value
‡ dP/dt is expressed as fraction of control (1800 ± 60 mm Hg/sec).

The failure of PGE_1 to cause phosphorylative events subsequent to protein kinase activation could be due to a generalized increase in phosphoprotein phosphatase activity by PGE_1. Thus, phosphorylative activation could occur but be counterpoised by dephosphorylation. To test this possibility, we assayed phosphatase activity in soluble fractions from frozen powders of control hearts and hearts treated with PGE_1 or INE, using phosphorylase a as substrate. Figure 3 shows that phosphorylase a activity was decreased at the same rate by extracts from all heart samples. The decrease was blocked by the inclusion of the phosphatase inhibitors NaF and β-glycerol phosphate; phosphorylase a activity was also restored by the addition of cyclic AMP and protein kinase to the phosphatase assay at t=60 minutes, (data not shown). Total (+AMP) phosphorylase activity did not change during the course of the assay (Fig. 3). These results indicate that the decreases in phosphorylase a activity depicted in Fig. 3 were due to phosphatase activity and that there were no differences in phosphorylase phosphatase activity due to drug treatments. Although our assay would likely have detected only stable changes in total phosphatase activity, the failure of protein kinase activation to be expressed in response to PGE_1 does not appear to result from an activation of phosphoprotein phosphatase.

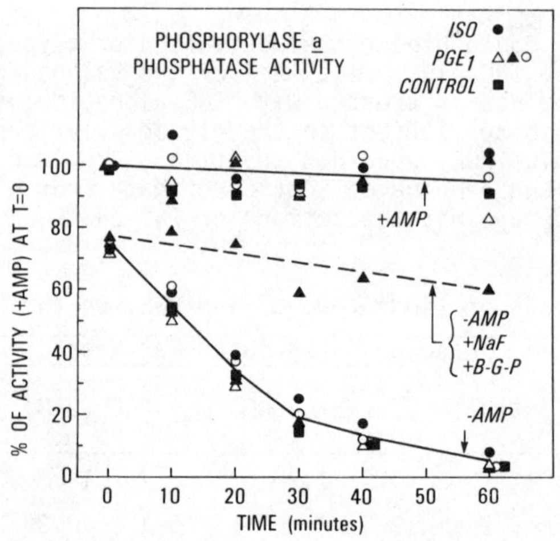

Fig. 3 Phosphorylase a phosphatase activity of perfused rat hearts

We are presently refining our experiments to see whether these striking differences persist with comparable activation (same -cAMP:+cAMP activity ratios) of protein kinase by INE and PGE_1. There are several intriguing explanations for the observations presented above. (1) INE supplies a necessary factor not supplied by PGE_1; Ca^{++} is a tempting factor but the apparent blockade is in protein kinase function, not a function thought to require Ca^{++}. (2) PGE_1 supplies an inhibitory factor (discussed above). (3) The protein kinases activated by INE and PGE_1 are comparted differently within the cell. However, our preliminary data indicate that at low protein kinase activity ratios (<0.3), there is no differential activation of soluble or particulate protein kinases by INE or PGE_1.

CHARACTERIZATION OF GUINEA PIG HEART PHOSPHORYLASE b KINASE

Phosphorylase b kinase from resting skeletal muscle or rat heart is present in a form that is essentially inactive at pH 7.0 and is only partially active at higher pH values. Under the influence of catecholamines the activity of phosphorylase b kinase increases; in skeletal muscle and perfused rat heart the change in the activation state of phosphorylase kinase associated with phosphorylation can be measured as an increase in the pH 6.8:8.2 activity ratio (18, 19). Langendorff perfused guinea pig hearts responded to a 2-minute exposure to isoproterenol (0.3μM) with increased intracellular cyclic AMP and phosphorylase activation (Table 3). This β-adrenergic effect occurred without activation of phosphorylase b kinase as assessed by the pH 6.0:8.2 or pH 6.8:8.2 activity ratios, but with a significant increase in activity measured at pH 8.2.

TABLE 3 Effects of Isoproterenol on the Perfused Guinea Pig Heart

	Cyclic AMP (pmol/mg wet wt)	Phosphorylase b Kinase Ratio (pH 6.0:8.2)	Phosphorylase b Kinase Activity at pH 8.2 (nmol/min/mg prot)	Phosphorylase (-AMP:+AMP)
Control	0.61 ± 0.17	0.14 ± 0.01	1.28 ± 0.10	0.12 ± 0.01
Isoproterenol	1.41 ± 0.38*	0.15 ± 0.01	3.11 ± 0.17*	0.60 ± 0.10*

Values are mean ± S.E.M. of determinations from 4 hearts
* = $p < 0.001$ compared to control values

The traditional measure of activation of phosphorylase b kinase is based on differential effects of pH on the phospho- and dephospho- forms of the enzyme. We therefore determined the pH profiles of the enzymes from guinea pig hearts perfused with isoproterenol (active) or control buffer (inactive). The activities of the two forms showed similar pH dependences, the active form having approximately twice the activity of the inactive form from pH 5 to pH 8.3 (Fig. 4). This is in contrast to the distinctly non-parallel profiles of the skeletal muscle enzymes (20). The active form of the skeletal muscle enzyme exhibits a 25-fold greater activity at pH 6.8 than does the inactive form, but nearly equivalent activity at pH 8.2. Thus, the pH 6.8:8.2 activity ratio can be used as a measure of the balance of the two forms of the skeletal muscle enzyme but does not distinguish between active and inactive states of guinea heart phosphorylase b kinase.

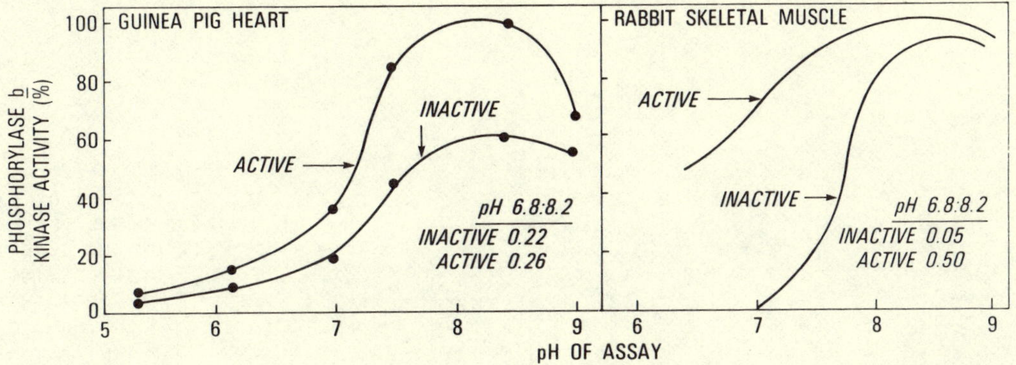

Fig. 4 The pH profiles for active and inactive phosphorylase b kinase

Activation of skeletal muscle phosphorylase b kinase increases the affinity of the enzyme for its substrate, as evidenced by 12- and 2-fold decreases in the K_m for phosphorylase b at pH 6.8 and pH 8.2, respectively (20). These differences can account for the observed increase in the pH 6.8:8.2 activity ratio accompanying activation of the enzyme. We have performed kinetic studies on soluble guinea pig heart phosphorylase b kinase activated or inactivated *in vitro* as described in Table 4. Enzymes prepared in this manner demonstrated pH profiles equivalent to those shown in Fig. 4 for enzymes activated *in vivo*. The data in Table 4 show that *in vitro* activation of guinea pig heart phosphorylase b kinase did not markedly reduce the K_m for phosphorylase b at pH 6.8.

TABLE 4 Kinetic Parameters for the Phosphorylation of Phosphorylase b by Guinea Pig Heart Phosphorylase Kinase

	Nonactivated		Activated	
pH	K_m (μM)	Vmax (nmol ^{32}P/min/mg prot)	K_m (μM)	Vmax (nmol ^{32}P/min/mg prot)
6.8	215 ± 51	0.97 ± 0.18	148 ± 29	1.10 ± 0.13
8.2	276 ± 45	4.40 ± 0.94	85 ± 11	2.7 ± 0.39

Values are means ± S.E.M. from 6 - 10 experiments. K_m is expressed as concentration of phosphorylase b monomer. Nonactivated enzyme was pre-incubated for 30 min. at 30° with the heat stable inhibitor of protein kinase and 5mM Mg^{2+}. The activated enzyme was pre-incubated with Mg^{2+}, ATP, cyclic AMP, and cyclic AMP-dependent protein kinase for 5 min. at 30° prior to assay.

If one substitutes the kinetic constants of Table 4 and the phosphorylase concentration used in the standard assay into the equation $v = V_m \cdot S/K_m + S$, the calculated pH 6.8:8.2 activity ratios for the active and inactive forms are identical (∿0.3). The kinetic changes effected by phosphorylation thus explain the absence of measurable increases in the pH 6.8:8.2 activity ratio of phosphorylase kinase in guinea pig hearts treated with INE (Table 4). Although the activity ratio is not a useful index of the activation state of the guinea pig heart enzyme, the effects of hormones can be assessed by measuring changes in total enzyme activity.

ACKNOWLEDGMENTS

We appreciate the technical assistance of Michael J. Branks and the support of NIH grants HLB12373 and 5F32 HL 05363.

REFERENCES

(1) R.W. Tsien, Cyclic AMP and contractile activity in heart, Adv. Cyclic Nucleotide Res. 8 (ed., P. Greengard and G.A. Robinson), Raven Press, New York (1977).

(2) D.H. Namm, S.E.Mayer and M. Maltbie, The role of potassium and calcium ions in the effect of epinephrine on cardiac cyclic adenosine 3'-5'-monophosphate,phosphorylase kinase and phosphorylase, Mol. Pharmacol. 4, 522 (1968).

(3) E.E. Largis, J. Ashmore and R.E. Toomey, Effect of varying K^+ concentration on phosphorylase activity, lipolysis, and cAMP levels in perfused rat heart, Proc. Soc. Exp. Biol. Med. 141, 31 (1972).

(4) S.E. Mayer, J.G. Dobson Jr., W.R. Ingebretsen Jr., E. Becker, J.H. Brown, W.F. Friedman and J. Ross Jr., Ionic regulation of signal transfer from adrenergic receptors in cardiac muscle in Adv. Cyclic Nucleotide Res. 9 (ed. W.J. George and L.J. Ignarro). Raven Press, New York (1978).

(5) L. Lundholm, T. Rall and N. Vamos, Influence of K-ions and adrenaline on the adenosine 3',-5'-monophosphate content in rat diaphragm, Acta Physiol. Scand. 70, 127 (1967).

(6) M. Huang, H. Shimizu and J. Daly, Regulation of adenosine cyclic 3', 5'-phosphate formation in cerebral cortical slices, Mol. Pharmacol. 7, 155 (1977).

(7) S.L. Keely, Activation of cAMP-dependent protein kinase without a corresponding increase in phosphorylase activity, Res. Commun. Chem. Path. Pharmacol. 18, 283 (1977).

(8) A.G. Gilman, A protein binding assay for adenosine 3':5'-cyclic monophosphate, Proc. Nat. Acad. Sci. U.S.A. 67, 305 (1970).

(9) S.L. Keely, J.D. Corbin and C.R. Park, Regulation of adenosine 3', 5'-monophosphate-dependent protein kinase: regulation of the heart enzyme by epinephrine, glucagon, insulin and methylisobutylxanthine, J. Biol. Chem. 250, 4832 (1975).

(10) E.M. Reimann, D.A. Walsh and E.G. Krebs, Purification and characterization of rabbit skeletal muscle adenosine 3', 5'-monophosphate-dependent kinase, J. Biol. Chem. 246, 1986 (1971).

(11) J.G. Hardman, S.E. Mayer and B. Clark, Cocaine potentiation of the cardiac inotropic and phosphorylase responses to catecholamines as related to the uptake of $[^3H]$-catecholamines, J. Pharmacol. Exp. Ther. 150, 341 (1965).

(12) J.A. Thomas, K.K. Schlender and J. Larner, A rapid filter paper assay for UDP-glucose-glycogen glucosyltransferase, including an improved biosynthesis of UDP-^{14}C-glucose, Anal. Biochem 25, 486 (1968).

(13) J.R. Neely and M.J. Rovetto, Techniques for perfusing isolated rat hearts, Methods in Enzymology 39 (ed., J.C. Hardman and B.W. O'Malley), Academic Press, New York (1975).

(14) R. Lindmar, K. Loffelholz and H. Pompetzki, Acetylcholine overflow during infusion of a high potassium-low sodium solution into the perfused chicken heart in the absence and presence of physostigmine, Naunyn-Schmied. Arch. Pharmacol. 299, 17 (1977).

(15) N. Vincent and S. Ellis, Inhibitory effect of acetylcholine on glycogenolysis in the isolated guinea-pig heart, J. Pharmacol. Exp. Ther. 139, 60 (1963).

(16) R.M. Gardner and D.O. Allen, The relationship between cyclic nucleotide levels and glycogen phosphorylase activity in isolated rat hearts perfused with epinephrine and acetylcholine, J. Pharmacol. Exp. Ther. 202, 346 (1977).

(17) J.H. Brown, B. Thompson and S.E. Mayer, Conversion of skeletal muscle glycogen synthase to multiple glucose-6-phosphate dependent forms by cyclic adenosine monophosphate dependent and independent protein kinases, Biochemistry 16, 5501 (1977).

(18) J.T. Stull and S.E. Mayer, Regulation of phosphorylase activation in skeletal muscle in vivo, J. Biol. Chem. 246, 5716 (1971).

(19) G.I. Drummond and L. Duncan, Activation of cardiac phosphorylase b kinase, J. Biol Chem. 241, 5893 (1966).

(20) E. G. Krebs, J.A. Beavo, P.J. Bechtel, P.J. England, T.S. Huang and J.T. Stull, Cyclic AMP and protein phosphorylation reactions in muscle in Exploratory Concepts in Muscular Dystrophy II (ed., A.T. Milhorat), Excerpta Medica, Amsterdam (1973).

Possible Regulation of the Calcium Permeability of Cardiac Cell Membranes by Cyclic Nucleotides

Harald Reuter

Department of Pharmacology, University of Bern, 3010 Bern, Switzerland

ABSTRACT

Evidence is summarized which indicates that in cardiac muscle the calcium permeability during membrane excitation is modified by beta-adrenergic and muscarinic cholinergic transmitters and drugs. While beta-adrenergic agonists increase the calcium permeability, muscarinic agonists decrease it. These effects can be mimicked by derivatives of cAMP (adrenergic effect) or cGMP (cholinergic effect). Therefore, it is suggested that the calcium permeability of cardiac cell membranes is regulated by neurotransmitters, possibly via cyclic nucleotide mediated phosphorylation and dephosphorylation reactions of membrane proteins which serve as calcium channels.

BETA-ADRENERGIC REGULATION OF CALCIUM PERMEABILITY

As first shown by Reuter (1) and Grossman and Furchgott (2), adrenaline and noradrenaline greatly increase ^{45}Ca influx in beating guinea pig auricles. Figure 1 illustrates several important points relating to this beta-adrenergic effect: (a) Isoproterenol has a higher potency than adrenaline, (b) the effect of adrenaline is stereospecific, and (c) ^{45}Ca influx is only increased during stimulation, but not in resting preparations.

Similar effects on ^{45}Ca influx have been demonstrated with the phosphodiesterase inhibitor theophylline (3) and with dibutyryl cyclic AMP (4). This may suggest that a change in intracellular cAMP is the common factor producing these permeability changes of the cardiac cell membrane (5).

Tracer experiments such as those shown in Fig. 1, however, do not give much information concerning the mechanisms by which the Ca permeability is regulated by beta-adrenergic agonists, phosphodiesterase inhibitors or cyclic AMP derivatives. One possibility to explore these mechanisms further is by using electrophysiological methods. These methods allow detailed investigations of the kinetics and voltage dependence of the Ca permeability change. This is exemplified in Fig. 2 which shows voltage clamp records obtained from a thin cow ventricular trabecula. The lower records show stepwise changes in membrane potential from -60 to -20 mV in the absence (control) and presence of adrenaline. The upper records show the corresponding membrane ionic current induced by the potential steps. Adrenaline increases an inward current (downward deflection) which is primarily carried by Ca ions (6). This Ca current, commonly called slow inward current, turns on rapidly

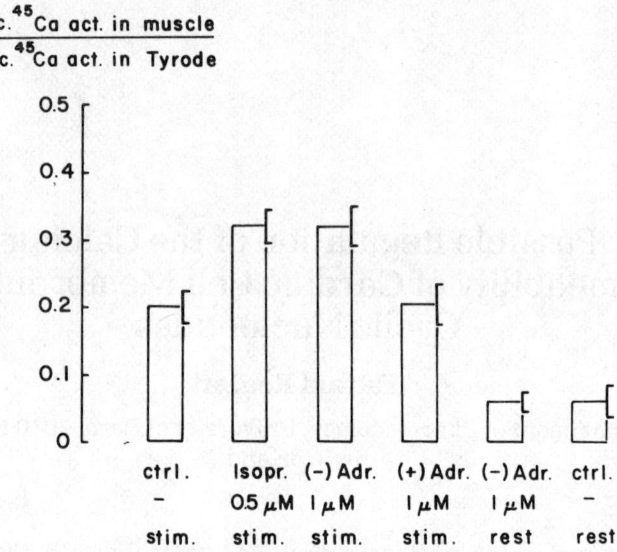

Fig. 1. Effect of isoproterenol, (–)-and (+)-adrenaline on ^{45}Ca uptake (10 minutes) in stimulated (1 Hz) and resting guinea pig auricles; ctrl. = controls.

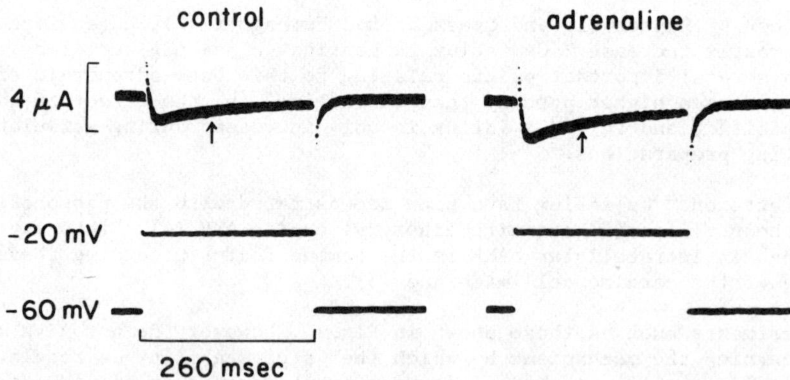

Fig. 2. Voltage clamp steps (lower traces) and corresponding membrane currents (upper traces) under control conditions and in the presence of adrenaline (0.2 μM); cow trabecula, single sucrose gap voltage clamp method (6). Note that the rates for activation and inactivation (time constants indicated by arrows) of the slow inward current are <u>unchanged</u> by adrenaline.

and turns off (inactivates) slowly. This membrane current is responsible for the plateau phase of the cardiac action potential. Corresponding to the increase of the slow inward current, the plateau phase is elevated by catecholamines (7,8,9), by iontophoretic injection of cAMP into the cell (10) or by exogeneous application of butyrate derivatives of cAMP (9,11).

A plot of peaks of inward currents, such as those illustrated in Fig. 2, at various clamp potentials is shown in Fig. 3. The holding potential, E_H, indicates the membrane potential from which voltage clamp steps were applied to different levels (abscissa). Peak inward current (negative current on the ordinate) or minimal outward current (positive current) at each potential has been plotted before (open circles), during (crosses) and after (closed circles) adrenaline application. In the voltage range -40 to +30 mV the slow inward current is greatly increased by adrenaline.

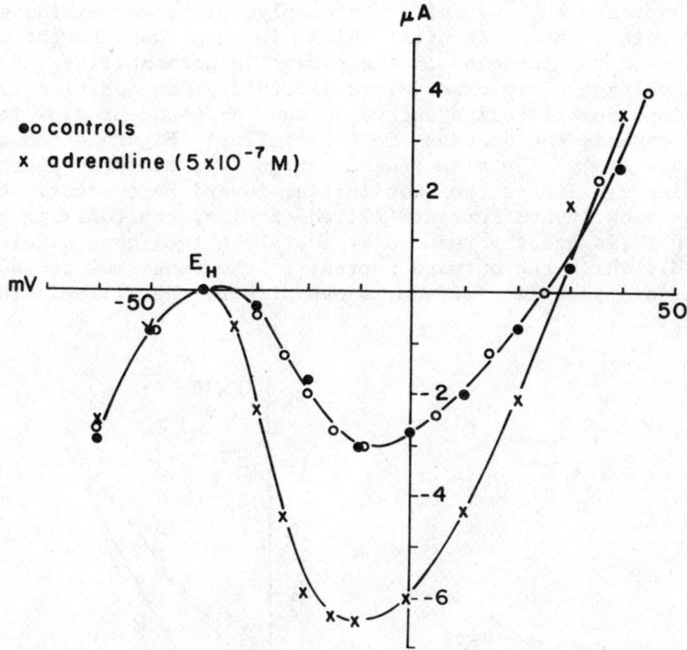

Fig. 3. Current-voltage relationships of the slow inward current (Ca current) in the absence and presence of adrenaline (0.5 μM); cat papillary muscle, single sucrose gap voltage clamp method (6).

Effects similar to those illustrated in Figs. 2 and 3 have also been obtained with theophylline (12) or butyrate derivatives of cAMP (9,11). Therefore, the electrophysiological experiments are in good agreement with the tracer experiments in showing a marked increase in Ca influx by substances which increase intracellular cyclic AMP levels. This Ca influx occurs during the plateau phase of the cardiac action potential.

However, the electrophysiological experiments also provided a clue about the possible site of regulation of the Ca permeability by catecholamines (or cAMP). In theory (13), the Ca permeability could be increased either by a change in the voltage dependent kinetic parameters determining the opening and closing of the Ca channels in the membrane, by an increase in the number of Ca ions transported per unit time through a Ca channel or by an increase in the number of functional Ca channels. We were able to exclude the first two possibilities (9,14) since neither the kinetics of the slow inward current were changed by adrenaline (see Fig. 2) nor was the selectivity of the Ca channels altered (14). This, however, would have to be expected if more Ca ions would be transported per unit time through a single channel (14,15). Therefore, we were left with the conclusion that the increase in the limiting Ca conductance which we observed with beta-adrenergic agonists or cAMP derivatives (9,14) is due to an increase in the number of functional Ca channels in the cardiac cell membrane.

MUSCARINIC CHOLINERGIC REGULATION OF CALCIUM PERMEABILITY

Acetylcholine decreases ^{45}Ca influx in beating guinea pig auricles (2,16), an effect opposite to that seen with beta-adrenergic drugs. Recently it has been shown (17) that this cholinergic effect could be mimicked by 8-bromo-cyclic GMP. Previously the decrease in ^{45}Ca influx by acetylcholine was explained by the shortening of the cardiac action potential which, in turn, was thought to be exclusively the result of the increase in the potassium permeability (18). However, as shown in recent voltage clamp experiments (19,20,21), in addition to the effect on the K permeability, acetylcholine decreases the Ca-dependent slow inward current. This could also explain the decrease in ^{45}Ca influx. Figure 4 shows an experiment by Giles and Noble (19). The slow inward current was measured in the presence of tetrodotoxin which eliminates the fast initial inward Na current. As can be seen from Fig. 4, the slow inward current (filled circles, controls) in the voltage range -30 to +20 mV is greatly reduced by acetylcholine (open circles, 0.03 μM; crosses, 0.12 μM), while the outward current in the range -80 to -40 mV is still unchanged. This acetylcholine effect is opposite to that illustrated in Fig. 3 for adrenaline.

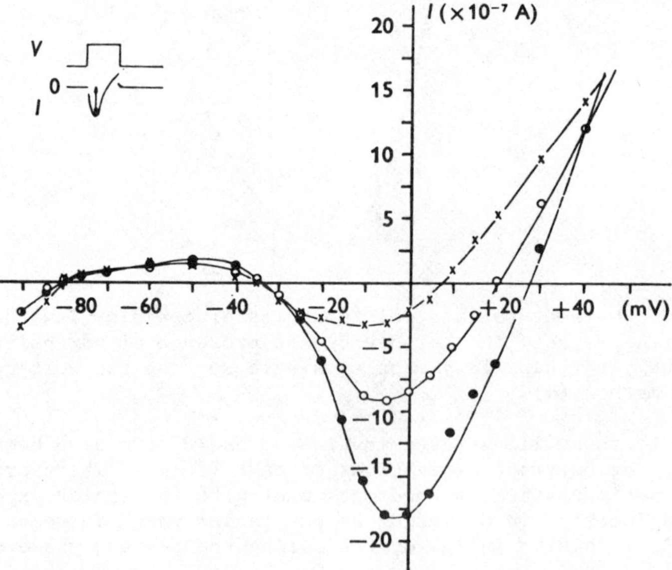

Fig. 4. Current-voltage relationships of the slow inward current (see inset) measured in the absence (•) and presence of acetylcholine (o,x). Bullfrog atrial trabecula double sucrose gap voltage clamp method. (Reproduced with permission from W. Giles and S.J. Noble (19)).

Although no detailed kinetic analysis of the acetylcholine effect on the slow inward current has been performed so far, it seems as if the decrease in this current is primarily due to a decrease in the limiting Ca conductance of the membrane (21). Atropine inhibits this acetylcholine effect (19), indicating a muscarinic action of the cholinergic transmitter.

No voltage clamp experiments have been reported so far in which the effect of 8-bromo-cyclic GMP on the slow inward current has been investigated. However, this cyclic GMP derivative, like acetylcholine, shortens the action potential and reduces the plateau height (17), an effect consistent with a depression of the slow inward current.

CONCLUSION AND HYPOTHESIS

There is increasing evidence that cyclic nucleotides may be involved in the regulation of the Ca permeability of the cardiac cell membrane by adrenergic and cholinergic neurotransmitters. The following cyclic nucleotide dependent biochemical reactions have been reported in the literature (see 22,23): (a) Intracellular cAMP and cGMP concentrations increase in the presence of beta-adrenergic and muscarinic cholinergic agonists, respectively; (b) in the heart these agonists interact and have opposing effects on a cAMP-dependent protein kinase (24); (c) phosphorylation of cardiac sarcolemma by a cAMP-dependent protein kinase reaction has been reported by Wollenberger and colleagues (25, this volume); (d) cGMP may cause dephosphorylation of membrane proteins, possibly by activating a phosphoprotein phosphatase (26).

On the basis of the electrophysiological experiments and biochemical reactions the following hypothesis may be advanced for the regulation of the Ca permeability of cardiac cell membranes by adrenergic and cholinergic neurotransmitters (see 22,23, 27,28):

Following the transmitter-receptor reaction at the outer surface of the membrane, adrenergic transmitters increase cAMP, while cholinergic transmitters increase cGMP. This may lead to cAMP dependent phosphorylation of Ca channels (adrenergic transmitter) or to cGMP dependent dephosphorylation (cholinergic transmitter).

Ca channels may be in the functional state and, hence, available for voltage dependent opening and closing (gating) reactions only when they are phosphorylated. Therefore, an increase in cellular cAMP by beta-adrenergic agonists could increase the number of phosphorylated, functional channels. This would result in the observed increase in the voltage dependent Ca conductance. Conversely, an increase in cellular cGMP by cholinergic agonists could reduce the number of functional channels by a dephosphorylation reaction and, hence, lead to a reduction in the Ca conductance of the cardiac cell membrane. This hypothesis implies that the number of functional, voltage-sensitive Ca channels is regulated by the ratio [cAMP]: [cGMP] in the cell. It is interesting to note that in frog hearts the contractile state not only depends mainly upon Ca influx during an action potential (28), but also upon this cyclic nucleotide ratio (29). The higher this ratio the larger is the force of contraction and, possibly, Ca influx.

This hypothesis, though consistent with biophysical and biochemical data, is still highly speculative. Further qualitative and quantitative tests will be required before it can be accepted or must be abandoned.

REFERENCES

(1) H. Reuter, Ueber die Wirkung von Adrenalin auf den cellulaeren Ca-Umsatz des Meerschweinchenvorhofs, Naunyn-Schmiedeberg's Arch. Pharmacol. 251, 401 (1965)

(2) A. Grossman and R.F. Furchgott, The effect of various drugs on calcium exchange in the isolated guinea-pig left auricle, J. Pharmacol. exp. Ther. 145, 162 (1964)

(3) H. Scholz, Ueber den Mechanismus der positiv inotropen Wirkung von Theophyllin am Warmblueterherzen. II. Wirkung von Theophyllin auf Aufnahme und Abgabe von ^{45}Ca, Naunyn-Schmiedeberg's Arch. Pharmacol. 271, 396 (1971)

(4) T. Meinertz, H. Nawrath and H. Scholz, Stimulatory effects of db-c-AMP and adrenaline on myocardial contraction and ^{45}Ca exchange. Experiments at reduced calcium concentration and low frequencies of stimulation, Naunyn-Schmiedeberg's Arch. Pharmacol. 279, 327 (1973)

(5) A.M. Watanabe and H.R. Besch, Cyclic adenosine monophosphate modulation of slow calcium influx channels in guinea pig heart, Circulation Res. 35, 316 (1974)

(6) H. Reuter and H. Scholz, A study of the ion selectivity and the kinetic properties of the calcium-dependent slow inward current in mammalian cardiac muscle, J. Physiol. 264, 17 (1977)

(7) H. Reuter, The dependence of slow inward current in Purkinje fibres on the extracellular calcium-concentration, J. Physiol. 192, 479 (1967)

(8) G. Vassort, O. Rougier, D. Garnier, M.P. Sauviat, E. Coraboeuf and Y.M. Gargouil, Effects of adrenaline on membrane inward currents during the cardiac action potential, Pfluegers Arch. 309, 70 (1969)

(9) H. Reuter, Localisation of beta adrenergic receptors and effects of noradrenaline and cyclic nucleotides on action potentials, ionic currents and tension in mammalian cardiac muscle, J. Physiol. 242, 429 (1974)

(10) R.W. Tsien, Adrenaline-like effects of intracellular iontophoresis of cyclic AMP in cardiac Purkinje fibres, Nature, New Biol. 245, 120 (1973)

(11) R.W. Tsien, W. Giles and P. Greengard, Cyclic AMP mediates the effects of adrenaline on cardiac Purkinje fibres, Nature, New Biol. 240, 181 (1972)

(12) H. Scholz and H. Reuter, Effect of theophylline on membrane currents in mammalian cardiac muscle, Naunyn-Schmiedeberg's Arch. Pharmacol. 293, R 19 (1976)

(13) A.L. Hodgkin and A.F. Huxley, A quantitative description of membrane current and its application to conduction and excitation in nerve, J. Physiol. 117, 500 (1952)

(14) H. Reuter and H. Scholz, The regulation of the Ca conductance of cardiac muscle by adrenaline, J. Physiol. 264, 49 (1977)

(15) W. Ulbricht, Ionic channels and gating currents in excitable membranes, Ann. Rev. Biophys. Bioeng. 6, 7 (1977)

(16) H. Hoditz and H. Lüllmann, Der Einfluss von Acetylcholin auf den Calciumumsatz ruhender und kontrahierender Vorhofmuskulatur in vitro, Experientia 20, 279 (1964)

(17) H. Nawrath, Does cyclic GMP mediate the negative inotropic effect of acetylcholine in the heart? Nature 267, 72 (1977)

(18) W. Trautwein and J. Dudel, Zum Mechanismus der Membranwirkung des Acetylcholin an der Herzmuskelfaser, Pfluegers Arch. 266, 324 (1958)

(19) W. Giles and S.J. Noble, Changes in membrane currents in bullfrog atrium produced by acetylcholine, J. Physiol. 261, 103 (1976)

(20) R. Ten Eick, H. Nawrath and W. Trautwein, On the mechanism of the negative inotropic effect of acetylcholine, Pfluegers Arch. 361, 207 (1976)

(21) Y. Ikemoto and M. Goto, Nature of the negative inotropic effect of acetylcholine on the myocardium, Proc. Japan. Acad. 51, 501 (1975)

(22) R.W. Tsien, Cyclic AMP and contractile activity in heart, Adv. Cycl. Nucl. Res. 8, 363 (1977)

(23) P. Greengard, Phosphorylated proteins as physiological effectors, Science 199, 146 (1978)

(24) S.L. Keely, T.M. Lincoln and J.D. Corbin, Interaction of acetylcholine and epinephrine on heart cyclic AMP-dependent protein kinase, Am. J. Physiol. 234, H 432 (1978)

(25) A. Wollenberger and H. Will, Protein kinase-catalyzed membrane phosphorylation and its possible relationship to the role of calcium in the adrenergic regulation of cardiac contraction, Life Sciences 22, 1159 (1978)

(26) I.V. Sandoval and P. Cuatrecasas, Opposing effects of cyclic AMP and cyclic GMP on protein phosphorylation in tubulin preparations, Nature 262, 511 (1976)

(27) H. Reuter, Properties of two inward membrane currents in the heart, Ann. Rev. Physiol. in press (1979)

(28) R. Niedergerke and S. Page, Analysis of catecholamine effects in single atrial trabeculae of the frog heart, Proc. R. Soc. Lond. B. 197, 333 (1977)

(29) F.W. Flitney, J.F. Lamb and J. Singh, Intracellular cyclic nucleotides and contractility of the hypodynamic frog ventricle, J. Physiol. 276, 38P (1978)

Some Characteristics of Low Molecular Weight Phosphoprotein Constituents of Cardiac Sarcoplasmic Reticulum and Sarcolemma

Horst Will, Hans-Jürgen Misselwitz [+], Tatyana S. Levchenko [++], and Albert Wollenberger

Division of Cellular and Molecular Cardiology, Central Institute of Heart and Circulatory Regulation Research, Academy of Sciences of the DDR, 1115 Berlin-Buch, German Democratic Republic

INTRODUCTION

Phosphorylation of proteins has become recognized as a major mechanism of cellular control (1). It constitutes the only well-documented mode of action of cyclic AMP in metazoan cells. In cardiac muscle both cytosolic, myofibrillar, and membrane-bound protein kinase-phosphoprotein systems have been identified, some of them cyclic AMP-dependent, others cyclic AMP-independent. The membrane-bound protein kinase substrates, some of which appear to be intrinsic membrane proteins, may play an important part in the regulation of transmembranal calcium ion transport in cardiac muscle (2). The study of these proteins can therefore be expected to contribute materially to the understanding of cardiac membrane regulation.

In our laboratory pigeon heart microsomes and their subfractions are presently being used in an attempt to isolate and characterize the cardiac membrane-bound phosphoproteins. To date at least 7 proteins have been identified in these membrane preparations as substrates of endogenous and/or exogenous cyclic AMP-dependent and independent protein kinases (3,4). Their molecular weights, as estimated by sodium dodecyl sulfate (SDS) polyacrylamide gel electrophoresis, range from 11,500 to greater than 140,000. On incubation of the membranes with Mg [γ-^{32}P]ATP by far the greatest part of protein-bound [^{32}P]phosphate is found in the three smallest proteins, with molecular weights of 11,500, 15,000, and 22,000 (3) (Fig. 1).

Isolation from the crude cardiac microsomes of purified sarcoplasmic reticular and sarcolemmal membrane fractions revealed that the three low molecular weight phosphoproteins are distinctly distributed

---

[+] Present address: Institute of Pharmacology, Humboldt University at Berlin, 108 Berlin, GDR.

[++] Present address: Laboratory of Cardiac Metabolism, All-Union Research Center of Cardiology, Academy of Medical Sciences of the USSR, Moscow, USSR.

among these membranes (3). Whereas the 22,000 and the 15,000 M_r proteins reside mainly in the sarcoplasmic reticular fraction, the 11,500 M_r protein is located predominantly in the sarcolemmal fraction (Fig. 1). In the following we outline the method employed for the isolation of purified sarcoplasmic reticular and sarcolemmal membrane fractions from pigeon myocardium, briefly summarize available data on the 22,000 and 15,000 M_r sarcoplasmic reticular phosphoproteins, and devote our main attention to the 11,500 M_r sarcolemmal phosphoprotein.

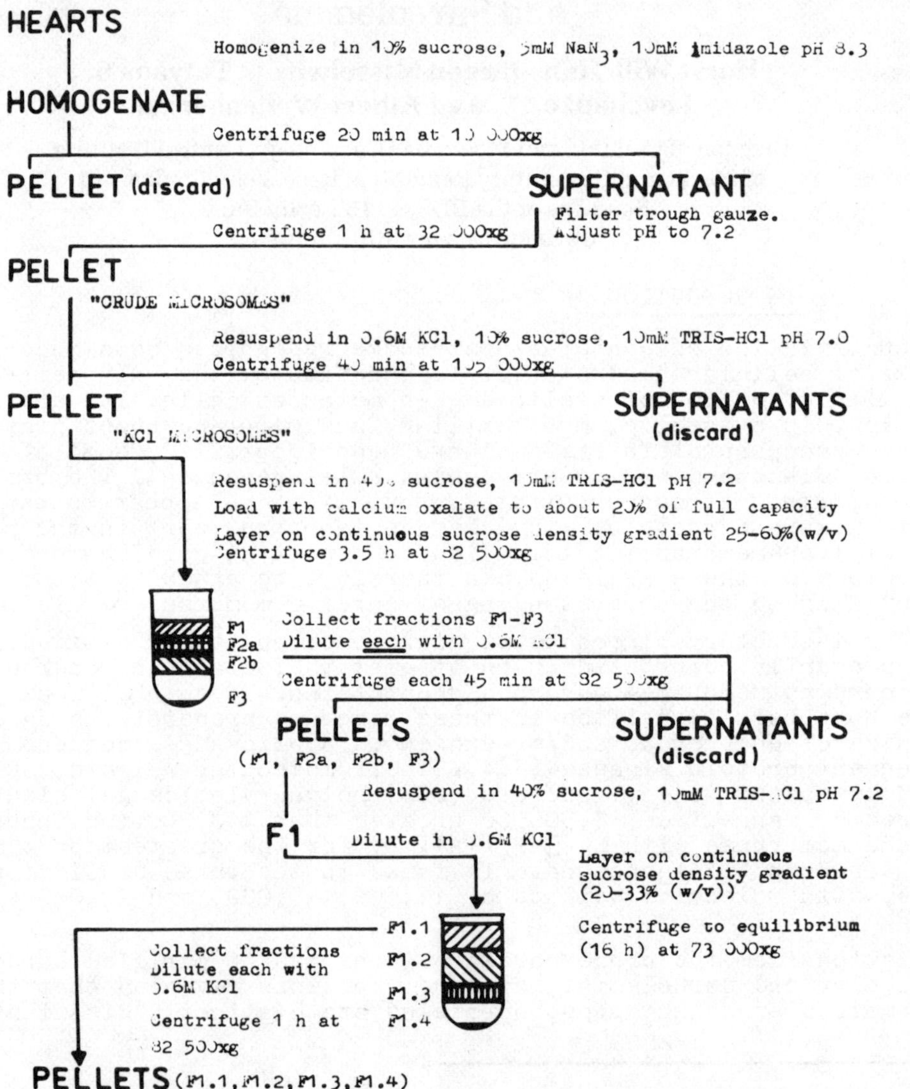

Fig. 1. Scheme for the isolation of sarcoplasmic reticulum and sarcolemma from pigeon hearts employing the methods of Levitsky et al. (5) and Misselwitz et al. (6).

ISOLATION OF SAROPLASMIC RETICULUM AND SARCOLEMMA FROM PIGEON HEART

The procedure used for preparing isolated cardiac sarcoplasmic reticulum and sarcolemma in high purity is illustrated in Fig. 1. Microsomes prepared in a conventional manner from pigeon hearts were washed free of myofibrillar proteins with a solution containing 0.6 M KCl ("KCl microsomes"), loaded with calcium oxalate to about 20 per cent of loading capacity, and subjected to rate zonal and isopycnic centrifugation (5,6). The pellet yielded by the rate zonal centrifugation is the sarcoplasmic reticulum fraction (F3 in Fig. 1). It consists chiefly of vesicles that are filled with precipitated calcium oxalate, clearly visible on electron micrographs, and that do not bind concanavalin A, and it exhibits high activity of Ca^{2+}, Mg^{2+}-ATPase (2.63 µmoles P_i/mg protein x min at 37°C) (5,6). The lightest fractions after the rate zonal and, even more so, after the isopycnic centrifugation steps (F1 and F1.1, respectively) are greatly enriched in sarcolemma, as evidenced, among other characteristics, by an ouabain-sensitive ATPase activity (unmasked) of 272 and 340 µmoles P_i, respectively, per mg protein x min at 37°C and by heavy concanavalin binding, made visible by a cytochemical procedure, to most (F1) or nearly all (F1.1) membrane fragments present (6).

The differential distribution of the 22,000, 15,000, and 11,500 M_r phosphoproteins between the sarcoplasmic reticular and the F1 sarcolemmal membrane fractions is shown in Fig. 2. The fractions enriched in mitochondrial membrane fragments, F2a and F2b, were virtually free of these proteins (3). There is thus little reason to doubt that, as mentioned above, the 22,000 and 15,000 M_r proteins are associated with the sarcoplasmic reticulum, while the 11,500 M_r protein is a constituent of the sarcolemma.

THE 22,000 AND 15,000 M_r SARCOPLASMIC RETICULAR PROTEINS

Cyclic AMP-dependent phosphorylation of a 20,000 - 22,000 M_r protein by both endogenous (7) and exogenous (8) protein kinases was observed several years ago in sarcoplasmic reticulum-enriched vesicular preparations from canine myocardium. Lately it was suggested (9) that this reaction involves two protein phosphorylation steps, the first catalyzed by a cyclic AMP-dependent and the second by a Ca^{2+}-dependent protein kinase. Phosphorylation of the 22,000 M_r protein, which was named phospholamban by Katz and coworkers (10), was closely correlated with parallel increase in Ca^{2+},Mg^{2+}-activity and in the rate of oxalate-supported Ca^{2+} uptake by the vesicles (11). As pointed out by Katz et al. (10), the consequence of this effect in contracting myocardium would be an acceleration of relaxation, which is a prominent feature of the action of cyclic AMP-raising and cyclic AMP-dependent protein kinase-activating ß-adrenergic catecholamines on the heart. For a convincing demonstration of a causal relationship between phosphorylation of the 22,000 M_r protein and stimulation of Ca^{2+}-activated ATPase, the 22,000 M_r protein would have to be isolated and recombined with the ATPase. However, good separation of the 22,000 M_r protein in the native state from the membrane has not yet been achieved. For more detailed information on research dealing with this protein we refer to a recent article by Tada and coworkers (12).

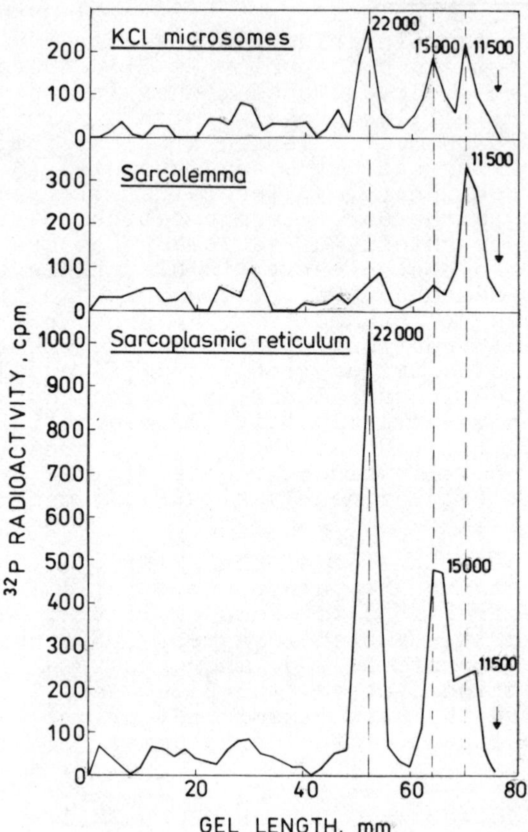

Fig. 2. SDS polyacrylamide gel electrophoresis of ^{32}P-labeled protein kinase substrates in KCl-washed microsomes and in purified sarcolemma and sarcoplasmic reticulum of pigeon heart. Membrane fractions were isolated as described elsewhere (3). Membrane protein, 0.5 to 2.0 mg per ml, was phosphorylated for 10 min at 30°C in 40 mM histidine-HCl buffer of pH 6.8 containing 0.12 M KCl, 2.5 mM $MgCl_2$, 4.0 mM [γ-^{32}P]ATP (49,000 counts/min/nmole ATP), 2.5 mM ethyleneglycol-bis-(ß-aminoethyl ether)-N,N'-tetraacetic acid (EGTA), 2.5 μM cyclic AMP, and 0.2 mg of soluble protein kinase per ml. For electrophoresis 80 μg of solubilized membrane protein were applied per gel. The arrows mark the position of the dye front. Modified from Will et al. (3).

Whereas phosphorylation of the 22,000 M_r protein of cardiac sarcoplasmic reticulum by both exogenous and endogenous cyclic AMP-dependent protein kinases occurs predominantly at a serine residue (3,7,10) - in distinction to the acyl phosphoprotein intermediate of sarcoplasmic reticular Ca^{2+}-activated ATPase -, the [^{32}P]phosphate transferred by endogenous protein kinase from [γ-^{32}P]ATP to the 15,000 M_r protein is recovered chiefly as [^{32}P]phosphothreonine (3). This phosphate transfer proceeds at a rate which is about twice as fast as the endogenous phosphorylation of the 22,000 M_r protein present in the same membrane preparation (3). In the presence of exogenous cyclic AMP-dependent protein kinase and cyclic AMP, [^{32}P] phosphate is incorporated also into serine residue(s) of the 15,000 M_r protein. Maximal phosphorylation of the protein in this case is 0.54 nmole P_i per mg membrane protein (3).

The 15,000 M_r protein can be extracted from sarcoplasmic reticular membranes with an acidic 2:1 chloroform-methanol mixture (3). Extraction was performed by the method proposed by MacLennan et al. (13) for solubilization of the proteolipid of skeletal muscle sarcoplasmic reticulum. However, the latter protein and the 15,000 M_r phosphoprotein are not identical. The possibility has not yet been explored that the 15,000 M_r protein of cardiac sarcoplasmic reticulum may have a function similar to that of the proteolipid described by MacLennan (13), which is thought to act as a coupling factor in skeletal muscle sarcoplasmic reticulum (14).

THE 11,500 M_r SARCOLEMMAL PROTEIN

The pigeon heart sarcolemmal protein of molecular weight 11,500 is phosphorylated mainly at serine residue(s) (3), distinguishing it from the acyl phosphoprotein intermediate of the Na^+,K^+-ATPase in the same membrane. In our purified sarcolemmal membrane fraction (Fig. 1) endogenous phosphorylation of this protein was found to be largely independent on cyclic AMP and insensitive to the soluble heat-stable protein kinase inhibitor. Addition of partly purified cyclic AMP-dependent protein kinase and cyclic AMP did not greatly enhance [^{32}P]phosphate incorporation into the protein. Under our incubation conditions phosphorylation was also unaffected by Ca^{2+} concentrations below 0.1 mM. Higher Ca^{2+} concentrations inhibited [^{32}P]phosphate incorporation. Moreover, most of the [^{32}P]phosphate incorporated into the 11,500 M_r protein is released from the membrane in the presence of 20 mM Ca^{2+} and part of the released [^{32}P] phosphate can be precipitated with 5 % trichloroacetic acid containing 0.25 % sodium tungstate and 0.06 N sulfuric acid (3). It was suggested (3) that 20 mM Ca^{2+} activated a membrane-bound protease or phospholipase which released the 11,500 M_r phosphorylated protein or a segment of it carrying the [^{32}P]phosphate from the membrane.

A phosphoprotein of molecular weight 11,700 has been found in a bovine heart Na^+,K^+-ATPase preparation (15). This protein, which may be identical with the 11,500 M_r protein of pigeon heart sarcolemma, was purified by repeated polyacrylamide gel electrophoresis in the presence of SDS. It was found to be rich in hydrophilic amino acid residues and to give a positive periodic acid-Schiff reaction (15).

In recent experiments the 11,500 M_r protein was solubilized with sodium deoxycholate (DOC) and purified by chromatography on DEAE cellulose and by gel filtration on Sephadex G-150. This is demonstrated in Figs. 3 and 4.

Two characteristics of the sarcolemmal phosphoprotein are immediately apparent from these experiments: (1) The phosphoprotein is a strongly polar protein which is eluted from DEAE-cellulose by NaCl only at concentrations above 0.5 M (Fig. 3). Its acid nature is suggested by the fact that it contains about 0.15 μmoles of sialic

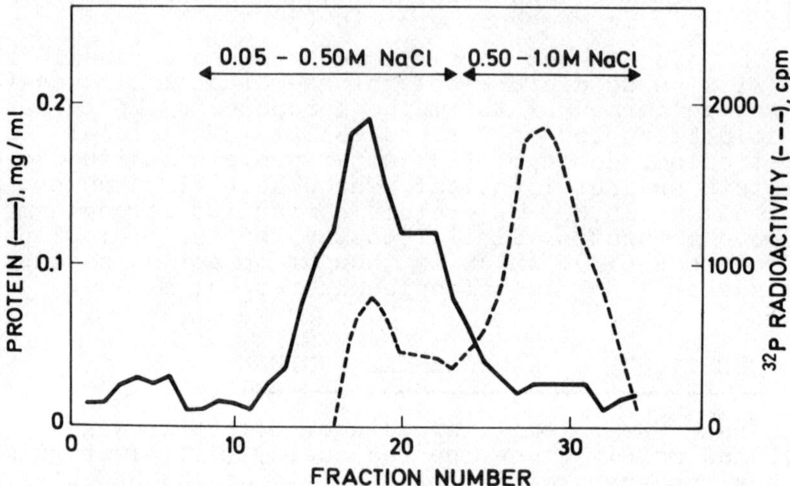

Fig. 3. Chromatography of deoxycholate-solubilized pigeon heart sarcolemma on DEAE-cellulose.
Sarcolemmal membranes were isolated from pigeon heart and phosphorylated for 5 min in a reaction catalyzed by endogenous protein kinas, as described previously (3). The membranes were solubilized with 0.1 % DOC in a solution also containing 1 M NaCl, 50 mM NaF, 10 mM EGTA, and 40 mM Tris-HCl buffer of pH 7.7 and were then dialyzed against 0.1 % DOC-50 mM Tris-HCl of pH 7.7 (16). The dialyzed extract containing about 10 mg protein and 3.5×10^4 counts /min of ^{32}P radioactivity was applied to a 1.5 x 16 cm DEAE-cellulose column which had been equilibrated with 0.1 % DOC-50 mM Tris-HCl buffer of pH 7.7. Protein was eluted with 90 ml linear gradient of NaCl (0.05 to 0.5 M) followed by a 70 ml linear 0.5 to 1.0 M gradient of NaCl. All gradient solutions contained also 0.1 % DOC and 50 mM Tris-HCl buffer of pH 7.7. Volume of each fraction was 6.5 ml.

acid residues per mg protein. (2) Whereas the phosphoprotein moves during SDS polyacrylamide gel electrophoresis like a protein of molecular weight 11,500, its apparent molecular weight estimated by gel filtration in the presence of 0.1 % DOC is 78,000 (Fig. 4).

A protein exhibiting similar properties was recently isolated by Feldman and Weinhold (16) from rat heart sarcolemma. The rat heart protein exhibits molecular weights of 12,300 and 71,400, respectively, on SDS gel electrophoresis and on gel filtration in the presence of DOC. It contains 0.12 µmoles of sialic acid residues and 7.35 µmoles of phospholipid per mg protein. The higher molecular weight obtained by gel filtration was explained by the presence of firmly bound phospholipids which became dissociated in 0.1 % SDS, but not in 0.1 % DOC. The 12,300 M_r protein was considered to be the apo-

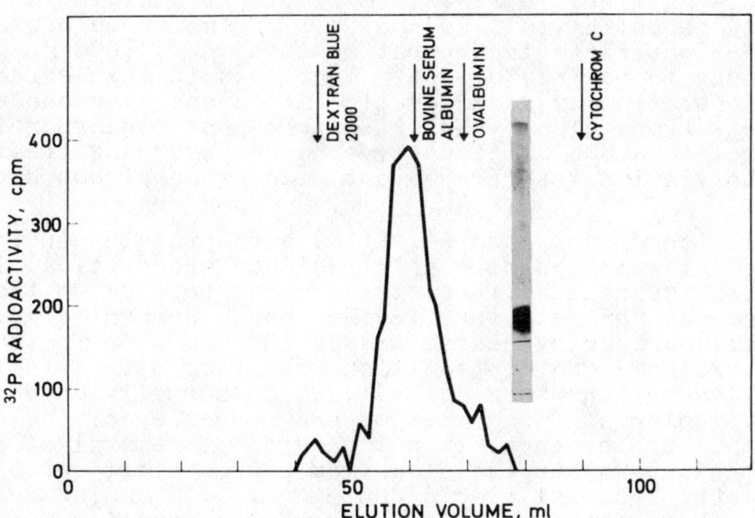

Fig. 4. Gel filtration of a sarcolemmal phosphoprotein on Sephadex G-150.
The main radioactivity peak eluted from DEAE-cellulose (Fig. 3) was concentrated by ultrafiltration, using an Amincon PM-10 filter. The concentrated solution was dialyzed against 1000 volumes of 0.1 % DOC-50mM Tris-HCl of pH 7.7 and applied to a 1.5 x 76 cm Sephadex column. Elution was carried out with 0.1 % DOC-50 mM Tris-HCl of pH 7.7. Inset: SDS polyacrylamide gel electrophoresis of the isolated sarcolemmal phosphoprotein. Electrophoresis was carried out with 35 ug protein as described previously (3). Protein in the gel was stained with Commassie brilliant blue. **Value of m_R = 0.91.**

protein moiety of a lipoprotein. Feldman and Weinhold (16) were not concerned with the question, whether or not the sarcolemmal lipoprotein is phosphorylated in a protein kinase-catalyzed reaction. They found, however, that the lipoprotein binds about 52 moles of Ca^{2+} per mole (K_d = 74 μM) and attributed to it a role in sarcolemmal Ca^{2+} storage.

Attention was recently called to the possibility that the 11,500 M_r phosphoprotein may be part of the electrogenic slow Ca^{2+} inward current channel in heart sarcolemma (2). On electrophysiological evidence it had been postulated that the availability of this slow channel for voltage-dependent activation may be regulated by cyclic AMP-dependent membrane protein phosphorylation (17). Agents that increase cellular cyclic AMP levels enhance electrogenic Ca^{2+} influx into the heart muscle fiber during sarcolemmal depolarization and restore excitability and contractions in hearts made inexcitable by blockade or inactivation of the fast sarcolemmal Na^+ channels (17,18). The enhanced Ca^{2+} influx seems to result from activation of "spare" channels rather than from a change in ion permeability properties of functional channels (19). As pointed out before, the 11,500 M_r protein is the chief site of protein kinase-catalyzed phoyphorylation in pigeon heart sarcolemma purified according to the procedure outlined in Fig. 1, maximal phosphorylation (in fraction F1) amounting to 0.65 nmoles P_i/mg membrane protein (3). Although [^{32}P]phosphate incorporation into the 11,500 M_r protein had been found to be only slightly increased in this sarcolemmal membrane preparation by added cyclic AMP-dependent protein kinase plus cyclic AMP and also by NaF, [^{32}P]phosphate incorporation into the same protein was significantly enhanced by these treatments when phosphorylation was carried out with crude pigeon heart microsomes (4).

According to Langer and coworkers (20) glycoproteins and, in particular, their terminal sialic acid residues are critical in the organization of the cell surface structure as related to the control of transmembrane Ca^{2+} exchange in the heart. Removal of sialic acid from the cell surface by neuraminidase increases Ca^{2+} exchange ability severalfold and, in addition, permits entry into the cell of competitive cations such as La^{3+}, which normally do not traverse the sarcolemma. These changes are rather specific, since K^+ permeability, for instance, is not altered by removal of siliac acid. The phosphoprotein isolated from pigeon heart sarcolemma is a glycoprotein. Its sialic acid content of 0.15 μmoles per mg is higher than that of the sarcolemmal membrane fraction, where it amounts to about 0.06 μmole per mg protein. It may well be that the saccharide part of the phosphoprotein, which can be assumed to be located on the external face of the membrane, provides superficial Ca^{2+} binding sites involved in transsarcolemmal Ca^{2+} influx.

Finally, a content of 96 amino acids which can be calculated for an 11,500 M_r protein on the basis of an average molecular weight of amino acid residues of 120, disregarding the presence of saccharide residues, is in excess of the number of amino acids needed to form an α-helical segment long enough to span the cell surface membrane. As was suggested in a recent review (2), the cardiac sarcolemmal slow cation channel may conceivably be formed by an assembly of several of such segments in a manner analogous to what has been proposed for channel formation by other glycoproteins (21). With this idea in mind, the ability of the 11,500 M_r protein to serve as substrate for cyclic AMP-dependent protein kinase (in crude cardiac microsomes, see above) can well be integrated into the notion (17, 19,22) that the cardiac sarcolemmal slow Ca^{2+} channel is regulated by cyclic AMP-dependent phosphorylation. This would provide for a biochemical mechanism by which cyclic AMP-raising agents such as the ß-adrenergic catecholamines increase the transsarcolemmal Ca^{2+} inward current during excitation, with increases in contractile force and automaticity as consequences.

Needless to say, the above suggestion of an involvement of the 11,500 M_r phosphoprotein in slow Ca^{2+} channel formation and function is at present purely hypothetical. But with the isolated protein having become available, the hypothesis is now open to experimental testing.

CONCLUSION

The still very rudimentary understanding of the function of the cardiac membrane phosphoproteins dealt with above makes further characterization of the phosphorylation reactions and the properties of the phosphoproteins highly desirable. The purification of the 11,500 M_r phosphoprotein of cardiac sarcolemma and the extraction of the 15,000 M_r protein from cardiac sarcoplasmic reticulum membranes with chloroform-methanol are promising steps. Reconstitution experiments to be carried out with isolated phosphoproteins may help to define the role of these proteins in cardiac membrane function. Knowledge of the molecular properties of the phosphoproteins will also be useful in forthcoming studies of membrane protein phosphorylation in contracting cardiac muscle.

REFERENCES

(1) P. Greengard, Phosphorylated proteins as physiological effectors, Science 199, 146 (1978).

(2) A. Wollenberger and H. Will, Protein kinase-catalyzed membrane phosphorylation and its possible relationship to the role of calcium in the adrenergic regulation of cardiac contraction, Life Sci. 22, 1159 (1978).

(3) H. Will, T. S. Levchenko, D. O. Levitsky, V. N. Smirnov and A. Wollenberger, Partial characterization of protein kinase--catalyzed phosphorylation of low molecular weight proteins in purified preparations of pigeon heart sarcolemma and sarcoplasmic reticulum, Biochim. Biophys. Acta (1978), in press.

(4) H. Will, Unpublished results.

(5) D. O. Levitsky, M. K. Aliev, A. V. Kuzmin, T. S. Levchenko, V. N. Smirnov and E. I. Chazov, Isolation of calcium pump system and purification of calcium ion-dependent ATPase from heart muscle, Biochim. Biophys. Acta 443, 468 (1976).

(6) H.-J. Misselwitz, H. Will and A. Wollenberger, Isolation of pigeon heart sarcolemma, To be published.

(7) P. J. La Raia and E. Morkin, Adenosine 3',5'-monophosphate--dependent membrane phosphorylation. A possible mechanism for the control of microsomal calcium transport in heart muscle, Circul. Res. 35, 298 (1974).

(8) M. Tada, M. A. Kirchberger and A. M. Katz, Phosphorylation of a 22 000 dalton component of the cardiac sarcoplasmic reticulum by adenosine 3':5'-monophosphate-dependent protein kinase, J. Biol. Chem. 250, 2640 (1975).

(9) H. L. Wray and R. R. Gray, Cyclic AMP stimulation of membrane phosphorylation and Ca^{2+} activated, Mg^{2+}-dependent ATPase in cardiac sarcoplasmic reticulum, Biochim. Biophys. Acta 461, 441 (1977).

(10) A. M. Katz, M. Tada and M. A. Kirchberger, Control of calcium transport in the myocardium by the cyclic AMP-protein kinase system, Adv. Cyclic Nucleotide Res. 5, 453 (1975).

(11) M. Tada, M. A. Kirchberger, D. I. Repke and A. M. Katz, The stimulation of calcium transport in cardiac sarcoplasmic reticulum by adenosine 3':5'-monophosphate-dependent protein kinase, J. Biol. Chem. 249, 6174 (1974).

(12) M. Tada, F. Ohmori and N. Kinoshita, Cyclic AMP regulation of active calcium transport across membranes of sarcoplasmic reticulum. Role of the 22,000 dalton phosphoprotein phospholamban, Adv. Cyclic Nucleotide Res. 9, (1978), in press.

(13) D. H. MacLennan, C. C. Yip, G. H. Iles and P. Seeman, Isolation of sarcoplasmic reticular proteins, Cold Spring Harbor Symp. Quant. Biol. 37, 469 (1972).

(14) E. Racker and E. Eyton, A coupling factor from sarcoplasmic reticulum required for the translocation of Ca^{2+} ions in a reconstituted Ca^{2+} ATPase pump, J. Biol. Chem. 250, 7533 (1975).
(15) P. J. Dowd, Jr., B. J. R. Pitts and A. Schwartz, Phosphorylation of a low molecular weight polypeptide in beef heart Na^+,K^+-ATPase preparations, Arch. Biochem. Biophys. 175, 321 (1976).
(16) D. A. Feldman and P. A. Weinhold, Calcium binding to rat heart plasma membranes: Isolation and purification of a lipoprotein component with a high calcium binding capacity, Biochemistry 16, 3470 (1977).
(17) N. Sperelakis and J. A. Schneider, A metabolic control mechanism for calcium ion influx that may protect the ventricular myocardial cell, Am. J. Cardiol. 37, 1079 (1976).
(18) A. M. Watanabe and H. B. Besch,Jr., Cyclic adenosine monophosphate modulation of slow calcium influx channels in guinea pig hearts, Circul. Res. 35, 316 (1974).
(19) H. Reuter and H. Scholz, A study of the ion selectivity and the kinetic properties of the calcium dependent slow inward current in mammalian cardiac muscle, J. Physiol. 264, 49 (1977).
(20) J. S. Frank, G. A. Langer, L. M. Nudd and K. Seraydarian, The myocardial cell surface, its histochemistry, and the effect of sialic acid and calcium removal on its structure and cellular ionic exchange, Circul. Res. 41, 702 (1977).
(21) R. C. Hughes, The complex carbohydrates of mammalian cell surfaces and their biological roles, Essays Biochem. 11, 1 (1975).
(22) H. Reuter, Possible regulation of calcium permeability of cardiac cell membranes by cyclic nucleotides, Abstr. 7th Internat. Congr. Pharmacol., Paris 1978 (in press).

Phosphorylation and Regulation of Contractile Proteins

James T. Stull, Donald K. Blumenthal, Primal deLanerolle, Charles W. High and David R. Manning

Department of Medicine, University of California, San Diego
La Jolla, California 92093, U.S.A.

SUMMARY

Phosphorylation of myofibrillar proteins may play an important role in regulating contractile events similar to the regulatory role of protein phosphorylation in metabolic processes. Specific proteins in thin and thick filaments have been purified from muscles and have been shown to exist as phosphoproteins. Purified cardiac troponin is readily phosphorylated by cAMP-dependent protein kinase and in perfused hearts in response to stimuli which increase cAMP. Although skeletal muscle troponin is a phosphoprotein, it is not readily phosphorylated by cAMP-dependent protein kinase nor is it phosphorylated in response to increases in cAMP or contractile activity. The phosphorylation of cardiac troponin appears to be restricted to beta-adrenergic stimulation and is associated with the positive inotropic effect. However, no biochemical effect of the phosphorylation has been provided to explain the positive inotropic response.

Myosin is also a phosphoprotein. A specific light chain subunit of myosin is phosphorylated by a myosin light chain kinase. The kinetic properties of three muscle kinases were determined using purified myosin light chain subunits and myosins from cardiac, skeletal and tracheal smooth muscle. Significant quantitative differences in the kinetic properties suggested that each of these muscles contains a catalytically distinct myosin light chain kinase. These myosin light chain kinases were all Ca^{2+}-dependent enzymes. The Ca^{2+} concentration required for half-maximal activation of myosin light chain kinase from heart is about 1 μM which is similar to the concentrations required for activation of myofibrillar ATPase activity. During electrical stimulation of skeletal muscle, myosin light chain is phosphorylated, presumably due to Ca^{2+} stimulation of the contractile and phosphorylation processes. Although it is not clear what role myosin phosphorylation may play in cardiac and skeletal muscle contraction, it has been suggested that phosphorylation of smooth muscle myosin may be essential for actin activation. These results would indicate that myosin phosphorylation may be essential for smooth muscle contraction.

CONTRACTILE PROTEINS

There are primarily four functional proteins involved in the contractile process in striated muscles. Myosin, located in the thick filament of the sarcomere, is a complex of two large polypeptides (heavy chains) and two to three smaller peptides (light chains). The light chain subunits are associated with the globular head regions which contain the ATPase and actin binding activities of myosin. Specific light chain subunits may be obligatory for enzymatic activity of myosin.

Actin is localized in the thin filament of the sarcomere in a polymerized form. The polymerized actin has the ability to activate myosin ATPase activity upon binding with the globular head region. Tropomyosin binds to 7 actin monomers and regulates the interaction of actin and myosin in the presence of troponin. Troponin is composed of three distinct subunits: TN-I, which binds to actin and inhibits actomyosin ATPase activity; TN-C, which binds Ca^{2+}; TN-T, which binds tightly to tropomyosin. At low Ca^{2+} concentrations (less than 0.1μM) troponin and tropomysin inhibit myosin binding to actin. At higher Ca^{2+} concentrations (10 μM) two molecules of Ca^{2+} bind to TN-C which leads to reversal of the inhibition. Actin then binds to myosin with a subsequent generation of force and shortening. A more detailed description of these processes may be found in recent reviews (1, 2, 3).

Troponin has not been isolated from smooth muscle. Many investigators feel that phosphorylation of myosin by a Ca^{2+}-dependent kinase may be essential for actomyosin interactions (see below). However recent evidence suggests that Ca^{2+} regulation occurs through an 80,000 dalton protein devoid of kinase activity. The reasons for conflicting results are not apparent at this time.

TROPONIN PHOSPHORYLATION

The original observations that purified TN-I from rabbit skeletal muscle was phosphorylated by cyclic AMP-dependent protein kinase (4) and the demonstration that troponin purified from skeletal muscle was a native phosphoprotein (5, 6) stimulated interest in the possible regulatory role of protein phosphorylation in myofibrils. Although purified troponin subunits from skeletal muscle were phosphorylated by cyclic AMP-dependent protein kinase and phosphorylase kinase, a role for these phosphorylation reactions in regulating the biochemical activity of skeletal muscle troponin has not been defined (See 7 for recent review). The capacity of phosphorylase kinase to phosphorylate purified TN-I may be related to the similarity in amino acid sequences of the phosphorylation sites in TN-I and phosphorylase. The maximal rate of phosphorylation of skeletal muscle troponin complex is only about 2% of the maximal rate of phosphorylation of histones (Table 1). However, cardiac troponin or the troponin·tropomyosin complex is phosphorylated at a rate greater than the phosphorylation of histones (Table 1). These results would suggest that cardiac troponin, but not skeletal muscle troponin, may be phosphorylated in vivo in response to stimuli which increase cyclic AMP.

TABLE 1 Phosphorylation of Proteins by Cyclic Nucleotide-Dependent Protein Kinases

Enzyme	Substrate	Vmax mol P/min/mg-enzyme
cAMP-dependent protein kinase	Histones IIA	3.30
	Cardiac Troponin	10.9 ± 1.3
	Skeletal Muscle Troponin	0.074 ± 0.002
	Myosin from cardiac, skeletal or tracheal smooth muscles	0
cGMP-dependent protein kinase	Histones 2b	1.37
	Cardiac Troponin	0.93 ± 0.27
	Skeletal Muscle Troponin	0
	Myosin from cardiac, skeletal or tracheal smooth muscles	0

No detectable phosphorylation of myosin from cardiac, skeletal or tracheal smooth muscles was observed, although a minor incorporation of ^{32}P was occassionally found with purified light chains. However the rate and extent of phosphorylation of these proteins were very low and may be related to the denaturing conditions used to purify light chains (8). The failure to observe phosphorylation could not be ascribed to saturation of phosphorylated sites because measurements of protein-bound phosphate (9) showed less than 0.1 mol phosphate per mol light chains.

The cGMP-dependent protein kinase also phosphorylated cardiac troponin (10, 11), but not skeletal muscle troponin nor myosin from three different types of muscle (Table 1). The biochemical properties of the reactions clearly indicated that cGMP-dependent protein kinase catalyzed the phosphorylation of cardiac troponin. The TN-I subunit was phosphorylated by the cGMP-dependent protein kinase, the same subunit phosphorylated by cAMP-dependent protein kinase (9). Considering the marked similarity of the two cyclic nucleotide dependent protein kinases, an analysis was made with respect to the sites phosphorylated in TN-I by the respective kinases (Table 2).

TABLE 2 Phosphorylation of Cardiac Troponin by cAMP and cGMP-dependent Protein Kinases

Kinase	Maximal Phosphate Incorporated
	mol P/mol cardiac troponin
cAMP-dependent protein kinase	1.10 ± 0.04
cGMP-dependent protein kinase	1.02 ± 0.03
Both protein kinases	1.01 ± 0.01

Each kinase alone catalyzed a maximal incorporation of 1 mol phosphate per mol troponin, suggesting phosphorylation of a single site. This value was not changed when cardiac troponin was incubated simultaneously with both enzymes. If distinct sites were phosphorylated by the respective kinases, a maximal incorporation of 2 mol phosphate per mol troponin would be expected. These results suggest that both cyclic nucleotide dependent protein kinases catalyze the phosphorylation of the same site in the TN-I subunit of cardiac troponin.

Would one expect cGMP-dependent protein kinase to catalyze the phosphorylation of cardiac troponin in vivo in response to an increase in cGMP levels? Previous results with perfused hearts indicated troponin phosphorylation and phosphorylase _a_ formation were not stimulated with increases in cGMP content (12). Although the rate of phosphorylation of cardiac troponin by cGMP-dependent protein kinase was similar to the rate of phosphorylation of histone, the rate of troponin phosphorylation by cGMP-dependent kinase was 12-fold lower than the maximal rate found with cAMP-dependent protein kinase. These reactions in vitro may reflect the relative rates of phosphorylation in vivo since the Km values for both enzymes (16-20 μM) are less than the concentration of cardiac troponin (50 μM) in the intact muscle (9). Furthermore it has been reported that cAMP-dependent protein kinase is present in an 11-fold greater amount than the cGMP-dependent protein kinase in heart (13). If both kinases were fully activated, the rate of phosphorylation of troponin by the cGMP-dependent protein kinase would be 130-fold less than the rate of phosphorylation by cAMP-dependent protein kinase. Because it takes about 25 seconds for cardiac TN-I to be phosphorylated in response to an increase in cAMP (12), the time required for full phosphorylation by cGMP-dependent protein kinase would be about 1 hour. Obviously this time is much too long to be important in the physiological regulation of cardiac performance by cGMP. Although kinetic information related to other common substrates for cAMP and cGMP dependent protein kinases are not available (14, 15), it seems likely that similar considerations could be evoked to explain why stimulation of cGMP formation is not associated with activation of cellular processes normally activated by cAMP. Phosphate incorporation into a particular protein (enzyme) which may even be associated with modification of its biochemical properties provide limited information. It must be emphasized that it is important to obtain critical information on the biochemical properties of phosphorylation reactions in order to evaluate regulatory processes at the cellular level.

Phosphorylation of the TN-I subunit of cardiac troponin has also been examined in the intact heart. A critical feature for these studies has been the purification of the subunit by affinity chromatography (16). England was the first to show that the stimulation of the inotropic state of cardiac muscle with epinephrine resulted in phosphorylation of TN-I (17). In a more detailed study, England (12) concluded that the correlation between phosphorylation of rat heart TN-I and stimulation of contractility by a beta-adrenergic agonist was more complex. The correlation was maintained during the onset of stimulation. However, upon removal of isoproterenol, TN-I remained phosphorylated while contractile activity returned to control values. The subsequent infusion of acetylcholine resulted in dephosphorylation of TN-I and was associated with an increase in cyclic GMP content in the cardiac muscle. These results show that TN-I phosphorylation may not always be associated with increased contractile activity. Phosphorylation of cardiac TN-I with beta-adrenergic stimulation was also found in hearts from rabbits (18) and guinea pigs (Stull and Beardsley, unpublished observations). Frearson et al. (19) recently reported that the positive inotropic response to a reduction of the sodium concentration or an increase in the Ca^{2+} concentration in the perfusion medium was not associated with a change in TN-I phosphorylation. In perfused cat hearts Ezrailson et al. (20) found that phosphorylation of TN-I occurred in the presence of isoproterenol, but not with other inotropic interventions such as ouabain, high calcium, treppe or the calcium ionophore X537A. Thus, it appears that TN-I phosphorylation may be specifically associated with increases in cAMP content in myocardium and not associated with ionic perturbations or other pharmacological stimuli which may increase contractile force by increasing the Ca^{2+} concentration within the myocardial cell.

What is the function of phosphorylation of cardiac troponin in regulating contractile performance of the myofibrils? It would be logical that phosphorylation of cardiac troponin alters Ca^{2+} regulation of actomyosin ATPase activity, but conflicting reports from several laboratories have prevented a definitive conclusion as to a possible biochemical effect. These effects include an increase in Ca^{2+} sensitivity of actomyosin ATPase activity (21), a decrease in Ca^{2+} sensitivity (23, 18) and no change in Ca^{2+} sensitivity but a decrease in the maximal ATPase activity (23). Studies on the calcium binding properties of cardiac troponin showed no change in the affinity or maximal amount of Ca^{2+} bound to troponin with phosphorylation (24) which is consistent with the report that cyclic AMP did not change the Ca^{2+} sensitivity of developed tension in skinned cardiac fibers (27). Thus, it appears that no consistent observations have been made which could describe a role of troponin phosphorylation in regulating the inotropic state of cardiac muscle.

MYOSIN PHOSPHORYLATION

A specific myosin light chain kinase (26) and light chain phosphatase (27) have been purified from rabbit skeletal muscle. It was shown in early reports that the kinase from skeletal muscle was dependent upon Ca^{2+} for activity which suggested that changes in Ca^{2+} concentration during a contraction cycle may regulate the phosphorylation of light chains. In this regard it has been recently shown that stimulation of contraction in skeletal muscle resulted in phosphorylation of the myosin light chain (28, 30). In rabbit skeletal muscle, tetanic electrical stimulation produced no change in cyclic AMP content or the activation state of phosphorylase kinase (Table 3). However, there was a marked increase in phosphorylase *a* formation along with an increase in the phosphate content in myosin light chain. Thus, both phosphorylase and myosin are phosphorylated with contraction, presumably due to the Ca^{2+} stimulation of the activity of non-activated phosphorylase kinase and of the Ca^{2+}-dependent light chain kinase, respectively.

It has been suggested recently that the same enzyme may catalyze a Ca^{2+}-dependent phosphorylation of phosphorylase kinase, myosin light chains and histone (30) although no kinetic properties for these reactions were presented. The phosphorylation of phosphorylase kinase would result in conversion to the activated state as reflected by the increase in the ratio of activity at pH 6.8/8.2. The results in Table 3 as well as earlier reports (31, 32, 33) show no conversion of phosphorylase kinase to the activated form with electrical stimulation. Thus, the Ca^{2+}-dependent light chain kinase may be relatively specific for myosin as a substrate in vivo.

TABLE 3 Phosphorylation of Myosin in Skeletal Muscle In Vivo

Condition	cAMP[a]	Phosphorylase Kinase[a]	Phosphorylase[a]	Myosin[a]
Control	0.18±0.05	0.039±0.006	0.06±0.01	0.50±0.03
Electrical Stimulation	0.21±0.06	0.030±0.007	0.41±0.06	0.87±0.04

[a] The results are presented as the means ± S.E.M.; cyclic AMP, µmol/kg; phosphorylase kinase, ratio of activity at pH 6.8/8.2; phosphorylase, ratio of activity -AMP/+AMP; myosin, mol phosphate per mol light chain.

The Ca^{2+} concentration required for 50% stimulation of maximal light chain kinase activity from skeletal and smooth muscles has been reported to be between 0.1 to 1 µM Ca^{2+}. These concentrations are very similar to those required for activation of contraction. In Fig. 1 the Ca^{2+} concentration required for stimulation of these processes in cardiac muscle are presented.

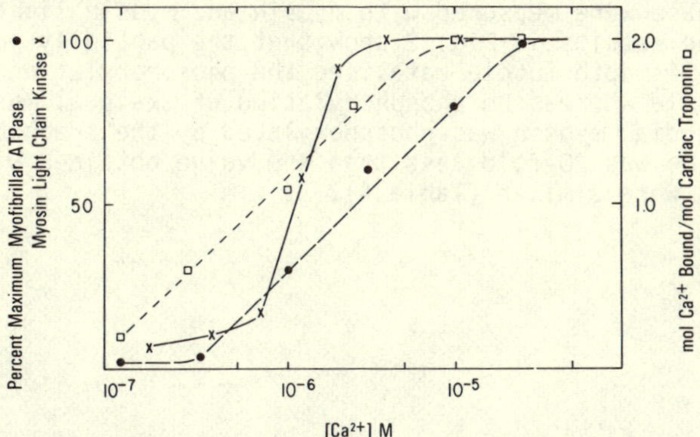

Fig. 1 Effect of Ca^{2+} concentration on three Ca^{2+}-dependent biochemical process in heart. Ca^{2+} binding to cardiac troponin (□) and Ca^{2+} activation of cardiac myofibrillar ATPase activity (•) are compared to Ca^{2+} activation of partially purified myocardial light chain kinase (x). Measurements of these reactions were made under similar conditions with the free Ca^{2+} concentration controlled with EGTA. The data for the ATPase activity (42) and Ca^{2+} binding to troponin (43) were previously reported.

The Ca^{2+} concentration required for 50% activation of cardiac light chain kinase is about 1 μM which is similar to the Ca^{2+} concentration for half saturation of the two high affinity sites in cardiac troponin. As the two sites in troponin are saturated, the ATPase activity is maximally activated. No cooperativity is indicated in the Ca^{2+} binding to troponin (slope of Hill plot = 1.0) whereas significant positive cooperative is found in the Ca^{2+} activation of the cardiac light chain kinase (slope of Hill plot = 3.8). At Ca^{2+} concentrations less than 0.5 μM both kinase activity and myofibrillar ATPase activity are inhibited. As the Ca^{2+} concentration exceeds 1 μM the myofibrillar ATPase is only partially activated while the light chain kinase is fully activated due to the positive cooperativity associated with the kinase. These results would suggest that in cardiac muscle the myosin light chain kinase may catalyze the phosphorylation of myosin during a contraction cycle. However this possibility needs careful examination because other important biochemical factors may modify the relative rates of these processes. In one communication it was reported that myosin light chain which was fully phosphorylated in perfused rabbit hearts was partially dephosphorylated with infusion of epinephrine (19). The reasons for these observations are not apparent from the biochemical properties of myosin light chain kinase or phosphatase.

Recent investigations have shed some light on the subunit structure of myosin light chain kinases, although the results are not yet unequivocal. One report suggests that myosin light chain kinase is a single polypeptide with an apparent molecular weight of 77,000 (26). Other reports suggest that the kinase from skeletal muscle may be composed of a Ca^{2+}-binding subunit and a catalytic subunit (30, 34) similar to the subunit structure of the light chain kinase from gizzard smooth muscle (35). In addition, evidence has been presented that the Ca^{2+} binding subunit of the myosin light chain kinases may be similar to calcium modulator proteins of cyclic nucleotide phosphodiesterase. Thus, these proteins may mediate the effects of Ca^{2+} on several cellular processes.

We have made comparisons of the enzymatic properties of myosin light chain kinases from heart, white skeletal muscle and tracheal smooth muscle. Myosin as well as the phosphorylatable light chain subunit was obtained from each of these muscles. All three enzymes were dependent upon Ca^{2+} for activity and had similar pH optima (pH 6.5 to 7.0) and Km values (0.15 to 0.2 mM) for ATP. The kinetic properties of a particular kinase were measured with myosin and myosin light chains from each type of muscle. The results in Fig. 2 show that the partially purified light chain kinase from tracheal smooth muscle catalyzed the phosphorylation of myosin from tracheal smooth muscle whereas no phosphorylation of skeletal muscle myosin was found. Although cardiac myosin was phosphorylated by the tracheal light chain kinase the Vmax value was 20-fold less than the value obtained with tracheal myosin. The Km values were similar (Table 4).

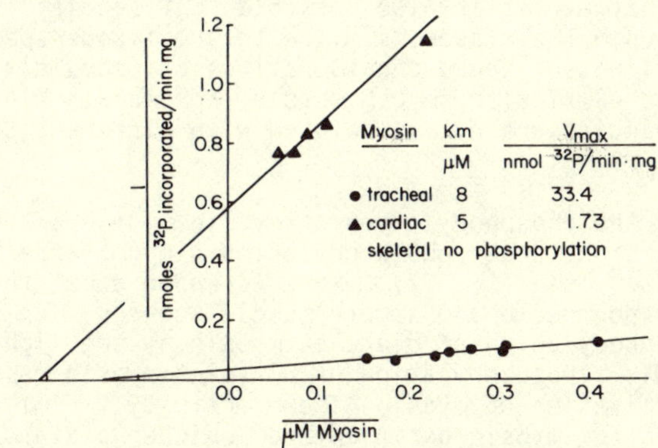

Fig. 2 Linweaver-Burk plot of the initial rates of phosphorylation of different myosins by partially purified myosin light chain kinase from tracheal smooth muscle.

A similar pattern for the phosphorylation of the purified light subunits was also found. Skeletal muscle myosin light chain was not phosphorylated by the tracheal light chain kinase at concentrations up to 100 μM light chain. Furthermore, the skeletal muscle light chain did not inhibit the phosphorylation of the tracheal light chain by the tracheal light chain kinase. A similar pattern of specificity was also found with skeletal muscle light chain kinase (Table 4). It phosphorylated skeletal and cardiac muscle myosins, but not tracheal smooth muscle myosin. Myosin and purified light chains from heart were phosphorylated by each of the other kinases. In addition, cardiac light chain kinase catalyzed the phosphorylation of purified light chains from all three types of myosin (data not shown). Thus the cardiac phosphorylation system does not show the same degree of specificity as found in skeletal and tracheal smooth muscles.

TABLE 4 Kinetic Characteristics of Partially Purified Light Chain Kinases from Skeletal and Tracheal Smooth Muscles.

Substrate	Skeletal Muscle Kinase		Tracheal Smooth Muscle Kinase	
	Km(μM)	Relative Vmax	Km(μM)	Relative Vmax
Skeletal Muscle Myosin	2	1.0	No Phosphorylation	
Cardiac Myosin	6	0.72	5	0.05
Tracheal Smooth Muscle Myosin	No Phosphorylation		8	1.0

These results emphasize that although the Ca^{2+} activation properties may be similar for myosin light chain kinases, the catalytic properties may be highly dependent upon the type of tissue in which the kinase is found. This is not surprising because the biochemical properties of the substrates (myosin) from cardiac, skeletal and smooth muscles are very different. Thus, it may be expected that there are a number of isozymic forms of myosin light chain kinase rather than a single species. What is not yet known is whether regulatory properties exist in addition to Ca^{2+} activation as found with the phosphorylation of phosphorylase kinase, another Ca^{2+}-dependent enzyme. It is also possible that specific regulatory mechanisms may be dependent upon the tissue, similar to the tissue specific phosphorylation of troponin. In light of these considerations one should also be cautious in generalizing about a particular smooth muscle light chain kinase until it is clear that unique properties are not associated with different types of smooth muscle.

A biochemical role for the phosphorylation of myosin from heart or skeletal muscle has not been defined. Light chain phosphorylation did not effect the ATPase activities of myosin or actomyosin (26, 27). More is known about the role of phosphorylation of myosin from nonmuscle and smooth muscle sources. An important biochemical role for the phosphorylation of platelet myosin by the light chain kinase has been described (36). Phosphorylation of platelet myosin resulted in an increase in the actin activation of myosin ATPase activity. Similar results have recently been described for myosin purified from chicken gizzard smooth muscle (37), pig stomach (38), and guinea pig vas deferens (39). Thus, the phosphorylation of myosin may be essential for actin activation of the ATPase activity, implying that phosphorylation of myosin may be essential for contraction in these cells. However, recent evidence reported by Ebashi and his colleagues suggests that phosphorylation of gizzard (40) and aortic (41) smooth muscle myosin may not be essential for actin activation of myosin. Instead, Ca^{2+} regulation was ascribed to tropomyosin and an 80,000 dalton protein factor which was devoid of kinase activity. The reasons for the conflicting reports are not apparent.

CONCLUSION

Although phosphorylation of myofibrillar proteins may be important in regulating contractile activity of different types of muscles, the evidence for such putative roles is not complete. Phosphorylation of specific myofibrillar proteins by purified protein kinases has been demonstrated. However, some of the major problems include identification of the specific phosphorylation reactions which occur in intact tissues in response to pharmacological or physiological stimuli and elucidation of biochemical mechanisms for the phosphorylation reactions which may be important in regulating contractile activity. The biochemical properties of distinct protein kinases should provide insights into the relationship between the phosphorylation reaction in vivo and the biochemical events initiated by specific pharmacological and physiological stimuli.

> Progress is rather slow, yet biochemistry must inevitably move in this direction because it must ultimately deal with integrated enzymatic activity at the cellular level of organization rather than with individual enzymatic reactions.
>
> C.F. Cori, 1956

ACKNOWLEDGEMENTS

The research presented in this paper was supported by grants from NIH (HL18170, GM02267), American Heart Association, Muscular Dystrophy Association of American and California Lung Association. This work was done during the tenure of an Established Investigatorship of the American Heart Association for J.T.S.

REFERENCES

1. J. Kendrick-Jones. Myosin-linked calcium regulation. In: Molecular Basis of Motility, Springer-Verlag, New York, p. 132 (1976).
2. H.G. Mannherz and R.S. Goody. Proteins of contractile systems. Ann. Rev. Biochem. 45, 428 (1976).
3. S.V. Perry. The contractile and regulatory proteins of the myocardium. In: Contraction and Relaxation in the Myocardium, Academic Press, London, p. 29

(1975).
4. C. Bailey and C. Villar-Palasi. Cyclic AMP-dependent phosphorylation of troponin. Fed. Proc. 30, 1147 (1971).
5. J.T. Stull, C.O. Brostrom and E.G. Krebs. Phosphorylation of the inhibitor component of troponin by phosphorylase kinase. J. Biol. Chem. 247, 5272 (1972).
6. P.J. England, J.T. Stull and E.G. Krebs. Dephosphorylation of the inhibitor component of troponin by phosphorylase phosphatase, J. Biol. Chem. 247, 5275 (1972).
7. J.T. Stull and S.E. Mayer. Biochemical mechanisms of adrenergic and cholinergic regulation of myocardial contractility. Handbook of Physiology: Circulation, in press.
8. D.B. Bylund and E.G. Krebs. Effect of denaturation on the susceptibility of proteins to enzyme phosphorylation. J. Biol. Chem. 250, 6355 (1975).
9. J.T. Stull and J.E. Buss. Phosphorylation of cardiac troponin by cAMP-dependent protein kinase. J. Biol. Chem. 252, 851 (1977).
10. D.K. Blumenthal, J.T. Stull and G.N. Gill. Phosphorylation of cardiac troponin by cGMP-dependent protein kinase. J. Biol. Chem. 253, 334 (1978).
11. T.M. Lincoln and J.D. Corbin. Purified c-GMP-dependent protein kinase catalyzes the phosphorylation of cardiac troponin inhibitory subunit (TN-I), J. Biol. Chem. 253, 337 (1978).
12. P.J. England. Studies on the phosphorylation of the inhibitory subunit of troponin during modification of contraction in perfused rat heart. Biochem. J. 160, 295 (1976).
13. T.M. Lincoln, C.L. Hall, C.R. Park and J.D. Corbin. Guanosine 3':5'-cyclic monophosphate binding proteins in rat tissues. Proc. Natl. Acad. Sci. USA 73, 2559 (1976).
14. J.C. Khoo, P.J. Sperry, G.N. Gill and D. Steinberg. Activation of hormone-sensitive lipase and phosphorylase kinase by purified cyclic GMP-dependent protein kinase. Proc. Natl. Acad. Sci. USA 74, 4843 (1977).
15. T.M. Lincoln and J.D. Corbin. Adenosine 3':5'-monophosphate and guanosine 3':5'-monophosphate-dependent protein kinases: possible homologous proteins. Proc. Natl. Acad. Sci. USA 74, 3239 (1977).
16. H. Syska, S.V. Perry and I.P. Trayer. A new method of preparation of troponin I using affinity chromatography. FEBS Lett. 40, 253 (1974).
17. P.J. England. Correlation between contraction and phosphorylation of the inhibitory subunit of troponin in perfused heart. FEBS Lett. 50, 57 (1975).
18. R.J. Solaro, A.J.G. Moir and S.V. Perry. Phosphorylation of troponin I and the inotropic effect of adrenaline in the perfused rabbit heart, Nature 262, 615 (1976).
19. N. Frearson, R.J. Solaro and S.V. Perry. Changes in phosphorylation of P light chain of myosin in perfused rabbit heart. Nature 264, 801 (1976).
20. E.G. Ezrailson, J.D. Potter, L. Michael and A. Schwartz. Positive inotropy induced by ouabain, by increased frequency, by X537A, by calcium and by isoproterenol: the lack of correlation with phosphorylation of TN-I. J. Mol. Cell. Cardiol. 9, 693 (1977).
21. R. Rubio, C. Bailey, and C. Villar-Palasi. Effects of cyclic AMP-dependent protein kinase on cardiac actomyosin: Increase in Ca^{2+} sensitivity and possible phosphorylation of TN-I. J. Cyclic Nucleotide Res. 1, 143 (1975).
22. K.P. Ray and P.J. England. Phosphorylation of the inhibitory subunit of troponin and its effect on the calcium dependence of cardiac myofibril adenosine triphosphatase. FEBS Lett. 70, 11 (1976).
23. Y.S. Reddy and L.E. Wyborny. Phosphorylation of guinea pig cardiac natural actomyosin and its effect on ATPase activity. Biochem. Biophys. Res. Commun. 83, 703 (1976).
24. J.E. Buss and J.T. Stull. Calcium binding to cardiac troponin and the effect of cyclic AMP-dependent protein kinase. FDBS Lett. 73, 101 (1977).

25. A. Fabiato and F. Fabiato. Relaxing and inotropic effects of cyclic AMP on skinned cardiac cells. Nature (London) 253, 556 (1975).
26. E.M.V. Pires and S.V. Perry. Purification and properties of myosin light-chain kinase from fast skeletal muscle. Biochem. J. 167, 137 (1977).
27. M. Morgan, S.V. Perry, and J. Ottaway. Myosin light chain phosphatase. Biochem J. 157, 687 (1976).
28. J.T. Stull and C.W. High. Phosphorylation of skeletal muscle contractile proteins in vivo. Biochem. Biophys. Res. Commun. 77, 1078 (1977).
29. K. Barany and M. Barany. Phosphorylation of the 19,000-dalton light chain of myosin during a single tetanus of frog muscle. J. Biol. Chem. 252, 4752 (1977).
30. D.M. Waisman, T.J. Singh and J.H. Wang. The modulator-dependent protein kinase. J. Biol. Chem. 253, 3387 (1978).
31. J.B. Posner, R. Stern and E.G. Krebs. Effects of electrical stimulation and epinephrine on muscle phosphorylase, phosphorylase kinase and adenosine 3':5'-phosphate. J. Biol. Chem. 240, 982 (1965).
32. G.I. Drummond, J.P. Harwood and C.A. Powell. Studies on the activation of phosphorylase in skeletal muscle by contraction and epinephrine. J. Biol. Chem. 244, 4235 (1969).
33. J.T. Stull and S.E. Mayer. Regulation of phosphorylase activation in skeletal muscle in vivo. J. Biol. Chem. 246, 5716 (1971).
34. K. Yagi, M. Yazawa, S. Ka Kiuchi, M. Ohshima and K. Uenishi. Identification of an activator protein for myosin light chain kinase as the Ca^{2+}-dependent modulator protein. J. Biol. Chem. 253, 1338 (1973).
35. R. Dabrowska, D. Aromatorio, J.M.F. Sherry and D.J. Hartshorne. Composition of the myosin light chain kinase from chicken gizzard. Biochem. Biophys. Res. Commun. 78, 1263 (1977).
36. R.S. Adelstein and M.A. Conti. Phosphorylation of platelet myosin increases actin-activated myosin ATPase activity. Nature 256, 597 (1975).
37. A. Gorecka, M.O. Aksoy and D.J. Hartshorne. The effect of phosphorylation of gizzard myosin on actin activation. Biochem. Biophys. Res. Commun. 71, 325 (1976).
38. J.V. Small and A. Sobieszek. Ca-regulation of mammalian smooth muscle actomyosin via a kinase-phosphatase-dependent phosphorylation and dephosphorylation of the 20,000 Mr light chain of myosin. Eur. J. Biochem. 76, 521 (1977).
39. S. Chacko, M.A. Conti and R.S. Adelstein. Effect of phosphorylation of smooth muscle myosin on actin activation and Ca^{2+} regulation. Proc. Natl. Acad. Sci. USA 74, 129 (1977).
40. T. Mikawa, Y. Nonomura and S. Ebashi. Does phosphorylation of myosin light chain have direct relation to regulation in smooth muscle? J. Biochem. 82, 1789 (1977).
41. M. Hirata, T. Mikawa, Y. Nonomura and S. Ebashi. Ca^{2+} regulation in vascular smooth muscle. J. Biochem. 82, 1793 (1977).
42. R. John Solaro and John S. Shiner. Modulation of Ca^{2+} control of dog and rabbit cardiac myofibrils by Mg^{2+}: Comparison with Rabbit Skeletal Myofibrils, Circ. Res. 39, 8 (1976).
43. J.T. Stull and Janice E. Buss. Calcium binding properties of beef cardiac troponin, (in press).

A Reassessment of the Criteria Used to Involve Cyclic Nucleotides in Hormone and Drug Mechanisms

Jean-Claude Stoclet

Laboratoire de Pharmacodynamie (C.N.R.S., E.R.A. 787), Faculté de Pharmacie, Université Louis Pasteur, 67083 Strasbourg, France

INTRODUCTION

During the past decade, the experimental strategy used to suggest a mediatory role of a cyclic nucleotide has been based on a series of recommendations initially proposed by Sutherland and collaborators (1) in the form of four criteria, which are paraphrased below. Before an agent is presumed to act through a change in the intra-cellular concentration of a cyclic nucleotide,

1) it should have a demonstrable effect on the activities of the enzymes involved in the regulation of the level of the cyclic nucleotide ;
2) it should change the cyclic nucleotide content of intact cells with a dose dependence and a time course consistent with a triggering action for the response ;
3) it should have pharmacological effects potentiated by phosphodiesterase inhibitors or, conversely, should potentiate the effects of hormones that elevate the cyclic nucleotide level ; and
4) an exogenous cyclic nucleotide should be shown to mimic its pharmacological effect(s).

Since these criteria were proposed in support of the hypothesis of a second-messenger role for 3',5'-AMP (cAMP), many publications on cyclic nucleotide research have accumulated. Other naturally occuring cyclic nucleotides, especially 3',5'-GMP (cGMP) have been found in many tissues. The criteria have been used by many investigators, who have sometimes reached opposite conclusions (2, 3). Because the methodology is critical, it may be useful to reexamine the criteria in the light of recent advances in the field of cyclic nucleotides. Extension of the criteria to cGMP will also be considered.

EFFECTS ON ENZYME ACTIVITIES

General Considerations

Fulfillment of the first criterion requires the demonstration of an effect on the enzyme systems synthesizing (cyclases) or hydrolyzing (phosphodiesterases) cyclic nucleotides. The major recent advances in this field bear on characterization, separation, and more or less thorough purification of various components of adenylyl cyclase (ADC), guanylyl cyclase (GC) and phosphodiesterase (PDE) systems, including modulator proteins. In at least some cases, the intracellular cyclic nucleotide level might be influenced by extrusion of the nucleotide from cells (4). The importance of this phenomenon may well have been underestimated in the past. However, it will not be considered here since its occurrence, role, and mechanism are still largely unknown.

Adenylyl Cyclase (ADC)

Various components of the ADC system. ADC catalyzes the reversible reaction

$$\text{ATP} \rightleftharpoons \text{cAMP + pyrophosphate}$$

probably towards the right, in view of the total cell contents of the components of the reaction.

As early as 1968, Robison, Butcher, and Sutherland (5) suggested that hormone receptor sites located at the outer surface of the plasma membrane are on regulatory subunits of the ADC system, the catalytic subunit of which functions at the inner surface of the membrane. The fact that receptor and catalytic subunits are indeed different entities was recently demonstrated by being physically separated (6) and recombined in cell-fusion experiments (7).

The mechanism of coupling between the binding of hormones to receptors and the activation of ADC remains largely unknown. It is well established that in many membrane preparations regulatory nucleotides (GTP, ITP, etc...) must be added to get hormones to stimulate ADC activity (8). In the case of the beta-adrenergic receptor, for instance, its occupation by an agonist is not a sufficient condition for response (9) : it has been suggested that a low-affinity state of the receptor is predominant in the presence of nucleotides and is responsible for enzyme activation, whereas a high-affinity state is predominant in the absence of GTP or other nucleotides and in then unable to activate ADC (9). Other factors have been shown to play a part in coupling between receptor and cyclase or to be able to modulate or alter the phenomenon. These factors include specific phospholipids (10), a "feedback regulator" characterized by Ho, and Sutherland in adipocyte preparations stimulated by adrenaline (11) and a Ca^{++}-regulated small protein which modulates both membrane adenylyl cyclase and PDE activity (see below). Recently it was reported that washing membrane preparations may uncouple beta-adrenergic receptors and ADC and that the preparations can be recoupled by a GTP-dependent cytosolic factor (12). It is too early to know the physiological significance of these various factors, which may play a role in refractoriness and supersensitivity to hormones, but their existence must obviously be taken into account when trying to relate pharmacological effects to ADC activity.

Evaluation of ADC activity. Absolute measurement of the ADC activity in intact cells is not possible because an unknown proportion of the cAMP formed from labelled ATP is hydrolyzed by PDE into 5'-AMP, which is also produced by the action of various other cell enzymes on ATP. However, the incorporation of a labelled precursor into cAMP may be used as an index of ADC activity (13). The preparation is treated with a PDE inhibitor (usually millimolar concentrations of theophylline) to decrease cAMP breakdown. As discussed below the specificity and efficacy of the available PDE inhibitors can be questioned : several or most of them alter many membrane properties. Recently sonication was proposed as a tool for the study of ADC activity in nondisrupted issolated cells (14). This procedure gives exogenously added ATP access to the catalytic site, thus permitting the use of the same assay techniques as with acellular preparations (see below). It may prove useful, but the nature and amounts of cellular damages produced by sonication need to be carefully investigated.

Techniques based on measurement of the pyrophosphate produced also have been proposed for cytochemical localization and estimation of hormone sensitive ADC. Their pitfalls have recently been discussed (15).

ADC from broken-cell preparations usually retains some ability to be stimulated by hormones. However this ability often seems markedly reduced from what is thought to occur in intact cells. Some preparations, especially those from brain, do not respond at all to hormones that stimulate ADC in intact cells (9). This loss of responsiveness may due to the loss of one or more of the coupling factors mentioned above.

Methods used for ADC assay in acellular preparations have been reviewed recently (16). The major difficulties arise from the presence of ATPase and cAMP-PDE activities in crude preparations and in purified membranes from eukaryotic cells. An ATP-regenerating system and a PDE inhibitor (or an excess cAMP) are added to the assay medium. For the reasons indicated below (see PDEs) it should be verified in each particular case that the PDE inhibitor does not alter the sensitivity of the ADC preparation to drug or hormone stimulation and that PDE activity is practically completely inhibited.

Purified membrane preparations provide a unique tool for studying the relationships between specific binding of a ligand and activation of ADC. An example is provided by studies performed in different laboratories on beta-adrenergic receptor (9). An extremely good correlation was found between the affinity constants calculated from binding studies and from activation of ADC by agonists or inhibition by antagonists of its hormonal activation. These data confirm the identity of binding sites and receptor sites. Such studies also offer the opportunity to analyze coupling between receptor occupancy and activation of ADC. In some "well coupled" preparations maximal activation of the enzyme resulted from fractional occupation of receptor sites. In some other "poorly coupled" preparations, occupation of receptors must be nearly complete before effects on ADC activity are seen. It is reasonable to assume that membrane displaying poor coupling were altered during their preparation.

In conclusion, studies performed on intact tissues or cells or on crude broken-cell preparations may provide preliminary evidence suggesting that activation of ADC may play a part in a pharmacological mechanism. Studies on purified membranes from homogenous cell preparations are necessary to demonstrate the specificity of binding to plasma membranes of target cells and the relationship between such binding and the activation of ADC.

Guanylyl Cyclase (GC)

GC catalyzes the formation of cGMP using GTP as substrate. Its properties have recently been reviewed (17). It is found in soluble and particulate fractions from most tissues, but, unlike ADC, is not directly sensitive to hormonal stimulation. Recent studies suggest that its soluble and particulate activities are structurally different since they are antigenically different.

In a series of recent papers Murad and coworkers (17) observed that all the agents and conditions activating GC appear to be related to the redox state of preparations and the formation of free reactive NO or hydroxyl radicals. They proposed a model that takes into account most of the observations regarding drug effects on GC activity and cGMP accumulation ; these effects include Ca^{++}-dependent and Ca^{++}-independent accumulation of cGMP, and the effects hydrazine, nitroglycerine, nitroprusside, and other compounds which can be converted to NO. According to this model, an "inactive" form of GC which is sensitive to Ca^{++}-stimulation is activated by free radicals to a more labile and less Ca^{++}-sensitive enzyme form which may also use ATP as a substrate and produce cAMP (5-15 % of cGMP formation). Elucidation of these mechanisms should provide a clue to the understanding of hormonal or pharmacological regulation of GC and help to distinguish between Ca^{++}-mediated and cGMP-mediated responses in tissues. At the present time the significance of the GC forms in regulating cGMP level is still unclear.

Cyclic Nucleotide Phosphodiesterases (PDEs)

Multiple forms. PDEs catalyze the hydrolysis of 3',5'-cyclic nucleotides to 5'-nucleosides. The properties of PDEs have been recently reviewed (18). PDE activity has been found in most particulate and soluble fractions from practically all mammalian tissues.

Multiple forms of PDE have been chromatographically or electrophoretically separated from practically every tissue examined. They differ in their kinetic properties, substrate specificity and sensitivity to endogenous or exogenous activators and inhibitors (18). For example we have separated three PDEs from bovine aorta (Table 1) : one (form A) displays no specificity for cAMP or cGMP as substrate and is activated by an endogenous Ca^{++}-regulated protein modulator (CRPM) ; the two others (forms B and C) are respectively specific for cGMP and cAMP and are not activated by CRPM (19).

TABLE 1 Cyclic Nucleotide Phosphodiesterases in Bovine Aorta and Lung (a)

		Phosphodiesterase A	Phosphodiesterase B	Phosphodiesterase C
Activation	(b)	+	-	-
Substrate(s)		cAMP cGMP	cGMP	cAMP
Km (μM) high	(c)	14ᵡ 7ᵡᵡ	100	70
Km (μM) low	(d)	2ᵡ 2ᵡᵡ	7	3
Aorta		+	+	+
Lung		-	+	+

(a) from the data reported by Ilien et al. (19)
(b) Ca^{++}-dependent activation by an endogenous protein modulator (CRPM)
(c) 0.5 to 5 μM substrate concentration range
(d) 10 to 50 μM concentration range
ᵡ no CRPM ᵡᵡ plus CRPM

It has been suggested that selective inhibition of the various PDEs may cause selective pharmacological effects in the tissues where they are differently distributed (20). Other authors objected that we do not know if the PDEs of the intact cell are identical to those which have been isolated or are artefacts resulting from the separation procedure (18). Our finding (19) that, in the same conditions, two PDEs only (one specific for cGMP, the other for cAMP) could be obtained from lung whereas an additional PDE (activated by CRPM but showing no substrate specificity) was obtained from the aorta of the same species is consistent with the idea that different tissues contain various proportions of the different PDEs. The fact that papaverine, which is 3-4 times more potent in the *in vitro* inhibition of the beakdown of cAMP than of cGMP (23) supports the hypothesis that selective inhibition of a PDE may induce selective accumulation of the corresponding cyclic nucleotide. It should be emphasized, however, that we do not know the respective role of the different PDEs in regulating cyclic nucleotide levels.

Practically all the PDEs which have been studied display nonlinear kinetics characterized by two ("high" and "low") K_m values depending on the substrate concentration range used. It has often been postulated that only the "low-K_m form" has physiological meaning at the estimated concentration of cyclic nucleotides in cells. However the high K_m PDEs may also play a part in the breakdown of cyclic nucleotides at these concentrations because they display higher V_{max} values than the low K_m PDEs. Various reasons have been proposed to explain the non Michaelian behaviour of the PDEs : the presence of isoenzymes with different K_ms (24), negative cooperativity (25), and/or interconversion by aggregation or disaggregation of subunits (26). While all these causes may play a part in the case of crude enzyme preparations, the purified forms also retain nonlinear kinetics with negative cooperativity (22) and/or two K_m values (19). Thus, the high and low K_m value represent the functional state of an enzyme preparation in a given experimental situation rather than a well defined form of PDE.

Evaluation of PDE activity. With respect to assay methods, the use of labelled substrates allow sensitive and accurate determination of PDE activity. The separation of products from the substrate remains the critical point. The 5'-nucleotide produced is generally transformed to nucleoside by snake venon 5'-nucleotidase in order to allow easier chromatographic or batch separation. As stated in a recent review (18), in crude preparations the metabolism may proceed beyond adenosine or guanosine because of tissue deaminases. Deaminated products may be less easily separated from residual substrate, especially when using anion-exchange batch or chromatographic techniques (27). It is therefore necessary to verify carefully in the case of each tissue preparation that recovery of labelled products is total. In the case of crude preparations, it may be preferable to use ^{32}P-substrate and to measure $^{32}PO_4^{3-}$ released by 5'-nucleotidase than to measure ^3H-nucleoside formed from ^3H-substrate. Another point to be stressed is that dilution of the PDE may cause interconversion from a low to a high K_m dissociated enzyme (25). Kinetic studies should thus be performed using one single enzyme concentration.

So far most studies on PDE activators or inhibitors have been performed on crude preparations. The presence in many crude tissue extracts of two or more PDEs with different K_m values for the same substrate makes it impossible to determine which PDE is affected by inhibitors and which type of inhibition is involved, unless the various PDE are separated. For the same reason it is not possible to determine true K_i values of inhibitors using crude preparations. Purification of the PDEs is thus a prerequisite to the search for specific inhibitors of the various PDEs. The finding of such compounds may in turn provide a clue to the respective roles of the PDEs.

Lung may be an interesting exception to what was said above, since it seems to contain only two PDEs (Table 1), which are respectively specific to cAMP and cGMP (19). The fact that Davis and Kuo (28) found identical results when studying the selectivity of inhibitors on crude or purified enzyme preparations from guinea pig lung is consistent with the view that, in that tissue, cAMP and cGMP are hydrolyzed by different enzymes which do not interfere with each other.

In their system 2'-deoxy derivatives of cAMP and cGMP show a high degree of selectivity to inhibit the hydrolysis of cAMP or cGMP, respectively. Inosine 3',5'-monophosphate (cIMP), too, selectively inhibits the hydrolysis of cGMP in lung, but in aorta it also inhibits the hydrolysis of cAMP by the activable PDE which has no substrate specificity (Table 2).

PDE inhibitors. The breakdown of cAMP is inhibited by a large variety of chemical compounds which have recently been reviewed exhaustively (29, 30). These compounds belong to different chemical groups with no obvious structural relationships and display a wide spectrum of pharmacological properties. The existence of various PDEs offers a wide diversity of mechanisms of interactions, which may perhaps explain the wide diversity in the chemical structures of inhibitors. In most cases, however, the type of inhibition and its selectivity for one or the other of the PDEs is unknown because it has not been studies on purified enzymes. Except in the case of the cyclic

TABLE 2 Inhibition of the PDEs by Cyclic Nucleotide Analogues

Analogue	I_{50} (μM) activatable PDE	cGMP-PDE	cAMP-PDE	Tissue	Reference
cAMP	–	10,000	–	aorta*	(19)
8-Br cAMP	–	150	150	lung**	(28)
N^6,2'-O-dibutyryl cAMP	–	200	470	lung**	(28)
N^6-monobutyryl cAMP	–	1,000	160	lung**	(28)
2'-deoxy cAMP	–	1,000	45	lung**	(28)
cGMP	–	–	10,000	aorta*	(19)
8-Br cGMP	–	1,000	1,000	lung**	(28)
2'-deoxy cGMP	–	4.5	1,000	lung**	(28)
cIMP	–	1.5	1,000	lung**	(28)
	4	1	165	aorta*	(19)

* from beef ** from guinea pig

TABLE 3 Active Concentration Range of Inhibitors of Cyclic Nucleotide Breakdown

	Concentration M cAMP-PDE	cGMP-PDE	Tissue	References
Methylxanthine				
Theophylline	10^{-4}	10^{-4}	Beef heart	(58)
MIX*	10^{-5}	10^{-6}	Pig coronary arteries	(22)
Isoquinoline				
Papaverine	10^{-5}	10^{-5}	Beef and Rat heart	(21) (33)
Flavonoid				
Quercetin	10^{-6}	–	Beef heart	(32)
Pyrazolopyridine				
SQ 20006	10^{-6}	10^{-5}	Rat brain	(60)
Imidazolidinone				
Ro 20-1724	10^{-7}	–	Rat erythrocytes	(31)
	10^{-5}	–	Rat brain	(61)
Phenanthroline				
M & B 22,948	10^{-5}	10^{-7}	Human lung	(62)

* 1-methyl 3-isobutylxanthine

nucleotide analogues (Table 2), the selectivity of inhibitors, when it has been studied, has generally been found to be relatively poor. This in the case of methylxanthines (22) and papaverine (28), for instance. Some new compounds seem more specific (62).

It should also be stressed that the apparent affinity of the presently available PDE inhibitors (Table 3) is lower than that of the substrate (with some possible exceptions, Ref. 31, 32, 62). Therefore very high concentrations of those drugs are necessary to inhibit PDE in intact cells. In some chemical series, the potencies of cAMP-PDE inhibition was correlated with the potencies of some pharmacological properties of various derivatives, for instance smooth muscle relaxation or vasodilatation (33, 34). At the pharmacologically active concentrations, however, virtually all presently available PDE inhibitors are not specific in that they also affect various other cell phenomena (35). For this reason, the link between PDE inhibition and pharmacological properties is not established in most cases.

In conclusion, some attempts to correlate the pharmacological effects of drugs with the drugs effects on PDE should be reexamined using purified PDEs. The use of the presently available nonspecific PDE inhibitors (including methylxanthines) to inhibit PDE in the assay of ADC (above) or to potentiate hormonal effects (below) should perhaps also be reconsidered. At least it should be verified that the PDE inhibitor does not interfere with the ADC of the system studied (theophylline, for instance, inhibits ADC activation by adenosine and related purines, Ref. 36)

Calcium-Regulated Protein Modulator (CRPM)

CRPM is a small molecular weight thermostable protein which has been found in all mammalian tissues and animal species so far examined (37). It binds Ca^{++} at physiological concentrations of the ion, and, in its Ca^{++}-bound form, activates both a PDE (see above and Ref. 38) and purified ADC (39). Interestingly enough, trifluophenazine and other phenothiazine derivatives bind to CRPM, thereby specifically inhibiting the activatable form(s) of PDE. Binding to CRPM seems the best way to inhibit the activatable PDEs since these enzymes do not discriminate cAMP from cGMP (19) or do so very partially (18). It was recently reported (41) that CRPM plays a regulatory role in the sensitivity of synaptic membranes to dopamine or noradrenaline stimulation : the hormone induced rise in cAMP content activates cytosol protein kinases which phosphorylate a specific membrane site, thereby triggering the release of the endogenous CRPM from the membrane ; CRPM in turn activates PDE and increases the hydrolysis of cAMP. It is too early to know if this self-regulatory mechanism takes place in other tissues, but if so, it should obviously be taken into account in the analysis of drug effects on the cyclic nucleotide system.

EFFECTS ON cAMP AND cGMP LEVELS

Fulfillment of the second criterion of Sutherland (1) requires that the intracellular concentration of cyclic nucleotides changes in a direction appropriate to induce the observed pharmacological property. There should be a quantitative correlation between the change in cyclic nucleotide content and the magnitude of the pharmacological response. Furthermore the change in cyclic nucleotide content should precede or at least coincide with the onset of the effect. An example of what should be done to establish a cause-and-effect relationship between cAMP and a pharmacological property may be found in the work of Mayer and co-workers on the inotropic effect of glucagon (42). Unwarranted conclusions have sometimes been drawn from experiments where the quantitative and temporal correlations have not been studied. For instance a mediatory role of cGMP in smooth muscle contraction and in cardiodepression produced by acetylcholine had been suggested. However, it was subsequently shown that the cGMP level increases after the onset of smooth muscle contraction (43) and that the concentrations of acetylcholine that elevate the cardiac cGMP level are 100 times higher than those that depress contraction (44). Obviously, the demonstration of temporal and quantitative relationships between cyclic nucleotide level and effect requires many cyclic nucleotide assays. The sensitivity and simplicity of the assay methods are thus critical. This point is discussed by Delaage et al. in this Methodological Seminar.

Cyclic-nucleotide-mediated effects are thought to be brought about in eukaryotic cells by activation of cyclic nucleotide-dependent protein kinases. In the activation process, the cyclic nucleotide is bound to a regulatory subunit of protein kinase. Thus, one should expect a better correlation between the cell content of protein-bound cyclic nucleotide or activated protein kinase and effect than between the cell content of total cyclic nucleotide and effect. An increase in the cyclic nucleotide concentration over that which maximally activates protein kinase would not bring about pharmacological consequences. Techniques have been developed to estimate kinase activity (45) or bound cAMP (46, 47) levels as they are in cells prior to homogenization. In the systems studied, a good correlation was found between bound cAMP and protein kinase activation while intense hormonal stimulation could increase total cAMP to above the saturation of its binding protein. Interestingly, it was also observed that small changes in the total cAMP concentration induced marked variations in the amount of cAMP bound to the regulatory subunit of protein kinase (47). Thus, measurement of bound cAMP in cells may prove very useful in investigating the mediatory effects of cyclic nucleotides.

It should be pointed out that the significance of cyclic nucleotide contents with respect to a pharmacological effect should be considered cautiously in tissues containing various cell populations with different functions. Homogeneity of the cell population tested is certainly a critical factor. It has also been pointed out by many authors that the concentration of cyclic nucleotides is not necessarily uniform in cells. The problem of cell compartimentalization is discussed in this seminar by Steiner.

POTENTIATION OF HORMONE EFFECTS BY PHOSPHODIESTERASE INHIBITORS

The third criterion requires that PDE inhibitors should potentiate the cAMP-mediated effect(s) of ADC-activating hormones. Obviously this criterion cannot be extended to cGMP since no hormonal stimulation of GC has been demonstrated. If we consider the second criterion (preceding section), potentiation of the effects thought to be mediated by cAMP should follow potentiation of the rise in cAMP level. It should also be pointed out that demonstration of a potentiation requires demonstration of a shift to the left of the concentration-effect curve.

Many examples of such potentiation have been reported in the literature (48-50, for instance). However an equal number of cases of lack of potentiation has been reported by others, sometime studying the same effect (compare Ref. 49 and 51). Thus, the evidence related to this criterion is in many cases controversial or unsatisfactory. As discussed above, the presently available PDE inhibitors are not specific. Their ability to raise the cAMP level has not always been verified in the choosen experimental conditions. Furthermore they may also increase the cGMP level (with unknown consequences) and act on many other cellular events. This criterion should therefore be reexamined if more specific and reliable PDE inhibitors are available in the future.

EFFECTS OF EXOGENOUS CYCLIC NUCLEOTIDES

The fourth criterion rests on the demonstration that an exogenously applied cyclic nucleotide mimics the pharmacological effect being studied. One cannot use cAMP or cGMP itself, since each is rapidly degraded in extracellular fluids and probably does not penetrate inside cells : at high concentrations (up to millimolar) they produce effects which may be ascribed to their metabolites, especially adenosine (52). Cyclic nucleotides analogues (8-Br and N^6,2'O-dibutyryl derivatives) have been extensively used in attempts to satisfy this criterion. They are thought to penetrate more easely into cells because they are more liposoluble and resist hydrolysis by PDE (53). They are actually moderately potent inhibitors of PDE (acting in the 10^{-5} M range), but they nonselectively inhibit the breakdown of cAMP and cGMP (22, 23).

It has been shown in HeLa cells that dibutyryl-cAMP is transformed into its N^6-monobutyryl derivative which activates cAMP-dependent protein kinase (52). Thus exogenously applied dibutyryl-cAMP can indeed reproduce the effects of endogenous cAMP. However it cannot be considered as specific evidence for a mediatory role of cAMP because, like other PDE inhibitors, it may also induce other nonspecific effects : besides possibly increasing the cGMP level, it may also present some effects which are indentical to those of other purine nucleotides and nucleosides (54, 55). A study to verify the absence of any effect of butyrate and a comparison with other noncyclic purine nucleotides and with adenosine should at least be systematically made.

CONCLUSION

It is clear that there are many methodological pitfalls in the studies required to satisfy the Sutherland's criteria. These pitfalls have appeared gradually as progress has been made in the increasingly complex field of cyclic nucleotides and have undoubtedly contributed to the many controversies in this area. The conclusions of some of the early studies should obviously be reexamined using improved methodology.

Of the four criteria initially proposed by Sutherland, the first and the second remain highly relevant. They should be studied on well defined purified enzymes or subcellular preparations or in sufficiently homogenous cell populations to establish the relevance of the results to the pharmacological effect. The choice of the system on which the study is performed is critical. Fulfillment of the third and fourth criteria cannot now be considered as specific evidence supporting a mediatory role of a cyclic nucleotide since the PDE inhibitors and cyclic nucleotide analogues which have been used are not specific. These criteria might be reconsidered when one may use specific inhibitors of the different PDEs. Some of the recently studied synthetic cyclic nucleotides may fulfill this need (28).

In addition the ultimate demonstration of a mediatory role of cAMP or cGMP requires the understanding of the relationships between activation of cyclic-nucleotide-dependent protein kinases and the pharmacological effect studied. The study of physiological events controlled by protein phosphorylation represents a rapidly expanding new field (56) and should perhaps be added to the criteria. The work of Katz and his coworkers (57) on the role cAMP-dependent phosphorylation of a sarcoplasmic reticulum protein in the mechanism of action of catecholamines on heart gives an example of this type of investigation *in vitro*. Cyclic-nucleotide-dependent phosphorylation of proteins involved in the studied cell function should also be demonstrated *in vivo*.

Acknowledgements. The experimental part of this work was partially supported by grants from D.G.R.S.T. (77-7-1399) and I.N.S.E.R.M. (C.R.L. 78-1-071-5).

REFERENCES

(1) E. W. Sutherland, G. A. Robison and R. W. Butcher, Some aspects of the biological role of adenosine 3',5'-monophosphate (cyclic AMP), Circulation 37, 279 (1968).

(2) D. H. Namm and J. Leader, Occurrence and function of cyclic nucleotides in blood vessels, Blood Vessels 13, 24 (1976).

(3) E. D. Jacobson and W. J. Thompson, Cyclic AMP and gastric secretion : the illusive second messenger, Adv. Cyclic Nucleotide Res. 7, 199 (1976).

(4) J. Penit, S. Jard and P. Benda, Probenecide sensitive 3',5'-cyclic AMP secretion by isoproterenol stimulated glial cells in culture, FEBS Letters 41, 156 (1974).

(5) G. A. Robison, R. W. Butcher and E. W. Sutherland, Adenyl cyclase as an adrenergic receptor, Ann. N. Y. Acad. Sci. 139, 703 (1967).

(6) L. E. Limbird and R. J. Lefkowitz, Resolution of β-adrenergic receptor and adenylate cyclase activity by gel exclusion chromatography, J. Biol. Chem. 252, 789 (1977).

(7) J. Orly and M. Schramm, Coupling of catecholamine receptor from one cell with adenylate cyclase from another cell by cell fusion, Proc. Nat. Acad. Sci. U.S.A. 73, 4410 (1976).

(8) M. Rodbell, M. C. Lin, Y. Salomon, C. Londos, J. P. Harwood, B. R. Martin, M. Rendell and M. Berman, Role of adenine and guanine nucleotides in the activity and response of adenylate cyclase systems to hormones : evidence for multisite transition states, Adv. Cyclic. Nucleotide Res. 5, 3 (1975).

(9) M. E. Maguire, E. M. Ross and A. G. Gilman, β-adrenergic receptor : ligand binding properties and the interaction with adenylyl cyclase, Adv. Cyclic. Nucleotide Res. 8, 1 (1977).

(10) G. S. Levey, The role of phospholipids in hormone activation of adenylate cyclase, Recent Prog. Horm. Res. 29, 361 (1973).

(11) R. J. Ho and E. W. Sutherland, Action of feedback regulator on adenylate cyclase, Proc. Nat. Acad. Sci. USA 72, 1773 (1975).

(12) F. Pecker and J. Hanoune, Uncoupled β-adrenergic receptors and adenylate cyclase can be recoupled by a GTP-dependent cytosolic factor, FEBS Letters 83, 93 (1977).

(13) J. F. Kuo and E. C. De Renzo, A comparison of the effects of lipolytic and antilipolytic agents on adenosine 3',5'-monophosphate levels in adipose cells as determined by prior labelling with adenine-8-^{14}C, J. Biol. Chem. 244, 2252 (1969).

(14) S. B. Achar, S. J. Strada, R. B. Sanders, W. J. Pledger, W. J. Thompson and G. A. Robison, Sonication as a tool for the study of adenylyl cyclase activity, J. Cyclic Nucleotide Res. 3, 189 (1977).

(15) A. Lemay and L. Jarrett, Pitfalls in the use of lead nitrate for the histochemical demonstration of adenylate cyclase activity, J. Cell. Biol. 65, 39 (1975).

(16) G. Schultz, General principles of assays for adenylate cyclase and guanylate cyclase activity, in : J. G. Hardman and B. W. O'Malley, Methods in Enzymology, 38, pp 115-125, Academic Press, New-York (1974).

(17) C. K. Mittal and F. Murad, Properties and oxidative regulation of guanylate cyclase, J. Cyclic Nucleotide Res. 3, 381 (1977).

(18) J. N. Wells and J. G. Hardman, Cyclic Nucleotide phosphodiesterases, Adv. Cyclic Nucleotide Res. 8, 119 (1977).

(19) B. Ilien, A. Stierle, C. Lugnier, J. C. Stoclet and Y. Landry, Separation of three cyclic nucleotide phosphodiesterases from bovine aorta, Biochem. Biophys. Res. Commun. (in press).

(20) B. Weiss and W. N. Hait, Selective phosphodiesterase inhibitors as potential therapeutic agents, Ann. Rev. Pharmacol Toxicol. 17, 441 (1977).

(21) C. Lugnier and J. C. Stoclet, Inhibition by papaverine of cAMP and cGMP phosphodiesterases from rat heart, Biochem. Pharmacol. 23, 3071 (1974).

(22) J. N. Wells, Y. J. Wu, C. E. Baird and J. G. Hardman, Phosphodiesterases from

porcine coronary arteries : inhibition of separated forms by xanthines, papaverine, and cyclic nucleotides, Mol. Pharmacol.11, 775 (1975).

(23) F. Demesy-Waeldele and J. C. Stoclet, Effect of papaverine on cyclic nucleotide levels in the isolated rat aorta, Europ. J. Pharmacol. 46, 63 (1977).

(24) Y. M. Lin and W. Y. Cheung, Cyclic 3',5'-nucleotide phosphodiesterase. A theoretical account of its anomalous kinetics in terms of multiple isoenzymes, Int. J. Biochem. 6, 271 (1975).

(25) M. M. Appleman and W. L. Terasaki, Regulation of cyclic nucleotide phosphodiesterase, Adv. Cyclic Nucleotide Res. 5, 153 (1975).

(26) A. L. Pichard and W. Y. Cheung, Cyclic 3',5'-nucleotide phosphodiesterases. Interconvertible multiple forms and their effects on enzyme activity and kinetics, J. Biol. Chem. 251, 5726 (1976).

(27) K. K. Ong and P.I.C. Rennie, Assay errors of cyclic 3',5'-nucleotide phosphodiesterase activity based on recovery of adenosine alone using an anion-exchange resin column, Anal. Biochem. 76, 53 (1976).

(28) C.W. Davis and J. F. Kuo, Differential effects of cyclic nucleotides and their analogs and various agents on cyclic GMP-specific and cyclic AMP-specific phosphodiesterases purified from guinea pig lung, Biochem. Pharmacol. 27, 89 (1978).

(29) S. M. Amer and W. E. Kreighbaum, Cyclic nucleotide phosphodiesterases : properties, activators, inhibitors, structure activity, J. Pharm. Sci. 64, 1 (1975).

(30) M. Chasin and D. N. Harris, Inhibitors and activators of cyclic nucleotide phosphodiesterase, Adv. Cyclic Nucleotide Res. 7, 225 (1976).

(31) H. Sheppard and G. Wiggan, Analogues of 4-(3,4-dimethoxybenzyl)-2-imidazolidinone as potent inhibitor of rat erythrocyte adenosine cyclic 3',5'-phosphate phosphodiesterase, Mol. Pharmacol. 7, 111 (1971).

(32) A. Beretz, R. Anton and J. C. Stoclet, Flavonoid compounds are potent inhibitors of cyclic AMP phosphodiesterase, Experientia (in press).

(33) C. Lemoulinier, J. M. Scheftel, G. Leclerc, C. G. Wermuth et J. C. Stoclet, Dialcoxy-6,7 isoquinoleines : activité inhibitrice de l'hydrolyse de l'AMP 3',5' cyclique par une phosphodiestérase artérielle, Eur. J. Med. Chem. (in press).

(34) B. Lesèche, J. Gilbert, C. Viel, A. Stierlé et J. C. Stoclet, Nouveaux analogues de la papavérine : benzylidène-4- et benzyl-4-dihydro- et tetrahydro-isoquinoléines, Eur. J. Med. Chem. 13, 183 (1978).

(35) J. C. Stoclet and C. Lugnier, Inhibitors of Cyclic AMP breakdown, in : P. Vanhoutte, Mechanisms of Vasodilatation, Karger, Basel (in press).

(36) A. Sattin, T. W. Rall and J. Zanella, Regulation of cyclic adenosine 3',5'-monophosphate levels in guinea-pig cerebral cortex by interaction of α-adrenergic and adenosine receptor activity, J. Pharmacol. Exp. Therap. 192, 22 (1975).

(37) J. H. Wang, Calcium-reglulated protein modulator in cyclic nucleotide systems, in : H. Cramer and J. Schultz, Cyclic 3',5'-nucleotides : Mechanisms of Action, pp 37-56, Willey, London (1977).

(38) W. Y. Cheung, Cyclic 3',5'-nucleotide phosphodiesterase : evidence for and properties of a protein activator, J. Biol. Chem. 246, 2859 (1971).

(39) C. O. Brostron, Y. C. Huang, B. Brechenbridge and D. J. Wolff, Identification of a calcium-binding protein as a calcium dependent regulator of brain adenylate cyclase, Proc. Nat. Acad. Sci. USA 72, 64 (1975).

(40) R. M. Levin and B. Weiss, Mechanism by which psychotropic drugs inhibit cyclic AMP phosphodiesterase of brain, Mol. Pharmacol. 12, 581 (1976).

(41) E. Costa, M. E. Gnegy and P. Uzunov, Regulation of dopamine receptor sensitivity by an endogenous protein activator of adenylate cyclase, Naunyn-Schmiedeberg's Arch. Pharmacol. 297, 547 (1977).

(42) J. G. Dobson, J. Ross and S. E. Mayer, The role of cyclic adenosine 3',5'-monophosphate and calcium in the regulation of contractility and glycogen phosphorylase activity in guinea pig papillary muscle, Circ. Res. 39, 388 (1976).

(43) G. Schultz and J. G. Hardman, Regulation of cyclic GMP levels in the ductus deferens of the rat, Adv. Cyclic Nucleotide Res. 5, 339 (1975).

(44) G. Brooker, Dissociation of cyclic GMP from the negative inotropic action of carbachol in guinea pig atria, J. Cyclic Nucleotide Res. 3, 407 (1977).

(45) J. D. Corbin, S. L. Keely, T. R. Soderling and C. R. Park, Hormonal regulation of adenosine 3',5'-monophosphate-dependent protein kinase, Adv. Cyclic Nucleotide Res. 5, 265 (1975).

(46) S. Harbon, L. Do Khac and M. F. Vesin, Cyclic AMP binding to intracellular receptor proteins in rat myometrium. Effect of epinephrine and prostaglandin E_1, Mol. Cell. Endocrinol. 6, 17 (1976).

(47) G. Schwoch and H. Hilz, Protein-bound adenosine 3',5'-monophosphate in liver of glucagon-treated rats, Eur. J. Biochem. 76, 269 (1977).

(48) G. A. Robison, R. W. Butcher and E. W. Sutherland, Fundamental Concepts in Drug-Receptor Interactions, Academic Press, New York (1970).

(49) G. Pöch and W. R. Kukovetz, Studies on the possible role of cyclic AMP in drug-induced coronary vasodilatation, Adv. Cyclic Nucleotide Res. 1, 195 (1972).

(50) P. Greengard and J. W. Kebabian, Role of cyclic AMP in synaptic transmission in the mammalian peripheral nervous system, Fed. Proc. 33, 1059 (1974).

(51) J. H. Mc Neill, J. B. Hook and R. S. Davis, Inhibition and enhancement of the noradrenaline constrictor response by methylxanthines, Can. J. Physiol. Pharmacol. 51, 553 (1973).

(52) E. Kaukel, K. Mundhenk and H. Hilz, N^6-monobutyryladenosine 3',5'-monophosphate as the biologically active derivative of dibutyryl-adenosine 3',5'-monophosphate in HeLa S3 cells, Europ. J. Biochem. 27, 197 (1972).

(53) T. Posternak, E. W. Sutherland and W. F. Henion, Derivatives of cyclic 3',5'-adenosine monophosphate, Biochem. Biophys. Acta 65, 558 (1962).

(54) E. Bülbring and J. H. Hardman, Effects on smooth muscle of nucleotides and dibutyryl analogs of cyclic nucleotides, in : Worcel and Vassort, Smooth Muscle Pharmacology and Physiology, pp 125-134, INSERM Paris (1976).

(55) V. A. W. Kreye and G. Schultz, Inhibition of norepinephrine-,angiotensin II - and vasopressin-induced contractions of smooth muscle by acyl derivatives of adenosine 3',5'-monophosphate, Europ. J. Pharmacol. 18, 297 (1972).

(56) H. G. Nimmo and P. Cohen, Hormonal control of protein phosphorylation, Adv. Cyclic Nucleotide Res. 8, 145 (1977).

(57) A. M. Katz, M. Tada and M. Kirchberger, Control of calcium transport in the myocardium by the cyclic AMP-protein kinase system, Adv. Cyclic. Nucleotide Res. 5, 453 (1975).

(58) E. W. Sutherland and T. W. Rall, Fractionation and characterization of a cyclic adenine ribonucleotide formed by tissue particles, J. Biol. Chem. 232, 1077 (1958).

(59) W. R. Kukovetz and G. Pöch, Inhibition of cyclic nucleotide phosphodiesterase as a possible mode of action of papaverine and similarly acting drugs, Arch. Exptl. Pathol. Pharmakol. 267, 189 (1970).

(60) M. Chasin, A potent new cyclic nucleotide phosphodiesterase (PDE) inhibitor, Fed. Proc. 30, 1268 (1971).

(61) I. Weinryb, M. Chasin, C. A. Free, D. N. Harris, H. Goldenberg, I. M. Michel, V. S. Paik, M. Phillips, S. Samaniego and S. M. Hess, Effects of therapeutic agents on cyclic AMP metabolism in vitro, J. Pharm. Sci. 61, 1556 (1972).

(62) H. Bergstram, J. Kristofferson, B. Lundquist and A. Schurmann, Effects of antiallergic agents, Compound 48/80, and some reference inhibitors on the activity of partially purified human lung tissue adenosine cyclic 3',5'-monophosphate and guanosine 3',5'-monophosphate phosphodiesterases, Mol. Pharmacol. 13, 38 (1977).

Radioimmunoassay of Cyclic AMP, Cyclic GMP, Cyclic CMP

M.A. Delaage, D. Roux and H.L. Cailla

Centre d'Immunologie INSERM-CNRS de Marseille-Luminy,
Case 906, 13288 Marseille Cedex 2 (France)

INTRODUCTION

The radioimmunoassay of cyclic nucleotides has brought about a profound change in our vision of the biological processes in which these molecules are involved. The basic methodology was established by A.L. Steiner et al. (1, 2), who obtained highly specific antibodies and synthesized iodinated analogs of cyclic nucleotides. Later, Cailla et al. (3, 4, 5) showed that cyclic nucleotides could be acylated in aqueous medium and that their succinylation in biological samples increased the sensitivity of the radioimmunoassay by two orders of magnitude (3). Within this general framework, improvements have been continuous (6) in order to confer on the radioimmunoassay, simplicity and reliability. This concerns the investment phase, preparation of antibodies and iodinated derivatives, and the exploitation phase of the radioimmunoassay, treatment of biological samples and incubation. These different questions will be successively examined, with the purpose of giving for each of them the current status or the most efficient protocol.

PREPARATION OF ANTIBODIES AND IODINATED DERIVATIVES

Preparation of 2'O succinyl cyclic nucleotides

Approximately 20 mg of succinyl cyclic nucleotide are needed to prepare sufficient quantities of immunogen and of tyrosylated derivative. Several methods of preparation have been described (1, 7). The following one is based on the fact that the succinylation of 2'OH position of ribose in 3'-5' cyclic nucleotides can be achieved in aqueous medium: 25 milligrams of cyclic nucleotide are dissolved in 2.5 ml of water. Then 0.500 ml of 4 M KOH and 0.112 g of succinic anhydride are added. The mixture is stirred for 15 minutes at room temperature. The incubation medium is then acidified with 140 µl perchloric acid (11.8 N), the final pH being around 2. The $KClO_4$ pellet is discarded and the supernatant is five fold extracted with 25 ml diethylether. In these conditions about 99% of initial succinic acid is eliminated. The solution of succinyl cyclic nucleotide is neutralized with KOH and lyophilized. In the beginning of the synthesis, it is recommended to add 100 µCi of tritiated cyclic nucleotide in order to facilitate the determination of the ratio hapten/carrier in the immunogen.

Preparation of tyrosylated derivatives

In the cold room 6 µmole of succinyl derivative are dissolved in 150 µl of freshly distilled dimethylformamide. A dilution (1/11) of triethylamine in dimethylformamide is made, 10 µl of this solution is added to the succinyl cyclic nucleotide solution. Then 10 µl of a dilution (1/16) of ethylchloroformate in dimethylformamide is added. The mixture is stirred for 10 minutes with a small magnetic bar. Then 100 µl of a solution of tyrosine methyl ester are added. This solution consists of 15 mg of D.L. tyrosine methylester, 600 µl dimethylformamide and 10 µl triethylamine. An other phenol-containing molecule can be used instead of tyrosine methyl ester, for example the peptide glycyl-L-tyrosine. In this case, the solution is made of 15 mg peptide, 500 µl dimethylformamide, 100 µl H_2O and 10 µl triethylamine. The reaction mixture is left for 5 minutes, diluted with 1 ml H_2O and loaded on a small QAE Sephadex column (0.9 x 20 cm) with chloride as counterion. Elution is obtained by a sodium chloride gradient (2 x 300 ml, 0.02 M to 0.4 M). The tyrosylated derivative is identified by its U.V. spectrum, and by the nitrosonaphtol reaction of phenols. The fractions corresponding to the conjugate succinyl cyclic nucleotide tyrosine methyl ester (or Gly.Tyr) are pooled, diluted in a 0.1 M sodium phosphate buffer, pH 7.5 to a final concentration of about 5×10^{-5} M, divided into 100 µl aliquots and stored at -20°C.

Iodination

The method is derived from the initial one of Hunter and Greenwood (8) adapted for repetitive iodinations. A solution of chloramine T in dioxane (2 mg/ml) is distributed in Eppendorf tubes into 5 µl aliquots which are allowed to dry. Each tube is convenient for one iodination with one mCi and can be stored for months.

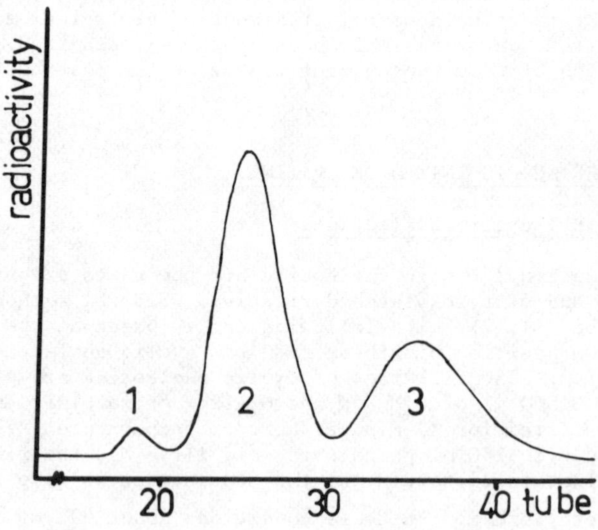

Fig. 1. Iodination of succinyl cyclic CMP tyrosine methyl ester. I Iodide. II Monoiodo-derivative. III Diiodo-derivative.

Iodination consists in adding 50 µl of tyrosylated derivative, 10 µl of ^{125}INa (carrier free, 100 mCi/ml) into the tube containing chloramine T. The tube is vortexed and left for 2 minutes. The reaction is stopped by adding 50 µl of a solution of 0.6 mg/ml sodium metabisulfite in phosphate buffer, and 500 µl of 0.1 M citrate buffer pH 6.2 with 0.5 g/l sodium azide. The same citrate buffer is used for chromatography of the iodinated derivative on a 0.9 x 60 cm G 25 Sephadex column which separates the diiodo-, the monoiodo- and the uniodinated derivatives. The elution profile of iodinated derivatives of cyclic CMP is shown in figure 1. The monoiodo derivative has the same specific radioactivity as iodine itself, approximately 1800 Ci/mmole (Radiochemical Center, Amersham). The monoiodo-derivative is diluted in 0.1 M citrate pH 6.2, 0.5 g/l sodium azide, 2 g/l albumin, up to approximately 3×10^5 cpm/ml. No loss of immunoreactivity was observed even after 5 months of storage.

Anti-cyclic nucleotide antibodies. Production and general properties

Immunogen. The initial protocol of Steiner and Oliver (1, 9) has never been questioned. 30 µmole of succinyl derivative are mixed with 20 mg serum albumin (bovine or human) in 2 ml of H_2O. The pH value is adjusted to 5.5. A total of 10 mg of 1 ethyl-3(3-dimethylaminopropyl) carbodiimide-HCl are added in five fractions over a period of one hour, in order to avoid excessive pH variations. After 2 hours, the reaction has practically stopped and the incubation mixture is loaded on a small Sephadex G.25 column (1 g) and eluted with 10^{-2} M NaCl. The ratio of fixation is determined at this stage by analysis of the U.V. spectrum, or better by the radioactivity content if a trace quantity of tritiated nucleotide has been added when preparing the succinylated derivative. After purification, the immunogen may be desalted by dialysis and lyophilized. Three to 12 moles cyclic nucleotide are fixed per mole of albumin, giving equally good results in the immunization step.

Immunization. Any classical immunization protocol can be used. We get good results with the one of Ross et al. (10). Each rabbit received a suspension of 0.25 mg to 0.5 mg immunogen with 2.50 mg lyophilized BCG in 1 ml of physiological serum, emulsified with 1 ml of complete Freund's adjuvant. This dose was injected at multiple (about 50) sites into the back of the animals. Booster injections were done at monthly intervals in an identical manner. The course of immunization has to be carefully followed in testing each rabbit once or twice a week. The titer of antiserum is estimated as the dilution able to bind fifty percent of diluted iodinated derivative.

Saturation of antibodies by endogeneous cyclic nucleotides. Antibodies produced against cyclic AMP and cyclic GMP are fully saturated by the corresponding cyclic nucleotides (6). This is due to the high concentration of circulating antibodies (up to 10^{-5} M in binding sites concentrations) as compared to the equilibrium dissociation constant for cyclic nucleotides (around 10^{-8} M), and to the capability of organism to produce cyclic AMP and cyclic GMP at a high rate. As shown in table 1 cyclic nucleotides are protected against phosphodiesterase by their antibodies. The self-saturation is without importance when using antibodies with iodinated derivatives which are succinylated and thus have a much better affinity for antibodies than cyclic AMP itself. This allows a high dilution of antiserum with about 10^{-10} M of binding sites, a concentration by which endogenous cyclic AMP is no longer bound and does not interfere in radioimmunoassay.

TABLE 1 Comparison of the hydrolysis of cAMP in anticyclic AMP immunserum and normal serum

At indicated times 200 µl of serum are taken off and acidified by 17 µl of 11.8 N $HClO_4$, then they are assayed for cAMP. Results expressed as 10^{-9} M.

Incubation time at 20°C	0	24 h	48 h	72 h
Antiserum	3000	-	-	2400
Normal serum added with 3 x 10^{-6} M cAMP	3151	11.9	4.5	0

TABLE 2 Specificities of anticyclic AMP, anticyclic GMP, anticyclic CMP antibodies

A = 2'O acetyl   S = 2'O succinyl

Antisera (Ours, except when cited)	Dissociation constant for the best ligand (M)	Cross-reactivity (a)	(b)	(c)	(d)
Anticyclic AMP	SCAMP	ACAMP	CAMP	SCGMP	
357	4 x 10^{-11}	1.4	100	10^3	
1-2	4 x 10^{-11}	1	40	10^4	
IV	4 x 10^{-11}	1	380	10^3	
Coll. Res.[12]	-	1	100	-	
Anticyclic GMP	SCGMP	ACGMP	CGMP	SCAMP	
I	4 x 10^{-11}	1.6	33	10^4	
II	1 x 10^{-10}	5.3	450	10^3	
III	7.5 x 10^{-11}	1.3	33	10^4	
IV	1.2 x 10^{-10}	1.7	63	2.5x10^3	
V	1 x 10^{-9}	-	50	4 x10^3	
Coll. Res.[12]	-	2	40	4 x10^4	
Coll. Res.[13]	-	-	70	2.5x10^3	
Anticyclic CMP	SCCMP	ACCMP	CCMP	SCUMP	SCAMP
1C	9.2 x 10^{-10}	1	1250	125	300
3C	1.5 x 10^{-10}	-	2600	6.3	7.0
4C	2.4 x 10^{-10}	-	110	32	-

Cross-reactivity is expressed as the ratio of concentrations giving the same displacement (generally 30% of binding), with the best ligand used as a reference.

Specificity of antibodies. Table 2 summarizes the main characteristics of affinity and specificity important for radioimmunology. Several conclusions emerge from this table:

- Antibodies exhibit the highest affinity for 2'O succinyl derivatives, the most closely related to the immunogen. Unmodified cyclic nucleotides are less recognized (column b).
- Acetylated derivatives (column a) are generally recognized as well as succinylated derivatives, except for anti-cyclic GMP antibodies which keep a slight preference for succinylated derivatives.
- Cross-reactivity between cyclic nucleotides has generally no practical importance, except when assaying cyclic CMP in the presence of excess of cyclic AMP (d). This problem will be examined further.
- Statistical analysis reveals a negative correlation between the gain provided by succinylation (column b) and the cross-reactivity with the most closely related succinyl cyclic nucleotide (column c). In other words a good recognition of the succinyl link slightly counteracts good recognition of the base (Fig. 2).

Heterogeneity of antibodies. Antiserum produced in rabbits against cyclic AMP exhibit a rather narrow heterogeneity. Some of them behave homogeneously, as judged by kinetics and thermodynamic properties (11). This is confirmed by electrofocusing (unpublished experiments) in which some antibodies appear nearly monoclonal.

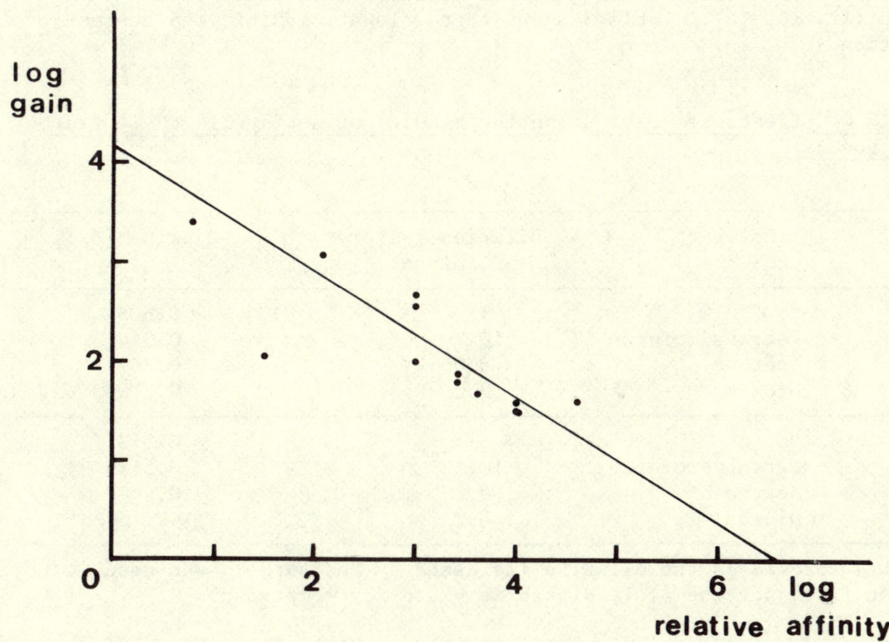

Fig. 2. Negative correlation between the recognitions of succinyl group and base. The double log scale reflects the free energy changes.

Factors affecting the hapten-antibody binding

In the following tables the effects of the most often encountered interfering substances are shown.

Cell culture media. Some of them contain interfering substances, such as HEPES buffer (table 3).

TABLE 3 Interference of some culture media on the radioimmunoassay of cyclic AMP

The interference is expressed as the apparent cyclic AMP concentration found in assaying pure medium.

Medium	Cyclic AMP equivalents (p mole/ml)
RPMI*	0.05
RPMI + 10% foetal calf serum	0.04
RPMI bicarbonate	0.14
RPMI bicarbonate + 10% human serum	0.91
Hanks HEPES	0.21
Hanks HEPES + 10% foetal calf serum	0.24
199 HEPES	1.68

*RPMI: Roswell Park Memorial Institute.

Effect of solvents. Some authors use solvents miscible with water in the acylation step (12, 13, 14). Table 3 shows that solvents inhibit the hapten-antibody reaction.

TABLE 4 Effects of solvents on the binding of anticyclic nucleotide antibodies

Antibody	Solvent	Dilution factor of solvent	Binding (B/T)
	No	-	0.603
Anticyclic AMP	Tetrahydrofuran	10.2	0.314
1-2	Acetone	9.0	0.287
	Dioxan	8.0	0.279
	No	-	0.673
Anticyclic GMP	Tetrahydrofuran	10.2	0.519
III	Acetone	9.0	0.464
	Dioxan	8.0	0.373

Each solvent was checked at the dilution indicated by the authors who used it: tetrahydrofuran (13), acetone (12), dioxan is given for comparison.

Effect of salts. We are essentially dealing with acetate and succinate of potassium or triethylammonium which are the salts introduced by the acylation procedures described in the next section. Salts have to be compared on the basis of their "normality" and not of their molarity since equal molar quantities of succinic anhydride and acetic anhydride result in twice as many acetate ions than succinate ions. In other words the molarity is different but the normality is the same, as are the concentrations of the counterions potassium or triethylammonium. Equal normalities correspond to equal dilutions of acylated sample. Figure 3 shows a typical salt effect on the hapten-antibody binding, for a given antibody.

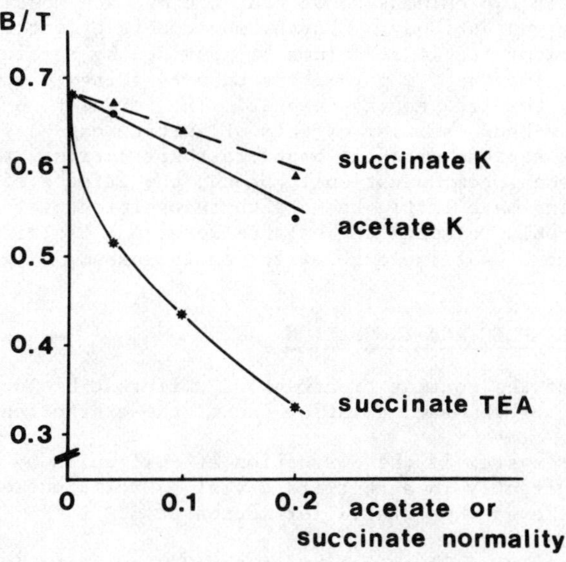

Fig. 3. Salt effect on anti-cyclic GMP antibody III.

The effect of salt is expressed by the initial slope of the curve, divided by the initial value of B/T; let S be this perturbation coefficient (table 5).

TABLE 5 Effects of ions on the binding of anti-cyclic nucleotide antibodies to their hapten

Antibody	Perturbation coefficients S (N^{-1}) TEA Succinate	K Succinate	K Acetate
Anti-cyclic AMP			
357	0.8	0.8	-
1-2	1.8	1.37	1.39
IV	-	5.05	5.05
Anti-cyclic GMP			
I	0.6	0.05	-
III	7.0	0.8	0.99
Anti-cyclic CMP			
1C	5.0	4.8	-
3C	4.8	4.0	2.81

These experiments performed by equilibrium dialysis give the exact values of binding. By contrast, when a precipitation or adsorption system is used, an additional salt effect on the separation system has to be taken into account.

The comparison between the first two columns indicates that potassium is less disturbing than triethylammonium, sometimes by a considerable factor. The comparison between the last two columns shows that acetate and succinate exert comparable effects on a given antibody. Furthermore, this effect is poorly correlated to the gain factor (table 2, column b) provided by succinylation. This can be shown in plotting the two parameters on a logarithmic scale, in order to reflect directly the free energy changes. The correlation is not significant. On the other hand, similar effects of acetate can be found for antibodies raised against haptens directly bound to their carrier, such as vasopressin (G. Rapuzzi, personal communication). Hence, the effects of acetate and succinate on hapten binding have little to do with recognition of the hapten-link by the antibody, and probably reflects the balance between a "salting in" effect on the phosphate group and a "salting out" effect on the base.

TREATMENT OF BIOLOGICAL SAMPLES AND INCUBATION

Table 6 gives a summary of the routine treatment of different biological samples, with complete integration of the succinylation and of the extraction steps.

Extraction. The simplest system is the disruption of cell walls by sonication in 1 N perchloric acid, directly in a centrifuge vial. Cyclic nucleotides are stable in perchloric acid even when stored for months at -20°C.

Acylation of the samples. The full acylation of the 2'O position in water is a special property of 3'5'-cyclic nucleotides, probably due to a favourable configuration of 2'O H not exhibited by other nucleotides. In the original method (3), we used succinic anhydride and triethylamine. We have now replaced triethylamine by potassium hydroxyde. On the other hand, Harper and Brooker (15) have proposed the replacement of succinic anhydride by acetic anhydride. These two points will be examined successively.

Acylation in the presence of KOH. The replacement of triethylamine with KOH results in a 96% yield of the succinylation of cyclic nucleotides in the biological sample. The loss of yield is negligible in radioimmunological applications. When KOH is replaced with an equivalent of potassium carbonate, calculated to give the same final pH, the initial pH jump is less and the yield of succinylation is only 60%. Intermediate situations were screened with various mixtures of KOH and potassium carbonate (Fig. 4). The fact that the yield was constant when over 50% KOH was used ascertains that a buffer capacity introduced by the biological sample is unlikely to reduce the succinylation yield. At least as good yields are obtained for acetylation with KOH, in the same conditions.

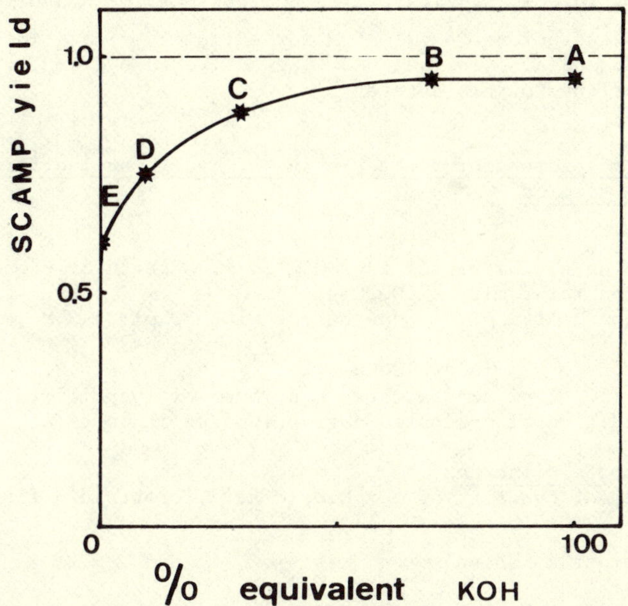

Fig. 4 Acylation of cyclic AMP with KOH in the presence of increasing buffering capacity.

There are four advantages in using KOH instead of triethylamine: 1) Diminution of salt effect on antibodies, as we have seen. 2) Easiness of handling and simplification of the protocol when perchloric acid is used. A single addition of KOH replaces the additions of potassium carbonate and triethylamine. 3) Better stability of succinylated derivatives during storage. We have observed that triethylamine, even at neutral pH, retains a sufficient nucleophilic power to slowly catalyze the hydrolysis of 2'O succinyl groups· 4) Absence of side reaction on the 6 OH position of guanine, when acetic anhydride is used. This reaction was found to occur (12) implying a considerable loss of immunoreactivity for the 2'O-, 6O-diacetyl cyclic GMP. With KOH, this side reaction (otherwise occurring with 30% yield) is practically avoided, as revealed by thin layer chromatography.

Succinic anhydride or acetic anhydride ? These two anhydrides are very similar in their ability to improve sensitivity (except for some anticyclic GMP antibodies) and in the secondary salt effects on antibodies. The choice between them is a purely technical matter, depending on the equipment of the laboratory. Acetic anhydride is ready for use but cannot be distributed in advance. Succinic anhydride is more convenient provided it is presented in lyophilized aliquots prepared in advance: a solution of succinic anhydride in dioxane (41 mg/ml) is distributed in 150 µl aliquots in Eppendorf tubes, with a Gilson automatic distributor. One hour lyophilization is sufficient to eliminate dioxane. Aliquots can be kept for several months at room temperature. This technique is easy to apply to very small samples (for example, 12 µl aliquots, 0.49 mg lyophilized succinic anhydride; those small quantities of anhydride are lyophilized for only 20 min) whereas pipetting precisely less than half a microliter acetic anhydride is much more difficult.

Interferences in the acylation step. During acylation substances susceptible to be acylated, and thus to compete with cyclic nucleotides for anhydride must not attain excessive concentrations (5×10^{-2} M for example). These include primary and secondary amines, and also molecules having activated hydroxyl groups. Tris buffer and citrate belong to this category.

TABLE 6 Procedures for the acylation of cyclic nucleotides in biological samples

I. Tissues
1. Homogeneization of the tissue in 200 µl of 1 N perchloric acid (final).
2. Centrifugation for 2 min at 10000 g.
3. Alkalinization of 160 µl of supernatant with 30 µl of 9 M potassium hydroxyde.
4. Centrifugation for 2 min at 10000 g,
5. Addition of 150 µl of supernatant to 6.15 mg of lyophilized anhydride.
6. Dilution, addition of iodinated derivative and incubation.

II. Fluids with deproteinization
1. Add a sufficient quantity of perchloric acid to obtain a final concentration of 1 N $HClO_4$.
2. Other operations as above.

III. Fluids without deproteinization (blood, urine)
Add at the same time 50 µl of the fluid and 100 µl of 1 M KOH to 6.15 mg succinic anhydride.

IV. Microsamples
Add 10 µl of sample and 2µl of 9 M KOH (if acidic) or 4 M KOH (if neutral) to 0.49 mg of lyophilized succinic anhydride.

V. Acetylation.
Replace 6.15 mg succinic anhydride with 5.74 µl acetic anhydride.

Incubation and separation of free and bound fractions.

The separation of free and antibody-bound radioactive ligand is the heart of any radioimmunoassay. From this point of view cyclic nucleotides present the most favorable case. Nearly every separation system works: precipitation with second antibody (1, 13, 16), filtration on millipore filter (14, 17) and equilibrium dialysis (18) give similar results. Ethanol precipitation and adsorption on charcoal (15, 19) lead to a slight overestimation of the binding while precipitation by ammonium sulfate or polyethylene glycol underestimates it. Equilibrium dialysis has become a practical tool, even for very small volumes (20 µl), as a result of the introduction of new dialysis devices (6, 18). Because of its intrinsic insensitivity to interference by salts, equilibrium dialysis is especially suited to the radioimmunoassay of cyclic nucleotides.

Special features of cyclic CMP radioimmunoassay.

The radioimmunoassay of cyclic CMP is especially difficult because of the very low level of this nucleotide in tissues and because of the characteristics of anticyclic CMP antibodies, which have a relatively low affinity, are sensitive to salt effects and are subjected to a strong cross-reactivity by cyclic AMP. The biological extracts (in perchloric acid, with subsequent neutralization with KOH) have to be concentrated by lyophilization and purified.

Purification of cyclic CMP from biological sample. The purification technique was selected to discard cyclic AMP which is the major contaminant of the cyclic CMP assay. Figure 5 shows the elution pattern of a mixture of ^{3}HcAMP and ^{3}HcCCMP from a Dowex 1 X 2 (1 x 10 cm) column, eluted with acidic gradient 100 ml 10^{-3}-M HCOOH-100 ml 0.5 M HCOOH. The profile was exactly the same when 4 ml of biological extract was loaded together with the tritiated derivatives and the recovery of cyclic CMP was always complete. The routine protocol consists of eluting with 5 x 10^{-2} M HCOOH in which only cyclic CMP is eluted. No ^{3}H-cCMP is added. The fractions containing cyclic CMP (50 ml) are pooled and lyophilized.

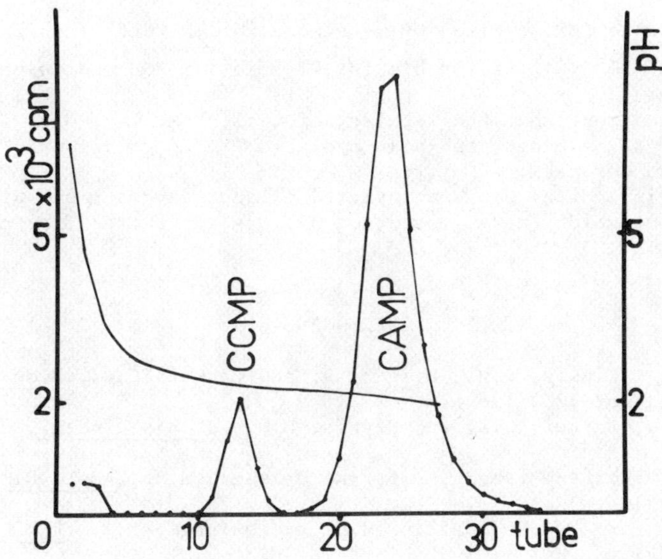

Fig. 5 Chromatography of a mixture of cAMP and cCMP on Dowex 1 X 2.

Cyclic CMP assay. The lyophilisates are redissolved in 1 or 2 ml H_2O. 300 µl of solution is alkalinized with 60 µl 4 M KOH. Potassium perchlorate is discarded. 150 µl of the alcaline supernatant is added to 6.15 mg lyophilized succinic anhydride, an other 150 µl is added to 7.2 mg of lyophilized succinic acid in order to serve as a control. Both are four fold diluted and incubated in dialysis with ^{125}I succinyl cyclic CMP tyrosine methyl ester and anticyclic CMP antibody. The amount of cyclic CMP is calculated from a reference curve in which standard solutions of cyclic CMP were incubated with the same potassium succinate concentration (.1 M) as the samples. Some results obtained from normal rat tissues are listed in table 7.

TABLE 7 Cyclic CMP content of normal rat tissues

Female rats (Sprague-Dowley) weighing 200 g were killed by decapitation. The tissues were immediately taken out by freeze-clamping and put in liquid nitrogen. The organs were weighed and allowed to thaw in 6 volumes of 1 N perchloric acid and homogeneized in a Potter-Elvejem apparatus. After centrifugation the pellet was discarded, the supernatant was neutralized by 9 M KOH. The potassium perchlorate was eliminated and the neutral supernatant was lyophilized.

Tissue	Cyclic CMP (10^{-12} mole/g wet weight)
Pancreas	45.75 ± 17.45 (N = 4)
Kidney	0.58 ± 0.03 (N = 4)
Spleen	0.87 ± 0.57 (N = 3)
Liver	1.02 ± 0.35 (N = 4)

Controls.

- Stability of cyclic CMP: cyclic CMP is stable in 1 N $HClO_4$ for 72 h at 4°C and for months at -20°C. It is stable for 72 h in 1 N KOH and in neutral medium except if non sterile.
- Controls of chromatography: the recovery of cyclic CMP is 100%. Negligible amounts of cyclic AMP are detectable in cyclic CMP. No cyclic CMP is detectable outside the cyclic CMP peak.
- Control of succinylation: non succinylated aliquot gave a small displacement, the contribution of which (never more than 5%) was subtracted from cyclic CMP results.

REFERENCES

(1) Steiner, A.L., Kipnis, D.M., Utiger, R. and Parker, C.W., Proc. Natl. Acad. Sci. USA, 64, 367-373 (1969).
(2) Steiner, A.L., Parker, C.W. and Kipnis, D.M., J. Biol. chem., 247, 1106-1113 (1972).
(3) Cailla, H.L. Racine-Weisbuch, M.S. and Delaage, M.A., Anal. Biochem., 56, 394-407 (1973).
(4) Goridis, C., Virmaux, N., Cailla, H.L. and Delaage, M.A., FEBS Letters, 49, 167-169 (1974).
(5) Cailla, H.L., Vannier, C.J. and Delaage, M.A., Anal. Biochem., 70, 195-202 (1976).
(6) Delaage, M.A., Roux, D. and Cailla, H.L. (1978) in Molecular Biology and Pharmacology of cyclic Nucleotides (G. Folco and R. Paoletti, Eds.), Elsevier/North-Holland, Amsterdam, pp. 155-171.
(7) Cailla, H.L. and Delaage, M.A., Anal. Biochem., 40, 62-72 (1972).
(8) Hunter, W.M. and Greenwood, F.C., Nature (London), 194, 495-496 (1962).
(9) Oliver, G.C., Jr., Parker, B.M., Brasfield, P.L. and Parker, C.W., J. Clin. Invest., 47, 1035 (1968).
(10) Ross, G.T., Vatukaitis, J.L. and Robbins, J.B. (1971) in Structure Activity Relationships of protein and polypeptide hormones (Marcoulies, M. and Greenwood, F.L., Eds.), vol. 1, p. 153, Excerpta Medica, Amsterdam.
(11) Denizot, F.C., Hirn, M.H. and Delaage, M.A., Immunochemistry (1978, in press).
(12) Frandsen, E.K. and Krishna, G., Life Sci., 18, 529-542 (1976).
(13) Zimmerman, T.P., Winston, M.S. and Chu, L.C., Anal. Biochem., 71, 79-95 (1976).

(14) Asakawa, T., Russel, T.R. and Ho, R.J., Biochem. Biophys. Res. Commun., 68, 682-690 (1976).
(15) Harper, J.F. and Brooker, G., J. Cyclic Nucleotide Res., 1, 207-218 (1975).
(16) Ferrendeli, J.A., Rubin, E.H., Orr, H.T., Kinscherf, D.A., and Lowry, O.H., Anal. Biochem., 45, 659-663 (1977).
(17) Weinryb, I., Michel, I.M. and Hess, S.M., Anal. Biochem., 45, 659-663 (1972).
(18) Cailla, H.L., Cros, G.S., Joly, E.J.P., Delaage, M.A. and Depieds, R.C., Anal. Biochem., 56, 383-393 (1973).
(19) Farmer, R.W., Harrington, C.A. and Brown, D.H., Anal. Biochem., 64, 455-460 (1975).

The Distribution of Cyclic Nucleotides and Their Protein Kinases in Tissues: An Immunocytochemical Approach

A.L. Steiner, Y. Koide, W.A. Spruill and J.A.Beavo*

Departments of Medicine and Pharmacology, University of North Carolina School of Medicine, Chapel Hill, North Carolina, and *Department of Pharmacology, University of Washington, School of Medicine, Seattle, Washington

INTRODUCTION

Immunocytochemistry is a powerful tool for studying cell function *in situ*. Cellular constituents can be localized by their reaction with specific antibodies that have been generated to intracellular antigens. In this paper we will review the immunocytochemical approach that we have utilized for localizing cyclic nucleotides (1-3) and more recently their protein kinases in tissues and cultured cells (4). With this immunocytochemical approach we attempt to identify the sites of action of the cyclic nucleotides and their protein kinases in tissues, and in that way provide clues for the physiological roles of the cyclic nucleotides in hormonal regulation. It seems likely that this immunocytochemical technique predominantly detects cyclic nucleotides that are tightly bound to tissue receptors, because the technique involves washing of the unfixed frozen tissue section, a procedure that probably results in the loss of free nucleotide and other freely soluble components. Since it is believed that protein-bound cyclic nucleotide is most likely the physiologically active form, it would appear that this technique may detect at least part of the functionally active cAMP and cGMP.

TECHNIQUE OF CYCLIC NUCLEOTIDE IMMUNOCYTOCHEMISTRY

Antibody Production

The immunocytochemical procedure for localization of cAMP or cGMP utilizes antibodies to the cyclic nucleotides produced in rabbits or goats (5,6). The cyclic nucleotides were rendered immunologic by conjugating them to carrier proteins through a 2'-O succinyl derivative of the cyclic nucleotide (5,6). The theoretical considerations involved in preparing the succinyl cyclic nucleotide derivatives have been discussed previously (3,5).

The most important criterion for selection of antibody for radioimmunoassay or for cyclic nucleotide immunocytochemistry is antibody specificity. Antibodies with virtually no cross-reactivity with ATP or 5' AMP are routinely found after immunization of several rabbits, and are particularly useful for cyclic nucleotide immunocytochemistry. These same antisera often will discriminate other cyclic nucleotides by a factor of 1000 to 10,000.

Tissue Preparation and Staining Procedure

The cytochemical procedure is routinely performed on 4 to 6 μm thick cryostat sections of unfixed frozen tissue. Alternatively, tissue sections can be fixed in 2% formaldehyde prior to the staining procedure. The staining patterns for both cyclic nucleotides are virtually identical from unfixed frozen or formaldehyde fixed tissue sections.

These tissue sections can be examined for cyclic nucleotide immunocytochemistry by either fluorescence or bright-field microscopy. For fluorescence microscopy, we prefer the indirect method because of the greater sensitivity and utilize a microscope equipped with episcopic illumination (the incident light is directed from above and is reflected off the surface of the specimen). The episcopic technique offers the advantage of increased staining intensity, as the exciting and emitting light is not adsorbed by passage through the tissue sections.

For bright-field microscopy we utilize the unlabeled antibody enzyme method "bridge technique" (7). This technique depends on the bivalence of IgG because the bridging species of IgG, if applied in excess, can both combine with the specific antibody to the antigen to be detected and still have binding sites free to combine with an antibody to horseradish peroxidase produced in the same species as the first antibody. For cyclic nucleotide immunocytochemistry, rabbit anti-cyclic nucleotide is applied to a tissue section followed by excess goat anti-rabbit IgG. This step is followed by the combination of purified rabbit anti-peroxidase with the excess combining sites on the goat anti-rabbit IgG. In the fourth step, peroxidase is added, followed by hydrogen peroxide and diaminobenzidine. The brown reaction product is visualized for light microscopy or can be made electron opaque for electron microscopy with osmium tetroxide. The technique is sensitive and specific, and the amount of first antibody employed is similar to that utilized in radioimmunoassay.

We use routinely Ig fractions of anti-cyclic nucleotide antiserum prepared by 40% $(NH_4)_2\ SO_4$ fractionation, and usually find negligible amounts of non-specific staining. In some instances non-specific staining is excessive. This can be reduced by preparing IgG fractions of anti-cyclic nucleotide antisera using chromatography on DEAE-cellulose in low ionic strength buffers at neutral pH (8). Commercial sources of fluorescein-labeled goat anti-rabbit IgG (2nd antibody) or peroxidase-conjugated goat anti-rabbit IgG do not usually require purification, but should be checked for specificity prior to use.

The selection of antibody dilution for immunocytochemistry is empirical. The process is designed to diminish non-specific staining as much as possible while maximizing the cyclic nucleotide staining patterns. Generally we incubate for 10 to 30 minutes with each antiserum or reagent, and wash for 10 minutes with .01 M phosphate, 0.15 M NaCl, pH 7.35 (PBS). Longer incubation times with cyclic nucleotide antibody (e.g., overnight) permit the use of more dilute antiserum. The procedures involved in cyclic nucleotide immunocytochemistry have been reviewed in detail (3).

Assessing Antibody Specificity

A commonly employed control for antisera specificity is the use of serum from an animal of the same species that has not been immunized with antigen. Thus, a preimmunized serum is obtained from all animals prior to the injection of the cyclic nucleotide protein conjugate, and is used in the same dilution as that of the cyclic nucleotide antiserum.

Another check of specificity is the application to tissue sections of antiserum that has been saturated by antigen. Cyclic nucleotides are incubated with antisera for a few hours (usually overnight). Other nucleotides such as ATP, 5'-AMP, or other cyclic nucleotides are incubated with the cyclic nucleotide antiserum in identical fashion. We have found that with certain cAMP or cGMP antisera, staining is inhibited by concentrations of the respective hapten at 10^{-6} M or less, and significantly higher concentrations of other nucleotides are required to eliminate the staining pattern. However, with other antisera, much higher concentrations of cyclic nucleotide are required to inhibit staining, and with a rare antiserum the staining pattern cannot be inhibited by 10^{-3} M concentrations of the respective cyclic nucleotide. Yet, when these antisera are utilized for immunocytochemistry, the staining pattern for the cyclic nucleotide is identical to that of an antiserum which is easily inhibited by hapten. We believe that while competitive inhibition of cyclic nucleotide staining is a helpful procedure for determining specificity, occasionally it is not definitive, and affinity chromatography for antiserum adsorption should be the final check. cAMP-Sepharose columns will selectively remove anti-cAMP antibody, but not cGMP antibody; conversely, antibody adsorption with cGMP-Sepharose has resulted in the selective removal of cGMP but not cAMP antisera, as determined by radioimmunoassay and immunocytochemistry.(2).

RECENT EXPERIENCE WITH CYCLIC NUCLEOTIDE IMMUNOCYTOCHEMISTRY

cAMP Immunocytochemistry

In a recent review we have summarized our cytochemical results utilizing this technique (3). We have been able to show in a variety of rat tissues enhanced fluorescence for cyclic AMP in selected cells within a heterogeneous tissue after hormonal stimulation, utilizing semi-quantitative photographic procedures. Increased fluorescence has been observed after hormonal stimulation in parotid (epinephrine) (1), thyroid (TSH) (2), kidney (PTH) (9), and liver (glucagon (3,10). In general the enhanced fluorescence is at the same sites as the basal staining pattern, and is usually in the area of the plasma membrane and adjacent cytoplasm. Nuclear membrane fluorescence has been observed in a number of tissues (3,11), and in certain cells fluorescence within nuclear elements is observed (12). This immunocytochemical procedure for localizing cyclic AMP has been particularly useful for identifying the cell type that is generating increased amounts of cyclic AMP within a heterogenous tissue.

We believe that cAMP immunocytochemistry is locating cAMP bound to receptor proteins because we have assumed that free cAMP would be washed away during the staining procedure. It will be important in future experiments to identify the species of "bound" cAMP that is identified by this procedure. While it is likely that the anti-cAMP antibodies are recognizing cAMP bound to receptors of protein kinase, it is clear, from preliminary immunocytochemical studies aimed at locating these receptor proteins, that the staining patterns of cAMP, these receptor proteins and the type I holoenzyme of cAMP dependent protein kinase are not entirely identical (see section V below). In addition, it seems quite likely that losses of cAMP bound to loosely bound receptor proteins will occur during the staining procedure. The extent of these losses need to be determined.

cGMP Immunocytochemistry

The staining patterns for cGMP in a variety of rat tissues have shown localization of cGMP antibody in the area of the plasma membrane, in cytoplasm, and in nuclear elements. These cytochemical observations have indicated a role for cGMP in membrane and nuclear regulation (10).

We have found that changes in the distribution of cGMP in rat liver during the process of regeneration following partial hepatectomy occurred without any change in the total tissue concentration of the nucleotide (13). The enhanced localization of cGMP antibody to the plasma membrane at 8 hours following hepatectomy and to the nucleus at 12 hours correlated with increased guanylate cyclase activity in purified particulate and nuclear fractions at these time points respectively (13). These observations provide evidence for compartmentation of cGMP in rat liver, and point out the need for developing techniques that help delineate compartmentation of the components of cyclic nucleotide metabolism in hormonal action (14).

cGMP antibody has been located to spermatocyte chromosomes in rat testis during the pachytene stage of meiosis (12). In Fig. 1 is shown cGMP on these chromosomes in squash preparations from rat testis. Because RNA synthesis is increased during pachytene, these results suggested that the nucleotide might be involved in nuclear synthetic activity and prompted studies utilizing polytene chromosomes from the salivary gland of Drosophila melanogaster (15). We have found that cGMP is found on these chromosomes in interbands and in puffs (Fig. 2). When the salivary glands are exposed to heat certain puffs occur on which there is increased RNA synthesis. When these chromosomes were stained for cGMP there was increased fluorescence on these puffs by 30 minutes (Fig. 2).

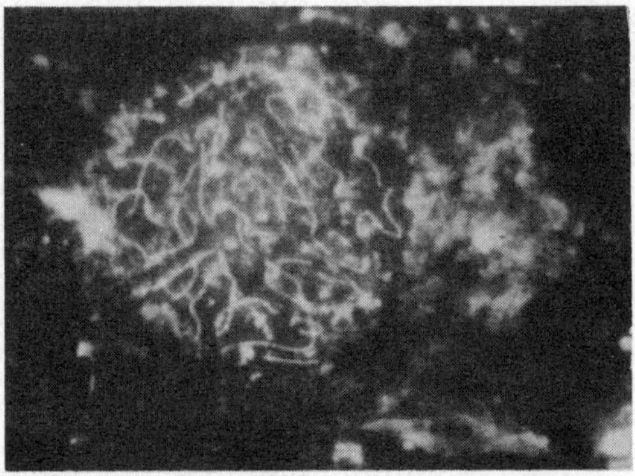

Fig. 1 Immunocytochemical localization of cyclic GMP on squash preparations of rat testis (X 2300)

Cyclic GMP fluorescence is observed associated with chromosomes of a pachytene spermatocyte. Note cGMP fluorescence associated with extra chromosomal material within the nucleus (X 2300).

The puff with the greatest uridine incorporation (93D) appeared to fluoresce the most brightly, suggesting a role for cGMP in the regulation of RNA synthesis. It is of interest that RNA polymerase has been immunocytochemically located to the same sites on these polytene chromosomes as cGMP (16). Future studies will be needed to determine if cGMP acts in association with RNA polymerase during transcription.

TECHNIQUE OF PROTEIN KINASE IMMUNOCYTOCHEMISTRY

Characterization of the Antisera for Cyclic Nucleotide Dependent Protein Kinase Cytochemistry

Type I cAMP dependent protein kinase from rabbit skeletal muscle, receptor subunit

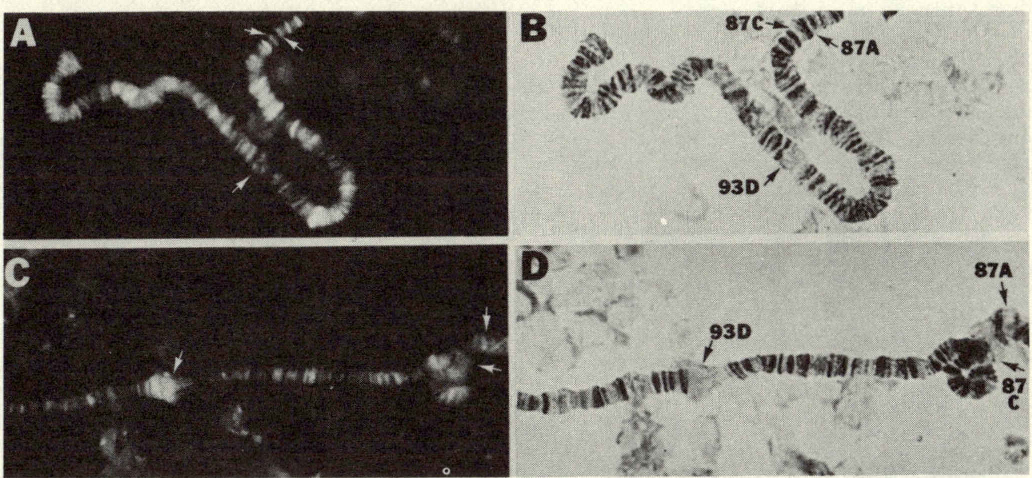

Fig. 2 (Left) cGMP fluorescence on control (A) and heat-shocked (C) chromosomes fixed with 2% formaldehyde. (Right) corresponding acetoorcein (B and D). Note enhanced fluorescence over locus 93D (C). (X 700). From Spruill et al (15). Reprinted by permission of the publisher.

of the type I kinase from bovine skeletal muscle, receptor subunit of the type II kinase from bovine heart, catalytic subunit of type I kinase from bovine skeletal muscle, and the soluble cGMP dependent protein kinase from bovine lung were purified as described previously (17,18). These antigens appeared to be homogeneous as determined by SDS acrylamide gel electrophoresis. They were utilized as immunogens in complete Freunds adjuvant. Antisera were generated in guinea pig (type I cAMP dependent holoenzyme, and catalytic subunit), rabbit (receptor subunit I, catalytic subunit, and cGMP dependent protein kinase), or goat (receptor subunit II). These antisera were characterized by the Ouchterolony double-diffusion technique. In addition, preincubation of the immunogen with its antiserum blocked the activity (either cAMP binding or phosphotransferase of the protein subunits). For immunocytochemical localization the antisera were competitively absorbed overnight, with their respective immunogen or other holoenzymes or subunits. An affinity column of receptor subunit II-Sepharose was utilized to ascertain the specificity of the receptor subunit II staining pattern.

Receptor Subunit I Antiserum

The antiserum was examined by the Ouchterolony double-diffusion technique. A single precipitin band was visualized which reacted both with cAMP receptor subunit I as well as with urea R (the subunit with cAMP removed). The antiserum did not cross-react with receptor subunit II, catalytic subunit, or the cGMP dependent protein kinase. The antiserum formed a line of identity with the type I cAMP dependent holoenzyme that fused with the line of receptor subunti I. Subsequent studies showed that this cross-reactivity was indeed with the holoenzyme as the holoenzyme did not dissociate in the agar during diffusion. The antiserum precipitated cAMP receptor subunit I and holoenzyme, but not cAMP receptor subunit II. In a number of rat tissues including liver, skeletal muscle, and testis, the staining pattern of the antigen could be competitively inhibited by absorption of the antiserum with its immunogen, but not by the other subunits. The results in rat liver are shown in Fig. 3d.

Catalytic Subunit Antiserum

By the Ouchterolony technique, the antiserum cross-reacted with catalytic subunit and type I holoenzyme, but not with cGMP dependent protein kinase, or the two cAMP dependent receptor subunits. Catalytic activity could be precipitated by the antiserum, but not by the other antisera, and its phosphotransferase activity was inhibited by the antiserum. Absorption of the antiserum by its immunogen prior to its fluorescent localization totally blocked the catalytic subunit staining pattern in a number of rat tissues. A representative example from rat liver is shown in Fig. 3b.

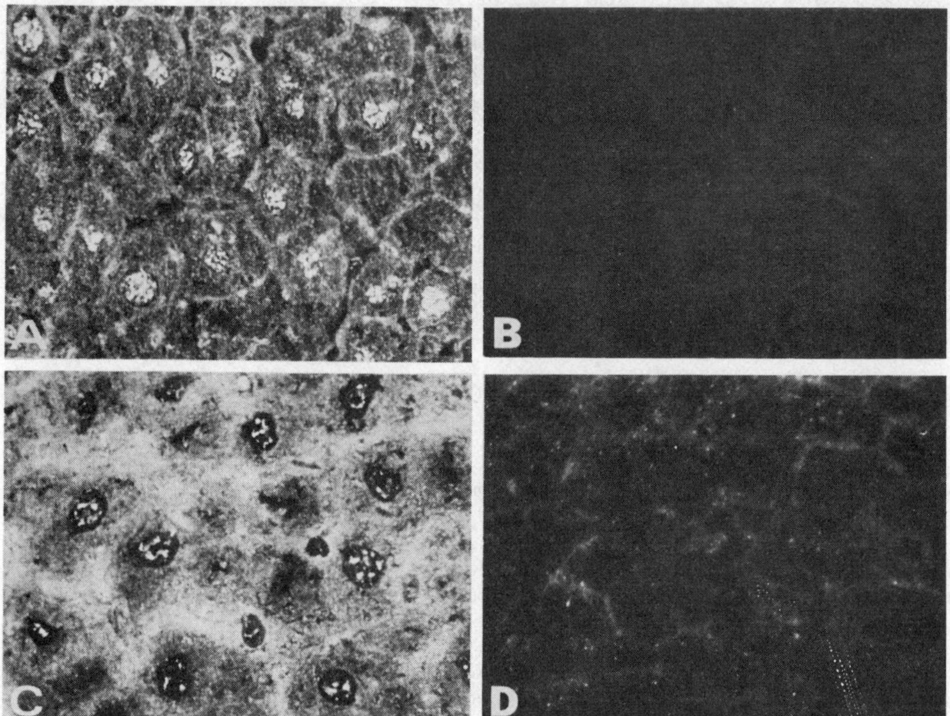

Fig. 3 Immunocytochemical localization of the catalytic subunit and receptor subunit I of cAMP dependent protein kinase in rat liver

In (A), catalytic subunit was localized in plasma membrane area, cytoplasm and in nuclei (X 580). In (B), fluorescence of catalytic subunit was totally eliminated when liver slice was stained with catalytic subunit antibody preincubated with catalytic subunit prior to the staining. In (C), receptor subunit I was located in cytoplasm and in nuclei (X 750). In (D), the receptor subunit I antiserum has been absorbed with receptor subunit I prior to the staining procedure.

Receptor Subunit II Antiserum

This antiserum formed a line of identity with its immunogen, but not with the other proteins (type I holoenzyme, receptor subunit I, cGMP dependent protein kinase, or catalytic subunit). Furthermore with progressive proteolytic degradation of receptor subunit II, a precipitin line was not detected. The antiserum precipitated receptor subunit II, but not receptor subunit I. Fluorescence in a number of tissues could be partially inhibited by the immunogen by liquid absorption, but affinity chromatography with receptor subunit II-sepharo e totally blocked the receptor subunit II fluorescence staining pattern (Fig. 4b).

Type I Holoenzyme Antiserum

Only one antiserum has been characterized. This antiserum did not form a precipitin band by the Ouchterolony technique. However, its fluorescence staining pattern was inhibited in a number of rat tissues by the immunogen (Fig. 4d). Furthermore, fluorescence intensity could be modified in rat liver by a physiological stimulus which increased cAMP concentrations (see section V below). We believe that this immunocytochemical data indicate that some determinants to the holoenzyme are present in this antiserum.

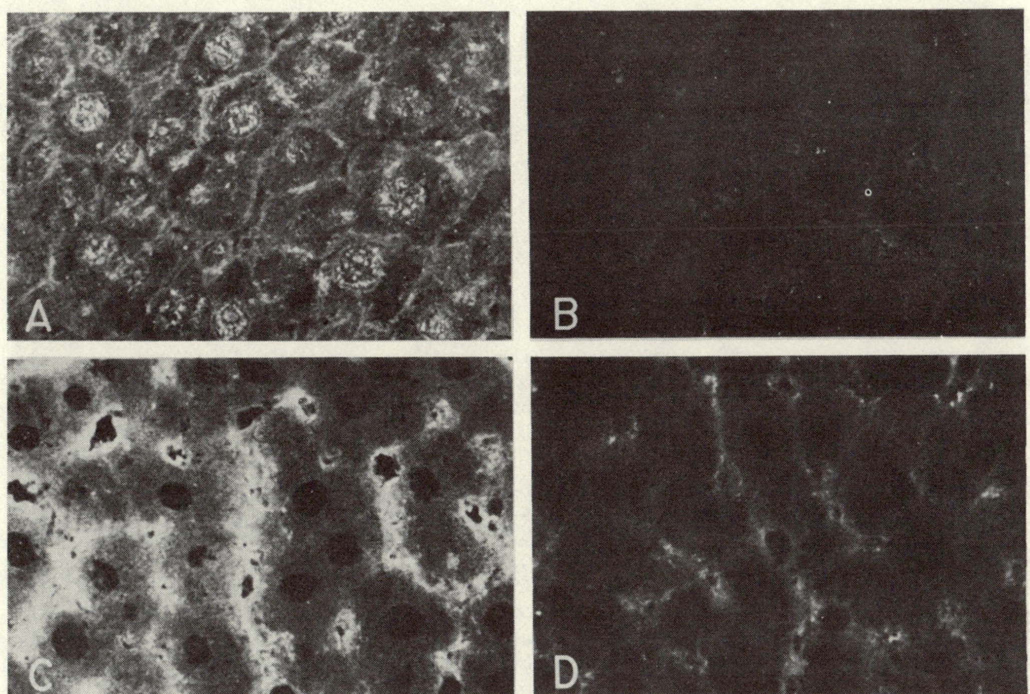

Fig. 4 Immunocytochemical localization of receptor subunit II and type I holoenzyme in rat liver (X 700).

In (A), receptor subunit II was localized in plasma membrane area, in cytoplasm and in nuclei. Nuclear membrane was stained in some cells. In (B) is depicted the staining pattern after the antiserum was passed over a receptor subunit II-Sepharose affinity column. In (C) type I holoenzyme was visualized along hepatic sinusoids. Only minimal fluorescence was observed in nuclei. In (D) the staining pattern of type I holoenzyme was virtually eliminated after competitive absorption of the antiserum with its immunogen.

cGMP Dependent Protein Kinase Antiserum

Two rabbit antisera have identical results on analysis by the Ouchterolony double-diffusion technique. The antisera cross-reacted with their immunogen, but not with the catalytic subunit of the cAMP dependent protein kinase, receptor subunit I or type I holoenzyme. Receptor subunit II formed a precipitin line that fused with the line of cGMP dependent protein kinase and its immunogen. To form approximately equal precipitin lines receptor subunit II was added in 5-fold excess as compared to cGMP dependent protein kinase. These results suggest possible homologeous regions for the two antigens as predicted by Corbin and Lincoln (19). Conversely, in these preliminary studies we have not totally ruled out minor contamination by receptor subunit II in the cGMP dependent protein kinase immunogen. The antisera precipitated their antigens and competitively inhibited their phosphotransferase activity. The fluorescence staining patterns of the immunogen could be diminished in a number of rat tissues.

Tissue Preparation and Staining Procedure

The cytochemical procedure is essentially the same as that utilized for cyclic nucleotide immunocytochemistry. Both unfixed frozen or paraformaldehyde fixed sections give comparable results. Cells in culture have been examined after washing and air-drying or after paraformaldehyde fixation. Tissue sections or cultured cells have been examined cytochemically by fluorescence or bright field microscopy with comparable results.

PRELIMINARY EXPERIENCE WITH PROTEIN KINASE IMMUNOCYTOCHEMISTRY

cAMP Dependent Protein Kinase and Subunits

Our major emphasis in this work during the past few years has been the characterization of the antisera, with emphasis on determining their specificity. We consider our immunocytochemical studies at this point quite preliminary.

Localization in control rat liver. The fluorescent staining patterns in control rat liver for receptor subunit I and catalytic subunit are shown in Fig. (3a and c). Both antigens were visualized in cytoplasm and nucleus. The pattern of nuclear localization appeared to be different for the two immunogens. Catalytic subunit had a more chromatin-reticular distribution and nucleoli fluoresced. Receptor subunit I had a more clumped distribution. In addition, catalytic subunit antibody localized to the area of the plasma membrane. Fluorescence staining patterns for both antigens were totally blocked by their respective antigens (Fig. 4b and d), but not by the other antigens. In rat skeletal muscle both catalytic subunit and receptor subunit I were visualized on periodic cross bands (whose identity have not been determined) and on what appeared to be sarcoplasmic reticulum (photographs not shown).

The staining patterns of receptor subunit II and holoenzyme in rat liver are shown in Fig. (4). Receptor subunit II was found in the area of the plasma membrane, in cytoplasm, and in nuclear elements, including the area of the nuclear membrane. Holoenzyme was concentrated along sinusoids, and minimal fluorescence was seen in nuclei. Both staining patterns were virtually eliminated after competitive absorption (holoenzyme, Fig. 3d), or after affinity chromatography (Receptor subunit II, Fig. 3b). In rat skeletal muscle both antigens were visualized on periodic cross bands, and the holoenzyme was particularly prominent in the area of the sarcollema (data not shown).

Localization in rat liver after glucagon administration. When rat liver was frozen at 5 minute intervals after administration of 0.5 mg of glucagon i.p. the subunits and the type I holoenzyme showed a change in their fluorescent staining patterns. The photographs of the catalytic subunit are shown in Fig. 5. Increased fluorescence was seen in cytoplasm and nucleus at the earliest time point examined (5 minutes (5b), peaked at 10 minutes (5c), and was decreasing by 15 minutes. The increased fluorescence in nucleus appeared in a chromatin-reticular distribution. Fluorescent staining patterns of receptor subunit I increased by 5 minutes in cytoplasm, but not convincingly in nucleus. The most prominent change with receptor subunit II antiserum was enhanced nuclear fluorescence starting at 10 and 15 minutes (photographs not shown). It is of note that while both receptor subunits and catalytic subunits were increased at early time points, fluorescence in the area of the hepatic sinusoids decreased with the holoenzyme antiserum (not shown).

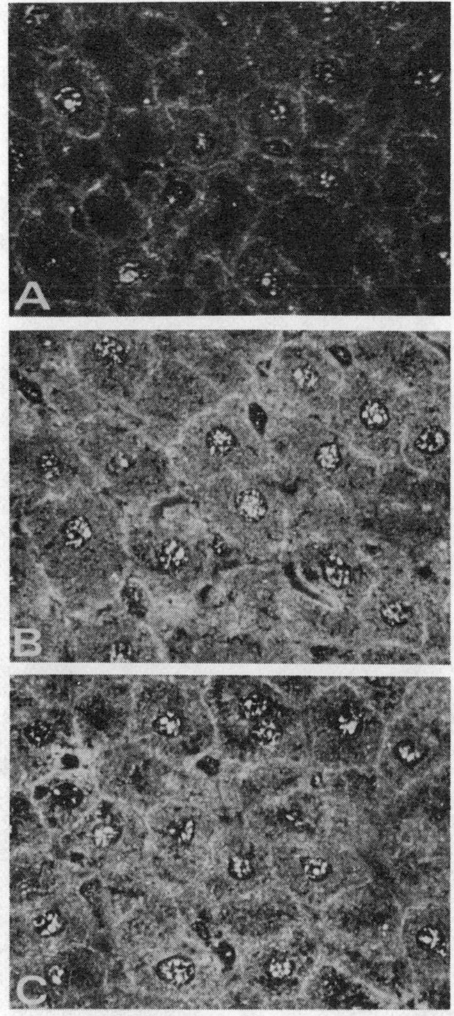

Fig. 5 Effect of glucagon on immunofluorescent staining pattern of the catalytic subunit of cAMP-dependent protein kinase in rat liver (X 700)

Glucagon, 0.5 mg i.p. was administered and sections were obtained at (A) 0 time, (B) 5 minutes, and (C) 10 minutes.

These results suggest that an antibody determinant in the holoenzyme antiserum is directed at the R_2C_2 combining site because holoenzyme fluorescence decreased as receptor and catalytic subunit fluorescence increased. It will be essential to characterize other holoenzyme antisera to determine the reproducibility of obtaining antibody determinants that distinguish holoenzyme from subunits. Furthermore, the enhanced fluorescence of catalytic subunit and receptor subunit II in nucleus after glucagon support the possibility of translocation of subunits from cytoplasm to nucleus. Biochemical evidence of translocation of both catalytic and receptor subunits has been provided in a variety of tissues after physiological stimulation (20-22), including in rat liver after glucagon administration (21).

cGMP Dependent Protein Kinase

The immunogen has a strikingly similar localization to cGMP in a number of rat tissues. Photographs comparing their localization in rat intestine is shown in Fig. 6. Note the bright fluorescence for cGMP (Fig. 6a), and its protein kinase (Fig. 6b) in the area of the brush border membrane. The antiserum will precipitate solubilized cGMP dependent protein kinase from these membranes and it will inhibit the activity of cGMP in phosphorylating a protein in the membrane (H. DeJonge, personal communication). These results provide evidence of the ability of this antiserum to cross react with cGMP dependent protein kinase of a different species because the immunogen was of bovine origin. Furthermore, the antiserum recognizes a particulate cGMP dependent protein kinase even though the immunogen was a soluble protein.

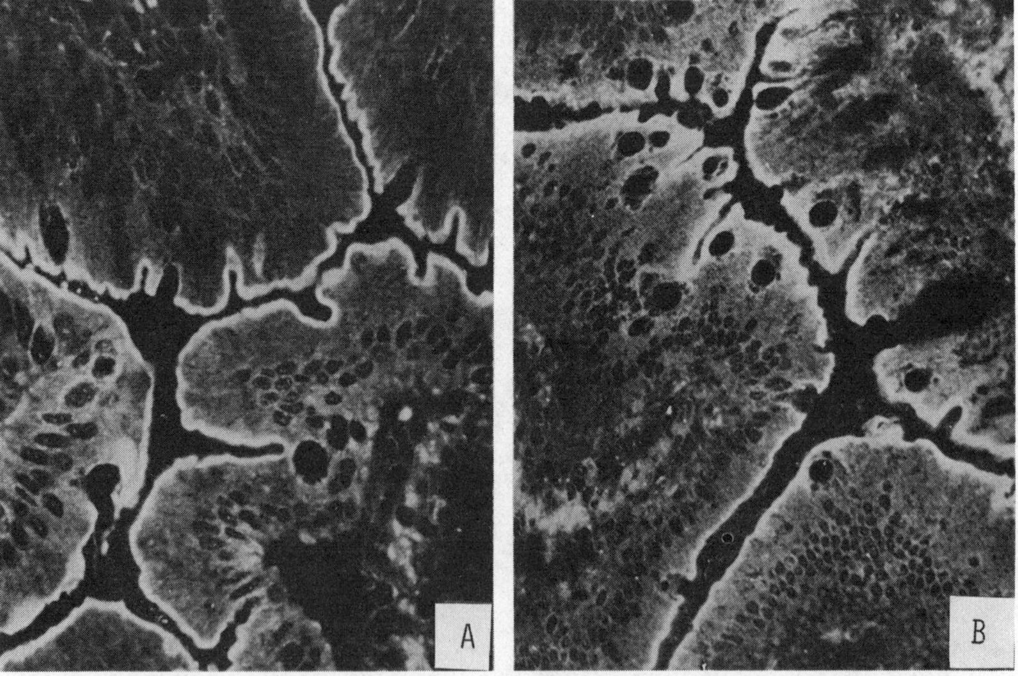

Fig. 6 Immunofluorescent localization of cGMP (A) and the cGMP dependent protein kinase (B) in rat jejunum (X 500). Note the similarity in distribution of both antisera in the area of the brush border.

CONCLUSION

The recent immunocytochemical utilization of antisera to the components of both cAMP and cGMP protein kinases complement our earlier work on cyclic nucleotide immunocytochemistry. These protein kinase antisera should be useful in providing clues for the sites of action of the cyclic nucleotides in hormonal action. Our preliminary cytochemical results provide some supportive evidence for the possible translocation of cAMP dependent protein kinase subunits from cytoplasm to nucleus. The adaption of this methodology to the ultrastructural level would further advance our understanding of cellular regulation by both cyclic nucleotides and their kinases. These rather preliminary results with the protein kinase antisera suggest that antisera can be produced which have determinants that are capable of distinguishing protein kinase subunits from holoenzyme. If this proves to be reproducible, it should be possible to obtain detailed cytochemical evidence of the sites of regulation of the protein kinases by cyclic nucleotides and determine the possible redistribution of protein kinases or their subunits after hormonal stimulation. More detailed immunocytochemical studies should provide information on the relative species cross-reactivity of these protein kinases and their subunits. The development of radioimmunoassays for the protein kinases and their subunits as has been described for receptor subunit II (23), would also be helpful in determining relative species cross-reactivity.

ACKNOWLEDGEMENT

This work was supported by USPHS grant AM19438 and a Grant-in-Aid from the American Heart Association. J.A.B. is an Established Investigator of the American Heart Association. We acknowledge the expert technical assistance of Hui-Lan Huang, and the help of Celeste Layton in preparing the manuscript.

REFERENCES

1. H.J. Wedner, B.J. Hoffer, E. Battenberg, A.L. Steiner, C.W. Parker, and F.E. Bloom, A method for detecting intracellular cyclic adenosine monophosphate by immunofluorescence, J. Histochem. Cytochem. 20, 293 (1972).

2. E.F. Fallon, R. Agrawal, E. Furth, A.L. Steiner, and R. Cowden, Cyclic guanosine and adenosine 3',5'-monophosphates in canine thyroid: Localization by immunofluorescence, Science. 184, 1089 (1974).

3. A.L. Steiner, S.H. Ong, and H.J. Wedner, Cyclic nucleotide immunocytochemistry, Adv. in Cyclic Nucleo. Res. 7, 115 (1976).

4. A.L. Steiner, Y. Koide, H.S. Earp, P.J. Bechtel, and J.A. Beavo, Compartmentalization of cyclic nucleotides and cyclic AMP-dependent protein kinases in rat liver: Immunocytochemical demonstration, Adv. in Cyclic Nucleo. Res. 9, 691 (1978).

5. A.L. Steiner, D.M. Kipnis, R. Utiger, and C.W. Parker, Radioimmunoassay for the measurement of adenosine 3',5'-cyclic phosphate, Proc. Natl. Acad. Sci. USA. 64, 367 (1976).

6. A.L. Steiner, C.W. Parker, and D.M. Kipnis, Radioimmunoassay for cyclic nucleotides. I. Preparation of antibodies and iodinated nucleotides, J. Biol. Chem. 247, 1106 (1976).

7. T.E. Mason, R.F. Phifer, S.S. Spicer, R.A. Swallow, and R.B. Dreskin, An immunoglobulin-enzyme bridge method for localizing tissue antigens, J. Histochem. Cytochem. 17, 563 (1969).

8. Fahey, J.L. (1967) Methods in Immunology and Immunochemistry, Williams and Chase, Academic Press, New York.

9. T.P. Dousa, L.D. Barnes, S.H. Ong, and A.L. Steiner, Immunohistochemical localization of 3':5'-cyclic AMP and 3':5'-cyclic GMP in rat renal cortex: Effect of parathyroid hormone, Proc. Natl. Acad. Sci. USA. 74, 3569 (1977).

10. S.H. Ong, T.H. Whitley, N.W. Stowe, and A.L. Steiner, Immunohistochemical localization of 3':5'-cyclic AMP and 3':5'-cyclic GMP in rat liver, intestine, and testis, Proc. Natl. Acad. Sci. USA. 72, 2022 (1975).

11. H.S. Earp, P. Smith, S.H. Ong, and A.L. Steiner, Regulation of hepatic nuclear guanylate cyclase, Proc. Natl. Acad. Sci. USA. 74, 946 (1977).

12. A. Spruill, and A. Steiner, Immunohistochemical localization of cyclic nucleotides during testicular development, J. Cyclic Nucleo. Res. 2, 225 (1976).

13. Y. Koide, H.S. Earp, S.H. Ong, and A.L. Steiner, Alterations in the intracellular distribution of cGMP and guanylate cyclase activity during rat liver regeneration, J. Biol. Chem. 253, 4439 (1978).

14. H.S. Earp, and A.L. Steiner, Compartmentalization of cyclic nucleotide-mediated hormone action, Ann. Rev. Pharmacol. Toxicol. 18, 431 (1978).

15. W.A. Spruill, D.R. Hurwitz, J.C. Lucchesi, and A.L. Steiner, Association of cyclic GMP with gene expression of polytene chromosomes of Drosophila melanogaster, Proc. Natl. Acad. Sci. USA. 75, 1480 (1978).

16. M. Jamrich, A.L. Greenleaf, and E.K.F. Bautz, Localization of RNA polymerase in polytene chromosomes of Drosophila melanogaster, Proc. Natl. Acad. Sci. USA. 74, 2079 (1977).

17. J.A. Beavo, P.J. Bechtel, and E.G. Krebs, Preparation of homogeneous cyclic AMP-dependent protein kinase(s) and its subunits from rabbit skeletal muscle, Methods Enzymol. 38, 299 (1975).

18. W.T. Dills, J.A. Beavo, P.J. Bechtel, and E.G. Krebs, Purification of rabbit skeletal muscle protein kinase regulatory subunit using cyclic adenosine 3':5'-monophosphate affinity chromatography, Biochem. Biophys. Res. Commun. 62, 70 (1975).

19. T.M. Lincoln, and J.D. Corbin, Adenosine 3':5'-cyclic monophosphate and guanosine 3':5'-cyclic monophosphate-dependent protein kinases: Possible homologous proteins, Proc. Natl. Acad. Sci. USA. 74, 3239 (1977).

20. R.A. Jungmann, P.C. Hiestand, and J.S. Schweppe, Mechanism of action of gonadotropin. IV. Cyclic adenosine monophosphate-dependent translocation of ovarian cytoplasmic cyclic adenosine monophosphate-binding protein and protein kinase to nuclear acceptor sites, Endocrinology. 94, 168 (1976).

21. W.K. Palmer, M. Castagna, and D.A. Walsh, Nuclear protein kinase activity in glucagon-stimulated perfused rat livers, Biochem. J. 143, 469 (1974).

22. E. Costa, A. Kurosawa, and A. Guidotti, Activation and nuclear translocation of protein kinase during transsynaptic induction of tyrosine 3-monooxygenase, Proc. Natl. Acad. Sci. USA. 73, 1058 (1976).

23. N. Fleischer, O.M. Rosen, M. Reichlin, Radioimmunoassay of bovine heart protein kinase, Proc. Natl. Acad. Sci. USA. 73, 54 (1976).

Contribution of Theoretical Simulation to the Study of the Cyclic AMP System

Stephane Swillens and Jacques E. Dumont

Institut de Recherche Interdisciplinaire, Université Libre de Bruxelles, B-1000 Brussels, Belgium

ABSTRACT

The purpose of this work is to show how the theoretical simulation of the cAMP system may give new informations by using known experimental data. In this study we have investigated on one hand the hormonal desensitization of adenylate cyclase system and on the other hand the control of protein kinase activation by cAMP. The biologist point of view on such a study is especially emphasized.

INTRODUCTION

The general subject of our works has been the theoretical study of the molecular interactions taking place in the biochemical reactions of the cyclic adenosine monophosphate (cAMP) system. In order to solve or at least to understand the problems issued from current data and knowledge, we have resorted to the computer simulation of the models describing the system. In this review, we should like to present a biologist personal point of view of such a work emphasizing the questions asked and the results obtained and deliberately leaving out the mathematics and calculations involved.

The role of simulation in our type of work is essentially to show the implications of conceptual models, i.e. the particular characteristics and the consequences of these, which must then be verified by some experiments for validating or rejecting the proposed models. It should be emphasized here that we all use models to define our hypothesis or to interpret our results, but few of us dare to draw models for fear of later disproval or of criticism for lack of experimental support. We therefore often prefer to rely on vague phrasing, thus avoiding to commit ourselves. On the other hand we often ignore that when a model is proposed, the predictions of the consequences of the model require as rigorous a proof as the validation of an hypothesis by experimental data.

The interest of simulation for the experimental researcher is at least threefold :

1) to demonstrate discrepancies, e.g. showing that a given model does not account for the experimental facts or that a given experimental result does not necessarily imply a model.

2) to provide an integrated and sometimes quantitative explanation of available data.

3) to predict consequences of a model and thus to propose experiments

to test the model or to distinguish between models.

It has been our experience that some of the consequences of a model are not generally accepted for a while but then, later, they become part of common knowledge and intuition and therefore no longer require simulation to be accepted or used : intuition often predicts the results of a simulation afterwards ... Therefore simulation work, although sometimes very useful in the solving of a problem, will often be forgotten when the concepts have been validated. The consideration of such a work mainly consists thus in providing a higher amount of informations deduced from a given set of data. However a theoretical analysis based on simulation is never used to prove the validity of any model or hypothesis but it can reject some model or hypothesis, implicitly or explicitly expressed and can propose how to test them in an experimental way.
We shall now consider a few examples of simulation work dealing with the cAMP system.

HORMONAL DESENSITIZATION OF ADENYLATE CYCLASE SYSTEM

Floating receptor concept was introduced to describe hormone receptor and adenylate cyclase as two distinct entities capable of diffusing independently in the plane of the plasma membrane (Ref. 1). It was proposed that the binding of a hormone molecule to a specific receptor promotes an interaction between receptor and adenylate cyclase in such a way that the enzymatic activity of adenylate cyclase is stimulated. As a result of this hormonal stimulation intracellular cAMP should accumulate up to a stationary level corresponding to the equalization of cAMP synthesis and degradation rates.

Significant evidence has now been gathered in support of the transient activation of adenylate cyclase in spite of the continuous presence of active hormone in the incubation medium. This phenomenon, referred to as the desensitization process, has been observed for an increasing number of tissues and for many different hormones. In all the cases, the desensitization process shares the common property that kinetics of cAMP accumulation resulting from hormone stimulation exhibit a fast increase, followed by a decline to a steady state somewhat higher than the basal level This hormone induced low sensitivity of adenylate cyclase leads also to a parallel decrease in the response to a further addition of hormone. The mechanism involved in such a process has been demonstrated to be not unique. Our study was dealing with the β-adrenergic desensitization as it was shown that it was the result of an apparent decrease of β-receptor number and that the recovery of full sensitivity after stopping hormone stimulation did not need new receptor synthesis (Ref. 2). Thus the decrease of receptor number would be caused by a temporary masking of the binding site. The question asked at this level of the study was : is it possible to elaborate a simple model based on the floating receptor concept which can account for the desensitization phenomenon as described for the β-adrenergic receptor ? To answer this question, we introduced the "locking receptor" concept (Ref. 3), which can be defined by four criteria :

1) the hormone has a double action as it induces both activation and inhibition of adenylate cyclase.

2) kinetics of activation are faster than kinetics of inhibition.

3) these different kinetic characteristics are a consequence of a conformational change of the receptor induced either by the hormone or by a direct or indirect action of adenylate cyclase.

4) the fall in the apparent number of binding sites induced by hormone stimulation is due to a persistent binding of the hormone to the receptors.

The two models described in Fig. 1 both fulfil the criteria and account for the floating receptor concept as receptor R and adenylate cyclase E can freely interact in the absence of hormone, E being active when bound to R (reaction 1). The double action of hormone (criterion 1 is a consequence that binding of H induces the formation of HRE (active complex) as well as the formation of HR (inactive complex).

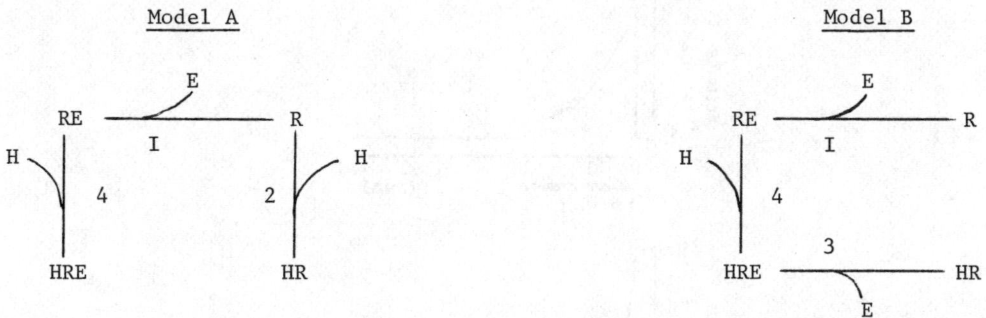

Fig. 1. Two simple models describing the locking receptor concept (H = hormone, R = receptor, E = enzyme (adenylate cyclase).

Criterion 2 requires that reactions 1 and 4 are faster than reaction 2 or 3 respectively for model A and model B. In model A, the kinetic characteristics of the activation and inhibition reactions (criterion 3) are dependent on the binding of E; the locked conformation of R consists of the locking of the binding site for E induced by H binding. On the other hand, in model B, H binding induces a change in the kinetics of E binding and the locked conformation of R consists of the locking of the binding site for H induced by E binding. In both cases, simulation of the locking receptor concept leads to theoretical results compatible with experimental observations (Fig. 2). The fall in the apparent number of receptor sites is due to the formation of the complex which cannot stimulate adenylate cyclase. Since this number is measured by antagonist binding to receptor, the occupancy of R by H prevents any binding of antagonist to the "masked"receptor. Though the biphasic kinetics of adenylate cyclase activation and thus of cAMP accumulation may be accounted for by both models, some distinguishing properties of these models should permit to reject one of them. For instance one can see that in the case of model B, the formation of the inactive complex HR requires the previous formation of HRE. Therefore, if the amount of E available for promoting RE and HRE formation is reduced, the kinetics of the decrease of the apparent receptor number would slow down. This phenomenon does not occur if model A applies since the complex HR is directly formed by reaction 2. The following experiment we propose on the basis of these theoretical considerations could lead to reject the wrong model : if we assume that two different hormones can stimulate cAMP synthesis, in one cell type, through the same adenylate cyclase molecules and that activation of adenylate cyclase requires the binding between R and E, then the simultaneous addition of the two types of hormone reduces the amount of available molecules of adenylate cyclase for each type of specific receptor corresponding respectively to the two types of hormone. As a consequence of this, if the kinetics of the number decrease of the receptor for one hormone type are dependent on the presence of the second type hormone, then model A should be rejected. Such an experiment could be designed in the case of the erythrocyte membranes which can be stimulated by β-agonist as isoproterenol and by prostaglandin, both hormone being able to induce the desensitization phenomenon.

In conclusion, we have shown that the floating receptor concept can be improved in such a way that it could account for the desensitization phenomenon. As experimentally observed, the desensitization is the consequence of the apparent decrease of

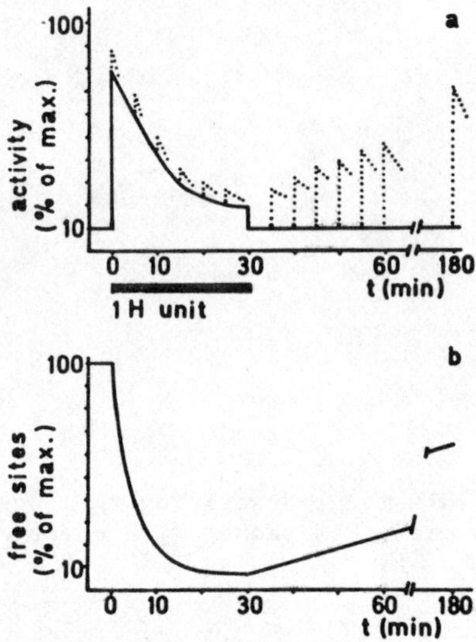

Fig. 2. Simulation of the desensitization phenomenon. Kinetics of adenylate cyclase activation (activity) and receptor sites availability (free sites) during and after a pulse of 30 min. of agonist addition. Interrupted lines : effect of further additions of agonist (Ref. 3).

hormone binding sites. We proposed that this decrease is the result of a persistent hormone binding to the receptor, the conformation of which cannot activate adenylate cyclase. This proposition could be tested experimentally. Theoretical predictions on the properties of the two models based upon the locking receptor concept led to propose a new experimental design which could give more insight into the mechanism of desensitization.

CONTROL OF PROTEIN KINASE ACTIVITY BY cAMP

The action of cAMP in hormone-sensitive cells is expressed through the interaction with an ubiquitous receptor protein R and through activation of a protein phosphorylating enzyme C. The combination of these two types of subunits defines the concept of cAMP dependent protein kinase. A close correlation in several systems between cAMP concentration and protein kinase activity in vivo supports the physiological significance of this mechanism. The analysis of such a mechanism should confirm the physiological role of protein kinase. In view of this, the two following criteria should be fulfilled by the cAMP-protein kinase system :

1) the affinity of cAMP for the protein kinase must be such that a variation of cAMP concentration around an intracellular cAMP level should significantly modulate the protein kinase activity.

2) the control of the activation of protein kinase by cAMP must be efficient, i.e. protein kinase activity must be sensitive to a slight relative variation of cAMP concentration between the basal and stimulated states of cAMP. This criteria is supported by some observations (hill coefficient higher than I, only small changes in total cAMP are required in vivo to activate protein kinase).

Though the first criterion has been now demonstrated in a number of systems, it is interesting to point out how it has led for a while to some misunderstanding and how theoretical considerations have successfully solved the discrepancies about the apparent high affinity of cAMP for protein kinases (Ref. 4). The starting point of the misunderstanding was that in vitro estimations of half activation concentration of cAMP referred to as the K_a constant, were gererally of the order of 10^{-8}M, while the cAMP concentration in unstimulated cells is generally about 10^{-6}M. Thus, at the concentration existing in the resting cell, the whole protein kinase activity should have been expressed. This discrepancy led some authors to postulate the existence of intracellular sequestration sites capable of binding most of cAMP and thus only a very low part of cAMP would have been available to protein kinase and would have induced only a small fraction of the total protein kinase activity. Much work, including in our laboratory, was carried out to identify this hypothetical sequestrating site. The interpretation of such observations resulted from the well-known theory of enzyme kinetics in which the so-defined Michaëlis-Menten constant K_m charaterizes the affinity of a substrate for an enzyme. Such a theory cannot be simply applied to the case of protein kinase because of the relatively high concentration of protein kinase compared to the cAMP concentration required to induce half maximal activation. If it is supposed that protein kinase activation parallels the occupancy level of the regulatory subunit by cAMP, it is clear that half-maximal activation level is obtained when half the receptor sites are occupied by cAMP. Thus, the total concentration of cAMP needed to obtain the half maximal activity, i.e. K_a, is higher than half the total concentration of receptor sites. This conclusion clearly does not depend on the mechanism of protein kinase activation. The parameter K_a is thus not a constant but depends on the protein kinase concentration. It was then realized that intracellular concentration of cAMP binding site was of the order of magnitude of basal cAMP concentration. On the other hand in in vitro conditions, the relatively low K_a value of approximatively 10^{-8}M has no physiological significance because of the dilution of the protein kinase in the assay. This theoretical prediction has been confirmed experimentally (Ref. 5). As a result of this, compartmentation of cAMP which has been suggested because of the discrepancy between the relatively low in vitro value of K_a and the high level of cAMP is not necessary to explain the effective physiological activation of protein kinase by cAMP as described here above by the first criterion.

It was our purpose to show if the models based upon the experimental observations on the protein kinase system are compatible with a physiological control by cAMP, i.e. the second criterion above-mentioned should be fulfilled.

The distinguishing feature of the mode of action of cAMP on protein kinase is the induced dissociation of the inactive holoenzyme : the free catalytic subunit C is then fully active. The first tentative description of this system was :

$$\text{cAMP} + \text{R-C} \rightleftharpoons \text{cAMP-R} + \text{C} \qquad (1)$$

Because of this particular mode of activation, the current methods for describing a biochemical system (Hill plot, Scatchard plot, Lineweaver-Burk plot) lead to erroneous conclusions in what concerns the characteristics of the system (type of apparent cooperativity). A new method has been elaborated which is based on the concept of the control efficiency (Ref. 6). By definition, a control is more efficient as the sensitivity of the effect to relative changes in the effector concentration is high and as this high sensitivity occurs in a wide range of values of the effect. The characteristic parameter which is called control efficiency E is higher than I if the control is efficient. The intuitive interpretation of the E value is just the same as that of value of the Hill coefficient ($n_H > I$ refers to an efficient control characterized by an apparent positive cooperativity).

In conclusion, the analysis of the protein kinase system is based on the determination of the control efficiency E of the protein kinase activity by cAMP. In order to fulfil the second criterion, the system should account for a control efficiency higher than I.

The computed efficiency E of the control of protein kinase by cAMP described by eq. 1 is equal to 0.68. Such a value refers to a poorly efficient control (apparent negative cooperativity) which is not compatible with the second criterion abovementioned. This low value is due to the dissociation of C leading to the concept of "retrooperativity"(Ref. 6). We have now to show if the different experimental observations on the protein kinase system can enhance the value of the control efficiency E.

The action of the heat stable inhibitor I occurs through the reversible binding to free C. The inhibitor would be capable of inhibiting in vivo about 20% of the total kinase activity, and exhibits a high affinity for free C. The analysis of eq. 1 combined with the action of the inhibitor leads to the conclusion that the control efficiency is slightly enhanced (E=0.78 instead of 0.68). However, in spite of this increase, E remains less than I.

The phosphorylation of specific substrates S by free C obeys probably the classic enzymatic mechanism i.e. a complex CS is formed for lowering the energy requirements for activation of the reaction. If the binding of the complex CS to R cannot occur, the equilibrium of eq 1 is displaced to the right (dissociation of C). Although such a substrate effect does not affect the control efficiency by cAMP, it enhances the affinity of cAMP for protein kinase. It is thus interesting to verify if this substrate effect actually takes place in the protein kinase system. The experimental design could consist in the comparison of cAMP binding with and without substrate for the protein kinase. If such an effect is operative, the cAMP affinity for protein kinase increases with substrate concentration because of the formation of the CS complex. Thus the activation of protein kinase cannot be under the cAMP control if the concentration of substrate is saturating, i.e. all the C subunits are bound to substrate molecules. We conclude therefore that the substrate concentration in vivo should be lower than the K_m of the reaction. In other words, the determination of the physiological value of this K_m gives an upper limit for the intracellular concentration of the considered substrate.

In several tissues the stoichiometry of the activation reaction of protein kinase by cAMP obeys the following schema :

$$2\ \text{cAMP} + R_2\text{-}C_2 \rightleftharpoons \text{cAMP}_2\text{-}R_2 + 2\ C \qquad (2)$$

The analysis of such a model comes to the conclusion that the maximal control efficiency is obtained when positive cooperativity occurs for the binding of cAMP to R as well as for the binding of C to R (E=1.05). This value, although it is higher than I is still too low : the positive cooperativity of cAMP binding efficiently counteracts the retrooperativity but cannot produce a very efficient control.

Protein kinase from bovine heart muscle can catalyse the phosphorylation of R. Such an autophosphorylation process is reversible and the binding of C to R is more rapid when R is dephosphorylated than when it is phosphorylated. The simulation shows that the control efficiency of this system may be as high as about 1.5 which is more compatible with the second criterion.

We can thus conclude that theoretical simulation may lead to a lot of informations only by taking into account the numerous experimental observations. By using the new method characterizing the efficiency of a biochemical control, the analysis leads to the conclusion that only the autophosphorylation of protein kinase combined with the actual stoichiometry of the reaction (eq.2) could account for a high control efficiency. However, as the exact description of the protein kinase system is still unknown, it is possible that other concepts not yet discovered could shed light on this unanswered question = which component of the protein kinase system could render the control by cAMP so highly efficient that only a slight increase of the intracellular cAMP concentration is sufficient to trigger the action of protein kinase ?

CONCLUSION

It is hoped that these few examples of simulation of the cAMP system have shown the interest of this approach in the integration of hypotheses into models, in defining the validity of experimental arguments with regard to models and in providing concepts and experimental guidelines for experimental work.

ACKNOWLEDGEMENTS

This work has been mostly carried out within the framework of the Association Euratom University of Brussels-University of Pisa, thanks to a grant of the Ministère de la Politique Scientifique (Actions Concertées).

REFERENCES

(1) P. Cuatrecasas, Ann. Rev. Biochem. 43, 169 (1974)

(2) R.J. Lefkowitz, L.E. Limbird, C. Mukherjee, M.G. Caron, Biochem. Biophys. Acta 457, 1 (1976).

(3) S. Swillens, J.E. Dumont, J. Cycl. Nucl. Res. 3, 1 (1977).

(4) S. Swillens, E. Van Cauter, J.E. Dumont, Biochim. Biophys. Acta, 364, 250 (1974).

(5) J.A. Beavo, P.J. Bechtel, E.G. Krebs, Proc. Nat. Acad. Sci. 71, 3580 (1974).

(6) S. Swillens, J.E. Dumont, J. Mol. Med., 1, 273 (1976).

Invited Lecture

Cyclic Nucleotides, Phosphorylated Proteins and Drug Actions in the Central Nervous System

Paul Greengard

Department of Pharmacology, Yale University School of Medicine, New Haven, Connecticut 06510, U.S.A.

A. CYCLIC AMP, CYCLIC GMP AND NEURONAL FUNCTION

In the past few years, a beginning has been made in understanding the molecular basis by which the activity of nerve cells is regulated. As in the case with hormones acting on non-neuronal tissues (Ref. 1), there is now evidence that cyclic AMP may act as an intracellular mediator for the actions of certain neurotransmitters on nerve cells (for recent reviews, see Refs. 2-8). There is also evidence that certain of the effects of some other neurotransmitters on their target cells may be mediated by cyclic GMP.

Neurotransmitter molecules, when released from the terminals of a presynaptic neuron, induce a change in the permeability properties of the membrane of the postsynaptic neuron by interacting with specific receptors located on the cell surface. The molecular nature of the events by which the binding of the neurotransmitter to its receptor is coupled with the alteration in the ionic properties of the postsynaptic cell membrane is not yet well-understood. However, there appear to be two general classes of mechanism by which this process may occur (Ref. 9).

The "receptor-ionophore" model: In one class the neurotransmitter receptor is thought to be coupled directly to an ionophore in such a manner that binding of the neurotransmitter to the receptor induces a conformational change in the ionophore, leading to a change in the permeability of the membrane to specific ions. One receptor that appears to work in this manner, and that has been intensively studied by Changeux, Heilbronn, Karlin, Raftery, Reich, and their colleagues, is the nicotinic cholinergic receptor.

The "receptor-second messenger" model: A second general class of mechanisms by which a neurotransmitter could affect the electrical properties of postsynaptic neurons is through a process in which the binding of the neurotransmitter to its receptor on the outside of the cell membrane leads to the production of another molecule, a second messenger, such as cyclic AMP, on the inside of the cell. This second messenger could then initiate a sequence of biochemical reactions that would ultimately result in a change in the electrophysiological properties of the neuronal membrane. Such electrophysiological changes might be caused by a biochemical modification of a membrane protein involved in regulating the permeability of the membrane to specific ions.

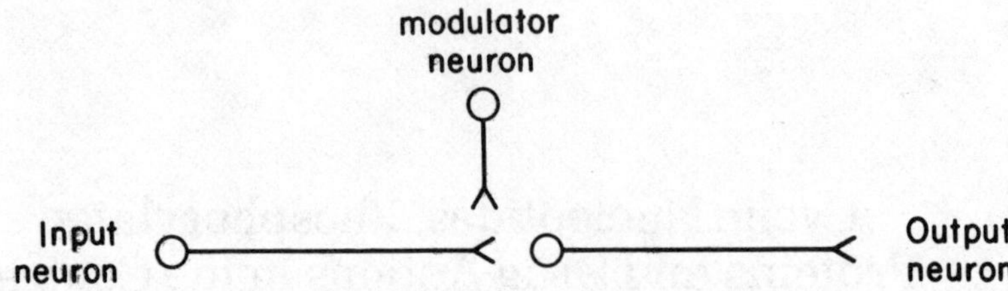

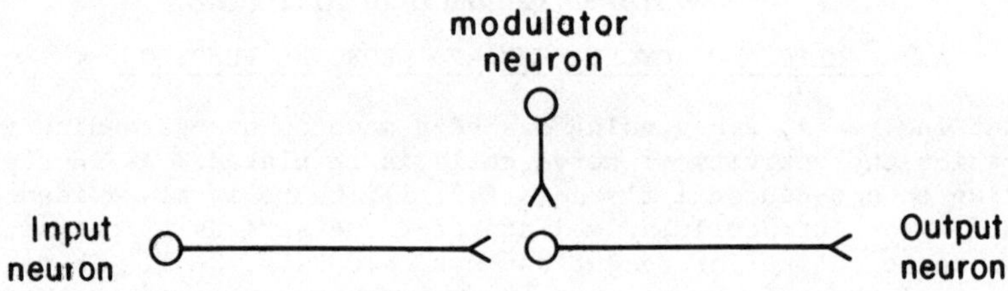

Fig. 1. Two types of modulation of synaptic transmission in which cyclic AMP has been implicated as a second messenger. (A) Presynaptic modulation. In this type of modulation, the postsynaptic cell of the receptor-second messenger synapse is in turn a presynaptic terminal, and is labeled "input neuron." (B) Postsynaptic modulation. In this type of modulation, the postsynaptic cell of the receptor-second messenger synapse is also the postsynaptic cell for another synapse, and is labeled "output neuron." (Taken from Ref. 8.)

One might anticipate that the physiological responses caused by activation of the two different classes of receptors would have different characteristics. In the first case, where the binding of the neurotransmitter to the receptor induces a conformational change in an ionophore, the permeability change resulting from this one-step, non-enzymatic process might be expected to have a very short latency and duration, on the order of several milliseconds. In contrast, it seems likely that the effects of cyclic nucleotides on the permeability properties of neuronal membranes are mediated by means of a sequence of biochemical reactions; therefore, the permeability changes generated in response to a cyclic nucleotide might be expected to have a much longer latency and duration.

There is now evidence for two types of synaptic transmission involving the receptor-second messenger system (Fig. 1). Thus, the postsynaptic cell of the receptor-second messenger modulatory synapse can be either the presynaptic or the postsynaptic cell of the synapse being modulated. In other words, the receptor-second messenger system can be utilized for either presynaptic (Fig. 1A) or postsynaptic

TABLE 1 Neurotransmitters Associated with Cyclic AMP System

Neurotransmitter	Antagonist
Dopamine	Chlorpromazine
Serotonin	LSD
Norepinephrine (β)	Propranolol
Histamine (H_2)	Metiamide
Octopamine	Phentolamine

(Taken from Ref. 8.)

(Fig. 1B) modulation of synaptic transmission. In addition to mediating the effects of modulator neurons on the permeability properties of input and output neurons, cyclic nucleotides appear to have many other roles in neuronal function (Refs. 4,7).

A variety of experimental criteria useful in evaluating whether a cyclic nucleotide mediates a postsynaptic permeability change have been presented elsewhere (Ref. 5). Failure to satisfy one or more of these criteria does not exclude the possibility that a cyclic nucleotide mediates a particular postsynaptic response. Some of the difficulties associated with testing certain of the criteria are discussed in detail elsewhere (Ref. 5).

The demonstration of the existence of neurotransmitter-sensitive adenylate cyclases in nervous tissue, their regional and subcellular distribution, and the properties of these enzymes suggest a role for cyclic AMP in the physiology of synaptic transmission. For each of the neurotransmitters listed in Table 1, an adenylate cyclase activated by low concentrations of the neurotransmitter in question has been found in nervous tissue. Moreover, compounds which act as physiological agonists (i.e. mimic the physiological actions) of each of these neurotransmitters stimulate the appropriate adenylate cyclase. Conversely, those substances which act as antagonists of the physiological effects of these neurotransmitters in general block the activation by the neurotransmitter of the particular adenylate cyclase. Representative antagonists for these neurotransmitters are also included in Table 1. For the group of neurotransmitters listed in Table 1, the close similarity between the properties of the neurotransmitter receptor, as characterized in physiological experiments, and the properties of the neurotransmitter-sensitive adenylate cyclase, using a variety of physiological and pharmacological agonists and antagonists, suggest an intimate association between neurotransmitter receptors and adenylate cyclase molecules.

Table 2 lists those neurotransmitters that cause increases in cyclic GMP in certain target cells. Each of these neurotransmitters is capable of causing increases in cyclic GMP in postsynaptic cells under conditions in which they produce a physiological response. In addition, substances that mimic the physiological effects of these neurotransmitters also mimic the neurotransmitter in causing increases in cyclic GMP in postsynaptic cells. Conversely, substances that act as antagonists of

TABLE 2 Neurotransmitters Associated with Cyclic GMP System

Neurotransmitter	Antagonist
Acetylcholine (muscarinic)	Atropine
Histamine (H_1)	Diphenhydramine
Norepinephrine (α)	Phentolamine
Glutamate	?

(Taken from Ref. 8.)

the physiological effects of these neurotransmitters block the increase in cyclic GMP brought about by the appropriate neurotransmitter in the postsynaptic cells. Representative antagonists for these neurotransmitters are included in Table 2. The available data suggest that cyclic GMP mediates some effects, and modulates other effects, of neurotransmitters on postsynaptic cells. The possible relationship of the neurotransmitter-induced increase in cyclic GMP to the physiological response of the postsynaptic cells is discussed in detail elsewhere (Ref. 8).

Studies of cyclic nucleotides in several neuronal preparations, including the superior cervical ganglion (Ref. 8) and the mammalian cerebellum (Ref. 2), leave little doubt that synaptic transmission under physiological conditions can result in an increase in the cyclic nucleotide content of certain postsynaptic neurons. The precise role(s) of the cyclic nucleotides formed in the postsynaptic cells, during physiological activity, is less well established. We have proposed that cyclic nucleotides formed in neurons in response to neurotransmitter action may have multiple roles including, but by no means limited to, the generation of certain slow postsynaptic potentials (Refs. 4,8,9). Two other possible roles for cyclic AMP in nervous tissue include regulation of microtubule function and of neurotransmitter biosynthesis. The increase in cyclic AMP during increased impulse flow may also serve to mobilize carbohydrate (Ref. 10) and lipid reserves, and thus help to meet the additional energy requirements associated with elevated functional activity in nervous tissue. In certain invertebrate nerve networks, an increase in cyclic AMP mediated by serotonin has been shown to be responsible for the increased neurotransmitter release that accompanies behavioral sensitization (Ref. 11). It is also of interest that, in various model systems, cyclic AMP may act at a transcriptional level to regulate the *de novo* synthesis of a number of proteins with specific functional roles (see Ref. 8 for review). It is possible that, in the nervous system, elevation of the cyclic AMP content of postsynaptic cells in response to increased presynaptic impulse flow may also result in altered synthesis of certain proteins having specific functional roles in synaptic physiology. This might constitute one component of the mechanism by which long-term information storage (*i.e.*, learning) and retrieval (*i.e.*, memory) can occur.

It is probably important to view the synaptic and nonsynaptic actions of cyclic AMP not so much in isolation but rather as part of an integrating system. It may be that enough synaptic stimulation not only raises cyclic AMP levels sufficiently to alter the permeability properties of neurons, but also initiates a logical sequence of events mediated by cyclic nucleotides. This sequence might include an

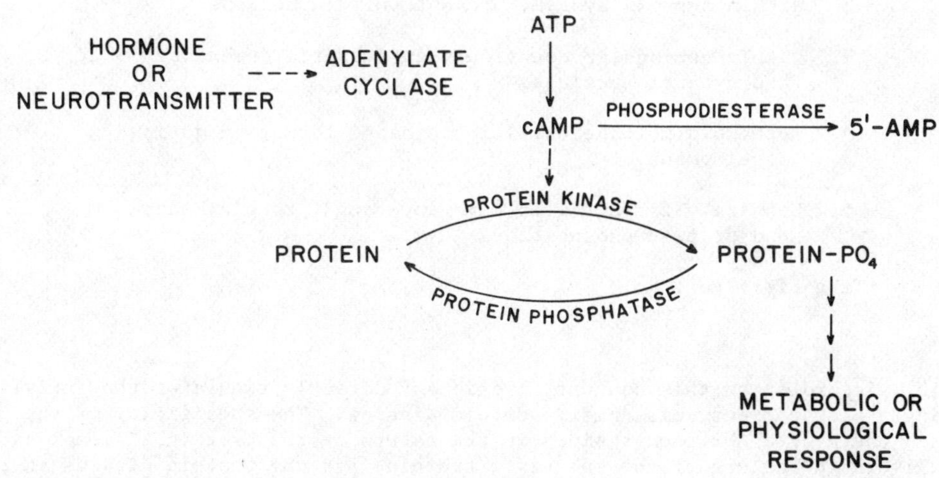

Fig. 2. Schematic diagram of apparent role played by protein phosphorylation in mediating the biological effects of those hormones and neurotransmitters that act through cyclic AMP. The appropriate hormone or neurotransmitter stimulates a specific adenylate cyclase present in the target tissue, causing an increased conversion of ATP to cyclic AMP; the newly formed cyclic AMP is either hydrolyzed by a phosphodiesterase to 5'-AMP or else it activates a protein kinase, leading to the phosphorylation of a substrate protein; the substrate protein, through one or more steps, controls the rate of some metabolic or physiologic response; the phosphorylation of this substrate protein leads to an increase or decrease in the rate of this response. (Taken from Ref. 7.)

increase or decrease in the synthesis or release of neurotransmitter in response to synaptic stimulation, an initiation of intracellular movements in order to transport newly synthesized products, the activation of carbohydrate metabolism to supply the necessary cellular energy requirements, and direct effects on the genetic material in the cell nucleus that may lead to long-term alterations of behavior.

B. Cyclic AMP-Dependent Protein Phosphorylation

I would like now to discuss a mechanism by which cyclic AMP may mediate the effects of neurotransmitters on postsynaptic cells. The protein kinase hypothesis, illustrated schematically in Fig. 2, provides a theoretical mechanism by which one simple molecule, cyclic AMP, could achieve a diversity of biological responses

TABLE 3 Some Biological Properties of Protein I

1. Present only in nervous system
2. Within nervous system, present only in neurons
3. Within neurons, present only in synaptic membranes and synaptic vesicles
4. Appears simultaneously with synapse formation during development
5. Substrate for specific membrane-bound protein kinase and protein phosphatase

(Taken from Ref. 8.)

(Ref. 12). According to this concept, cyclic AMP directly regulates the activity of a single class of enzymes, namely protein kinases. The specificity of the response to the cyclic AMP then resides in the nature of the protein kinases, and especially in the nature of the substrate proteins for the protein kinases in the various tissues.

An increasing amount of evidence supports the protein kinase hypothesis. Thus, numerous examples have now been found (for reviews, see Refs. 4,7,8,13-16) of systems in which the effects of cyclic AMP on various metabolic and physiological responses appear to be mediated through protein phosphorylation. At the present time, it seems likely that in eukaryotic organisms most, and conceivably all, of the metabolic and physiological effects of cyclic AMP will prove to be mediated through regulation of protein phosphorylation.

The literature on the subject of protein phosphorylation, even when restricted to the nervous system, is now vast. In this presentation, I shall review one study of this subject from our own laboratory (Refs. 17-21). In view of the apparent importance of cyclic AMP-dependent protein phosphorylation in the nervous system, my colleagues and I have searched for substrate proteins for cyclic AMP-dependent protein kinase in nervous tissue. The ability of endogenous protein kinase in a synaptic membrane fraction from the rat caudate nucleus to phosphorylate endogenous substrate proteins in this fraction is demonstrated by the experiment shown in Fig. 3. As shown by the protein staining pattern in this figure, a large number of distinct bands were found upon gel electrophoresis of the solubilized proteins from this synaptic membrane fraction. Of these many bands, the phosphorylation of three was markedly stimulated by cyclic AMP. We refer to these three bands as Proteins Ia, Ib and II. Recently, it has been possible to resolve Protein II into two bands (Ref. 20), one of which has been identified as the regulatory subunit of a protein kinase present in synaptic membranes (Ref. 19); this regulatory subunit is phosphorylated by the catalytic subunit in an autophosphorylation reaction (Ref. 19). Proteins Ia plus Ib, collectively designated Protein I, are present only in nervous tissue (Ref. 18). In view of the possibility that Protein I might be involved in the physiology of synaptic transmission, we have investigated its biological and biochemical properties. The results of those studies indicate that Protein I has a number of unique properties.

Some of the biological properties of Protein I are summarized in Table 3. Not only does Protein I appear to be confined to the nervous system (Refs. 17,18), but within the nervous system Protein I appears to be present only in neurons

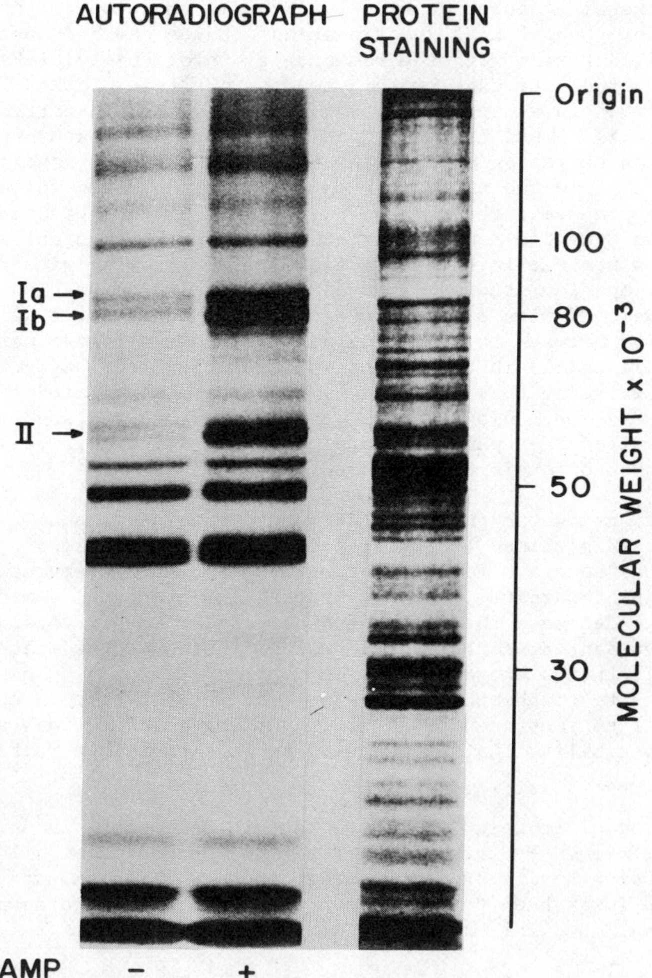

Fig. 3. Effect of cyclic AMP on endogenous protein phosphorylation in a synaptic membrane fraction from rat caudate nucleus. The synaptic membrane fraction was incubated with 7 μM [γ-^{32}P]ATP for 15 seconds at 30°C, in the absence (-) or presence (+) of 10 μM cyclic AMP. The reaction was terminated by the addition of the detergent sodium dodecyl sulfate, which also solubilized the membrane proteins. An aliquot of the reaction product was then subjected to sodium dodecyl sulfate-polyacrylamide gel electrophoresis in order to separate the membrane proteins from one another. The separated proteins were located in the gel by standard procedures of protein staining. Autoradiography was then carried out to determine those protein bands into which radioactive phosphate had become incorporated. Left, phosphorylation of endogenous protein in the absence or presence of cyclic AMP; right, protein staining. (Modified from Ref. 21.)

(Ref. 21). This latter conclusion is based on observations that injection, into brain, of kainic acid, a substance which selectively destroys neurons (Refs. 22-24), causes an extensive depletion of several brain phosphoproteins, including Proteins Ia and Ib. Subcellular fractionation studies carried out by Dr. Tetsufumi Ueda, in collaboration with the research group of Philip Siekevitz at Rockefeller University, indicate that, within neurons, Proteins Ia and Ib are present in high concentration in synaptic membrane fractions, and in particularly high concentration in fractions enriched in synaptic vesicles and postsynaptic densities (Ref. 25). The concentration of Proteins Ia and Ib in postsynaptic densities is much higher than that in the synaptic membrane fractions from which the postsynaptic densities are prepared; therefore, it is possible that the content of Proteins Ia and Ib in synaptic membrane fractions may be attributable to the postsynaptic densities present in these fractions. In agreement with the results of subcellular fractionation studies, immunocytochemical studies also indicate that Protein I is associated with synaptic vesicles and postsynaptic densities (Ref. 26). A correlation between the time course of synapse formation and the appearance of Proteins Ia and Ib in the brains of developing rats and guinea pigs has provided further evidence that Proteins Ia and Ib are associated with synaptic structures (Ref. 27). Protein I serves as a substrate for a specific Type II membrane-bound cyclic AMP-dependent protein kinase (Refs. 17,20,28) and a specific membrane-bound phosphoprotein phosphatase (Ref. 19).

Recently, alterations in the phosphorylation of Proteins Ia and Ib have been observed in intact brain slices (Ref. 29) and in whole animals (Ref. 30) in response to various chemical agents. For instance, several central nervous system depressants, including pentobarbitol, chloral hydrate and urethane, administered to mice intraperitoneally, decrease the state of phosphorylation of Protein I and, conversely, the convulsant drug pentylenetetrazole increases the state of phosphorylation of Protein I, in whole brains of mice *in vivo* (Ref. 30). These results indicate that the state of phosphorylation of Protein I, a protein confined to the synaptic region of neurons, can be altered by changes in neuronal activity, supporting the possibility that it may play an important role in the physiology of the synapse.

Biochemical studies of Protein I indicate that this molecule is unusual from a chemical point of view. Protein I is present in the brains of all mammalian species examined. Using bovine brain, in order to have large amounts of starting material, Protein I has been purified several thousand-fold to apparent homogeneity. Some of the unusual chemical properties of purified Protein I (Ref. 18) are summarized in Table 4.

C. Regulation of Protein Phosphorylation by Calcium, Cyclic GMP and Steroid Hormones

Finally, I would like to discuss protein phosphorylation in a broader context than that of its association with the cyclic AMP system. Some of the studies in our laboratory in the past few years were designed to test the possibility that phosphorylated proteins may play a much wider role in biological regulation than one limited only to mediating the effects of cyclic AMP. The results obtained thus far suggest that phosphorylated proteins probably mediate not only the effects of cyclic AMP and those regulatory substances that act through cyclic AMP, but also certain of the effects of a variety of other regulatory agents as well. Thus, it seems that protein phosphorylation may be a final common pathway for the actions of many types of biological regulatory agents. This concept is illustrated schematically in Fig. 4. Experimental data obtained in our own and other laboratories suggest that all of the regulatory agents listed in Fig. 4 achieve at least certain of their effects through protein phosphorylation. This subject was reviewed in some detail

TABLE 4 Some Chemical Properties of Protein I Purified from Bovine Brain

	Protein IA	Protein IB
1. Molar proportion	1	2
2. Molecular weight	86,000	80,000
3. Isoelectric point	10.3	10.2
4. Stokes radius	59 Å	59 Å
5. Shape	highly elongated	highly elongated
6. Amino acid composition	rich in glycine and proline	rich in glycine and proline
7. Other structural features	a globular region (phosphorylated) and an elongated collagenous tail (not phosphorylated)	

(Taken from Ref. 8.)

recently (Refs. 7,8), and therefore I shall limit my discussion to a brief summary of three of the specific regulatory substances which have been studied in our laboratory, namely Ca^{2+}, cyclic GMP and the steroid hormones.

1. Regulation by Ca^{2+} of Protein Phosphorylation

When an impulse travels down a nerve axon to the terminal, the wave of depolarization causes an influx of calcium that acts as the immediate trigger for the release of neurotransmitter from the terminal (Refs. 31-34). In addition, there is evidence that calcium influx controls the rate of neurotransmitter biosynthesis in the presynaptic terminal (Refs. 35,36). It seemed possible that one or more of the effects of calcium in the presynaptic nerve terminal might be mediated or modulated through the phosphorylation of specific proteins. For this reason, Bruce Krueger and Javier Forn studied the effect of calcium influx on protein phosphorylation in intact synaptosomes. They found (Ref. 37) that agents known to increase Ca^{2+} transport across the plasma membrane of nerve terminals altered the phosphorylation of specific endogenous proteins in intact synaptosomes from rat brain. Synaptosome preparations, preincubated *in vitro* with $^{32}P_i$, incorporated ^{32}P into a variety of specific proteins. Veratridine and high (60 mM) K^+, which increase Ca^{2+} transport across membranes, through a mechanism involving membrane depolarization, as well as the calcium ionophore A23187, each markedly altered the incorporation of ^{32}P into several specific proteins as determined by sodium dodecyl sulfate-polyacrylamide gel electrophoresis and autoradiography. All three agents failed to alter protein phosphorylation in calcium-free medium containing ethylene glycol bis(β-aminoethyl ether)N,N'-tetraacetic acid (EGTA). The effect of veratridine and of high K^+ on the phosphorylation of proteins in intact synaptosomes is illustrated in Fig. 5. Cyclic nucleotides and several putative neurotransmitters were without effect on protein phosphorylation in these intact synaptosome preparations.

The data suggested that conditions which cause an accumulation of calcium by synaptosomes lead to a calcium-dependent increase in phosphorylation of specific endoge-

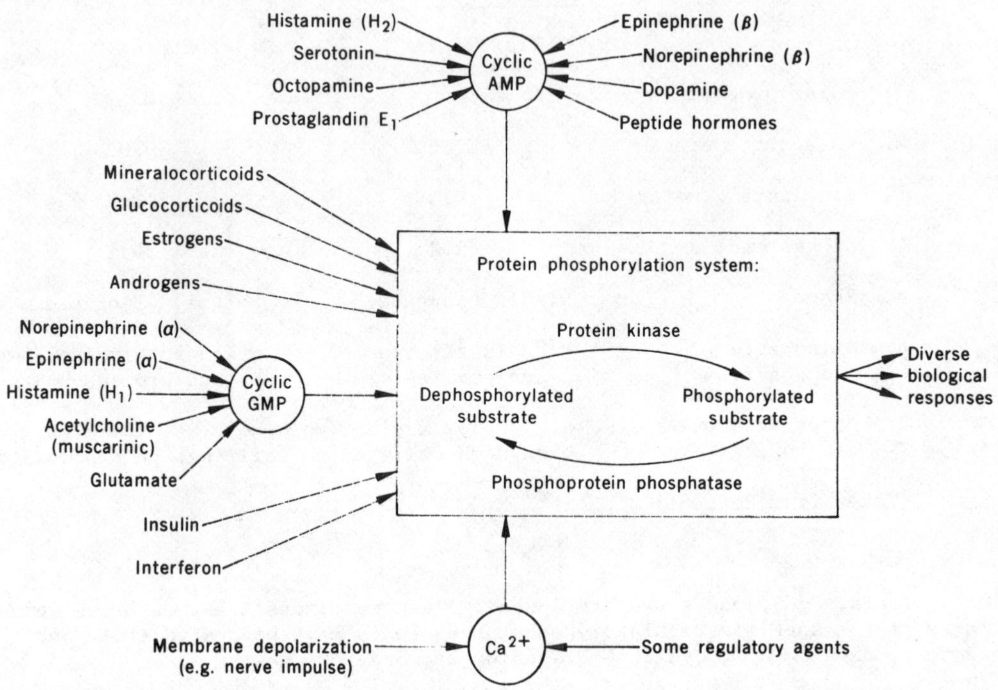

Fig. 4. Schematic diagram of postulated role played by protein phosphorylation in mediating some of the biological effects of a variety of regulatory agents. The diagram gives examples of regulatory agents some of whose effects may be mediated through regulation of the phosphorylation of specific proteins, and is not intended to be complete. In addition to cyclic AMP and a variety of neurotransmitters and hormones whose effects are mediated through cyclic AMP, these regulatory agents include cyclic GMP, a variety of neurotransmitters and hormones whose effects are mediated through cyclic GMP, Ca^{2+}, and agents whose effects are mediated through Ca^{2+}, as well as several classes of steroid hormones, insulin, and interferon. For brevity, the numerous peptide hormones whose effects are known to be mediated through cyclic AMP, and the various regulatory agents believed to act through translocation of Ca^{2+}, are not listed individually. It seems likely that some, but not necessarily all, of the biological responses elicited by any given one of the regulatory agents shown are mediated through the protein phosphorylation system; for simplicity, pathways from regulatory agent to biological response that do not involve protein phosphorylation are not shown. (Taken from Ref. 7.)

nous proteins. The results support the possibility that these phosphoproteins may be involved in the regulation of certain calcium-dependent nerve terminal functions.

Evidence concerning the mechanism by which calcium entry might stimulate the phosphorylation of synaptosomal proteins has come recently from experiments of Howard Schulman using lysed synaptosomes (Ref. 38). In these experiments, synaptosomes

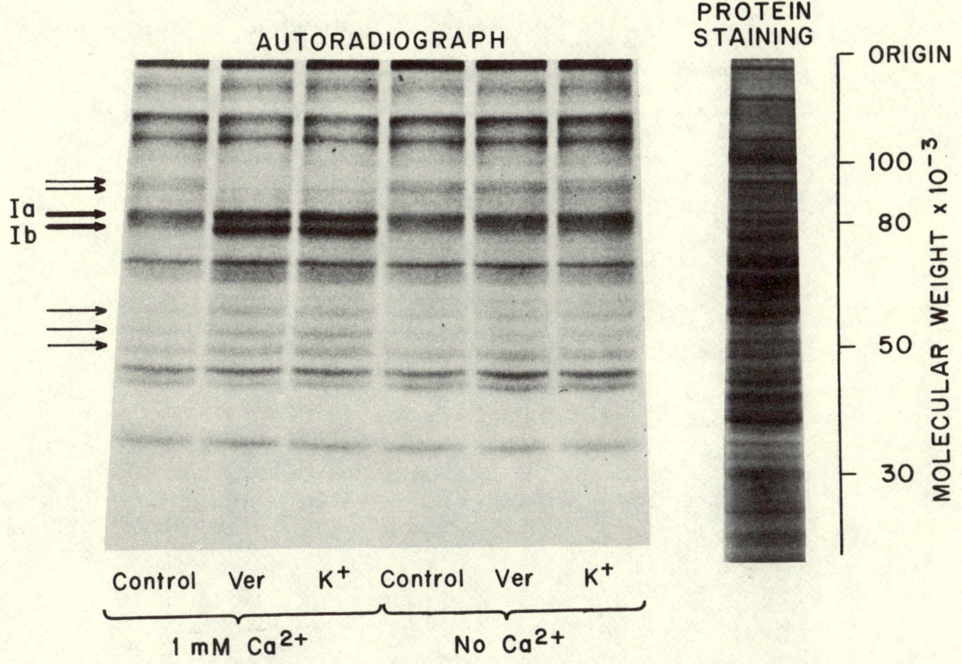

Fig. 5. Effect of veratridine and of high K^+, in the absence and presence of Ca^{2+}, on the phosphorylation of endogenous proteins in crude synaptosomal preparation from rat cerebral cortex. The synaptosome fraction was preincubated with $^{32}P_i$ for 30 minutes in the absence of Ca^{2+}. Aliquots of this suspension were then incubated for 30 seconds in the absence (Control) or presence of 100 μM veratridine (Ver) or 60 mM K^+; 1 mM Ca^{2+} was present where indicated. The incubation was terminated by the addition of sodium dodecyl sulfate and the samples were subjected to sodium dodecyl sulfate-polyacrylamide gel electrophoresis. The molecular weight scale was generated by determining the position of marker proteins of known molecular weight. The positions of Proteins Ia and Ib, whose phosphorylation was stimulated by veratridine or high K^+, are indicated by bold arrows. Light arrows indicate the positions of other protein bands whose phosphorylation was inhibited (upper two arrows) or stimulated (lower three arrows) by veratridine or high K^+. (Taken from Ref. 37.)

obtained from rat cerebral cortex were lysed by hypo-osmotic shock and assayed for incorporation of [^{32}P]-phosphate from [γ-^{32}P]ATP into protein. As shown in Fig. 6 (Lanes 1 and 2), calcium markedly stimulated [^{32}P]-phosphate incorporation into many protein bands in the lysed synaptosome preparation, which contained membranes (synaptic membranes plus synaptic vesicles) and synaptosomal cytoplasm. The calcium-dependent phosphorylation observed in lysed synaptosomes was lost upon preparation of cytoplasm-free membranes (Fig. 6, Lanes 3 and 4). This phenomenon was

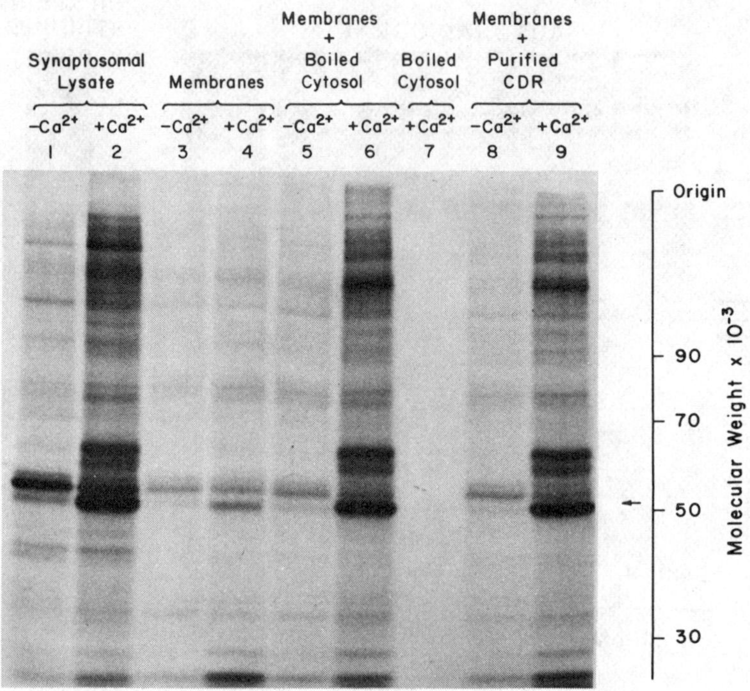

Fig. 6. Effect of Ca^{2+}, synaptosomal cytoplasm, and purified calcium-dependent regulator (CDR) on endogenous protein phosphorylation of brain membranes. Synaptosomes obtained from rat cerebral cortex were subjected to osmotic shock by homogenization in 10 volumes of ice-cold water (15 ml) and kept on ice for 30 minutes to optimize lysis. Total membranes were obtained from lysed synaptosomes by centrifugation at 150,000 x g for 30 minutes. The released synaptosomal cytosol was removed, heated at 95-100°C for 5 minutes, and used as the source of activator (boiled cytosol). The pellet was resuspended in 15 ml of 5 mM Tris-HCl (pH 7.0), centrifuged as above, and again resuspended in 15 ml of 5 mM Tris-HCl (pH 7.0). This resuspended pellet served as the source of activator-deficient membranes. The reaction mixture for assay of endogenous protein phosphorylation (final volume, 100 μl) contained: 50 mM PIPES buffer (pH 7.0), 10 mM $MgCl_2$, 1 mM dithioerythritol, 0.2 mM EGTA (minus calcium) or 0.2 mM EGTA + 0.5 mM $CaCl_2$ (plus calcium), 5 μM [γ-^{32}P]ATP (5.2 x 10^4 cpm/pmol), and, where indicated, lysed synaptosomes (50 μg protein), membranes (34 μg protein), boiled cytosol (14 μg protein) or purified CDR (0.25 μg protein). Incubation was carried out for 10 seconds at 30°C, the reaction terminated by addition of sodium dodecyl sulfate, and an aliquot of the sample analyzed for protein phosphorylation by sodium dodecyl sulfate-polyacrylamide gel electrophoresis and autoradiography. (Taken from Ref. 38.)

not due to inactivation of the calcium-dependent protein kinase activity during experimental manipulations, as it could be regained by addition of an amount of boiled (Fig. 6, Lanes 5 and 6) or unboiled synaptosomal cytoplasm commensurate with the amount of membrane protein present. Boiled cytosol, containing the heat-stable factor which restored calcium-dependent phosphorylation to washed membranes, did not exhibit any endogenous phosphorylation (Fig. 6, Lane 7), nor did it manifest any protein kinase activity using exogenous substrates.

It had been demonstrated earlier that Ca^{2+} plus a calcium-binding protein (calcium-dependent regulator or CDR) activate cyclic nucleotide phosphodiesterase (Refs. 39,40) and adenylate cyclase (Ref. 41) from mammalian brain. The heat stability of the factor in the synaptosomal cytosol that stimulated Ca^{2+}-dependent protein phosphorylation suggested a possible relationship to CDR. Therefore, CDR was purified to homogeneity from bovine cerebral cortex by the method of Teo _et al_. (Ref. 42), in order to test its effect in the calcium-dependent protein phosphorylation system. Addition of this protein to washed synaptosomal membrane fractions restored calcium-dependent protein phosphorylation (Fig. 6, Lanes 8 and 9). The pattern of phosphorylation obtained upon addition of heat-treated synaptosomal cytoplasm was indistinguishable from that obtained upon addition of CDR, suggesting that the two factors are functionally equivalent.

Stimulation of protein phosphorylation was dependent on the presence of both calcium and either the cytosol factor or CDR. Thus, calcium in the absence of boiled cytosol and boiled cytosol in the absence of calcium were ineffective, whereas addition of both calcium and boiled cytosol restored phosphorylation (Fig. 6, compare Lanes 3, 4, 5 and 6). Analogous results were obtained with CDR (compare Lanes 3, 4, 8 and 9). The endogenous activator, like CDR, was extremely heat stable, non-dialyzable, resistant to DNase and RNase, and sensitive to trypsin (Ref. 38).

We have recently purified a protein from bovine cerebral cortex based on its ability to stimulate Ca^{2+}-dependent protein phosphorylation in activator-depleted membranes (Ref. 43). Only a single peak of activity was present throughout the purification procedure. This activity co-purified with the phoshodiesterase activator, suggesting that the endogenous activator of the calcium-dependent protein phosphorylation system is CDR. Thus it would appear that regulation by calcium of calcium-dependent protein kinase activity in nervous tissue is mediated by the same calcium-binding protein that regulates cyclic nucleotide phosphodiesterase and adenylate cyclase activities.

These studies of nervous tissue demonstrated that calcium is a regulatory agent some of whose effects may be mediated by protein phosphorylation. In addition, recent studies indicate that many other tissues possess a protein kinase that requires both calcium and CDR for activity, suggesting that this enzyme may be of general importance as a mediator of the regulatory actions of calcium (Ref. 43).

2. Regulation by Cyclic GMP of Protein Phosphorylation

Several lines of evidence indicate that cyclic GMP plays an important role as an intracellular regulatory substance (Refs. 7,44). Moreover, there are some highly suggestive findings concerning the molecular basis for its actions in various biological systems. A number of years ago, Dr. J. F. Kuo and I undertook a study of the mechanism by which cyclic GMP might exert its biological effects. By analogy with cyclic AMP-dependent protein kinase, it seemed reasonable to postulate the existence of cyclic GMP-dependent protein kinase that might mediate the biological effects of cyclic GMP. Through the use of column chromatography, it was possible to demonstrate a cyclic GMP-dependent protein kinase in lobster muscle and to separate it from a cyclic AMP-dependent protein kinase present in the same tissue (Ref. 45). Subsequently, cyclic GMP-dependent protein kinases have been demonstrated in a variety of other invertebrate (Refs. 46,47) and vertebrate (Refs. 46,48-51) tissues. However, in contrast to the situation with the cyclic AMP-dependent protein kinases, where many substrates, some with known and some with unknown function, have been found (see Refs. 7,8,13-16 for reviews), efforts to find endogenous proteins that are specific substrates for these cyclic GMP-dependent protein kinases proved difficult. Nevertheless, endogenous substrates for cyclic GMP-dependent protein kinases have been identified in three locations, namely in membranes of smooth muscle (Ref. 52), in membranes of intestinal brush border epithelium (Ref. 53) and in cerebellar cytosol (Ref. 54). I shall briefly describe our experiments with the cerebellum.

When the cytosol fraction of rat cerebellum was incubated in the presence of [γ-^{32}P]ATP, in the absence or presence of low concentrations of cyclic GMP or cyclic AMP, and then subjected to polyacrylamide gel electrophoresis and autoradiography, results of the type shown in Fig. 7 were obtained. Cyclic GMP stimulated the phosphorylation of a specific protein band with an apparent molecular weight of 23,000 (dark arrow, Fig. 7). An equal concentration of cyclic AMP failed to stimulate the incorporation of [^{32}P]-phosphate into this protein. The observation that cyclic AMP stimulated phosphate incorporation into other proteins of lower and higher molecular weight (light arrows, Fig. 7) indicated that cyclic AMP-dependent protein kinase was active in the cytosol preparation.

The results shown in Fig. 7 indicated that the 23,000 dalton protein was a substrate for the cyclic GMP-dependent protein kinase rather than for the cyclic AMP-dependent protein kinase that was also present in cerebellar cytosol. This conclusion was supported by studies in which cyclic GMP-dependent protein kinase and cyclic AMP-dependent protein kinase were partially purified from rabbit cerebellum and added to cerebellar cytosol. The effect of various concentrations of cyclic GMP and cyclic AMP on the phosphorylation of the 23,000 dalton protein in the presence of added cyclic GMP-dependent protein kinase is shown in Fig. 8. Cyclic GMP caused a half-maximal phosphorylation of the 23,000 dalton protein at a concentration of 60 nM. When an equal number of units (see Ref. 54) of cyclic AMP-dependent protein kinase was added to the cytosol along with cyclic AMP or cyclic GMP, the increase in the phosphorylation of the 23,000 dalton protein was much less.

We are currently attempting to purify and characterize this soluble 23,000 dalton substrate protein for the cyclic GMP-dependent protein kinase from mammalian cerebellum. Hopefully, studies of the biological and chemical properties of this molecule will contribute to an understanding of the role of cyclic GMP and of cyclic GMP-dependent protein phosphorylation in nervous tissue.

3. Regulation by Steroid Hormones of Protein Phosphorylation

Steroid hormones have multiple actions on their target tissues. Many of these actions are either synergistic with, or antagonistic to, the effects of those hormones whose actions are known to be mediated through cyclic AMP. Although steroid hormones have been found to affect cyclic AMP levels and the enzymes (adenylate cyclase and phosphodiesterase) involved in cyclic AMP metabolism in a variety of tissues (Refs. 55-58), only a few aspects of steroid-hormone action can be accounted for by the effects of these hormones on cyclic AMP levels (Ref. 59). It is of significance, therefore, that representatives of all of the major classes of steroid hormones, including the mineralocorticoid, glucocorticoid, estrogen, and androgen classes, affect protein phosphorylation systems in their target tissues (Refs. 60,61). In every instance, administration *in vivo* of the appropriate steroid hormone affects the autophosphorylation of the regulatory subunit of a cyclic AMP-dependent protein kinase by the catalytic subunit of this enzyme. The effect is specific for the respective target tissue, occurs with low doses of the steroid hormones, can be demonstrated within 1 hour of steroid hormone administration, and

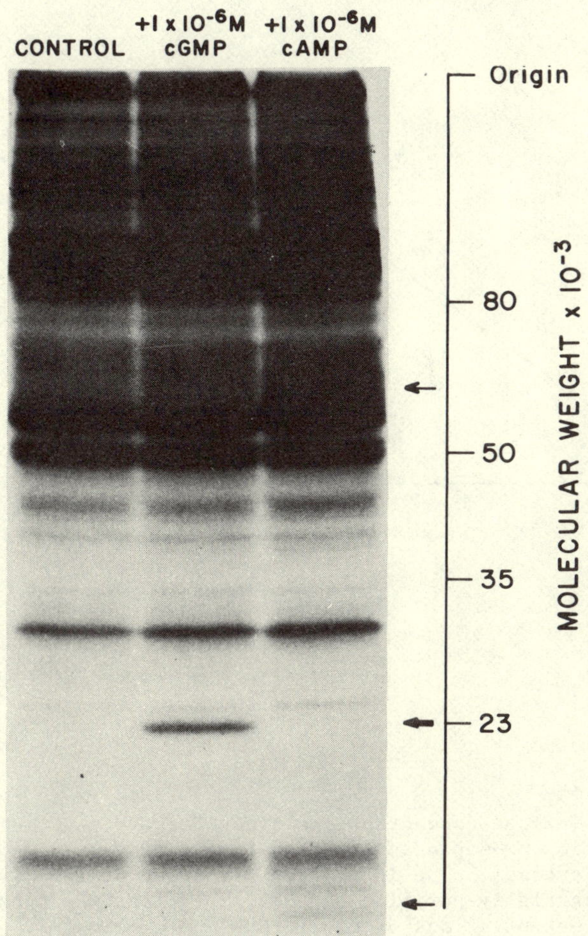

Fig. 7. Cyclic GMP-dependent endogenous protein phosphorylation in cytosol from rabbit cerebellum. The cytosol fraction was passed over a Sephadex G-25 column to remove endogenous ATP and cyclic nucleotides, and was then incubated with [γ-^{32}P]-ATP for 1.5 minutes at 30°C, in the absence or presence of cyclic GMP or cyclic AMP as indicated. The reaction was terminated by the addition of sodium dodecyl sulfate and the samples heated at 100°C for 1 minute. The entire sample was then subjected to sodium dodecyl sulfate-polyacrylamide gel electrophoresis and autoradiography. Cyclic GMP, but not cyclic AMP, stimulated the phosphorylation of a protein with an apparent molecular weight of 23,000 daltons, indicated by the dark arrow. No specific stimulation by cyclic GMP of phosphate incorporation into other proteins was observed, even when electrophoresis was carried out on a 7.5% polyacrylamide gel, which increased the separation of higher molecular weight proteins. (Taken from Ref. 54.)

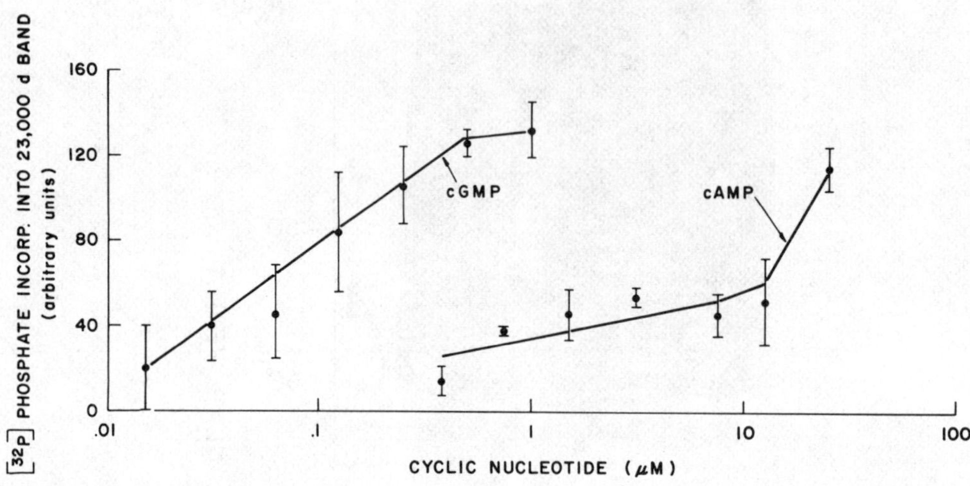

Fig. 8. Effect of various concentrations of cyclic GMP and cyclic AMP on the phosphorylation by cyclic GMP-dependent protein kinase of the 23,000 dalton band present in cerebellar cytosol. The incubation mixture contained [γ-^{32}P]ATP, cerebellar cytosol, a partially-purified cerebellar cyclic GMP-dependent protein kinase and the indicated concentrations of cyclic GMP and cyclic AMP. Incubation was for 1 minute at 30°C. Other experimental procedures were as described in the legend to Fig. 7. Incorporation of radioactive phosphate was measured quantitatively by densitometric analysis of the autoradiograms. Phosphorylation of the 23,000 dalton protein was calculated as the increase above that observed in the absence of added nucleotide, which amounted to 87 arbitrary units. (Taken from Ref. 54.)

is associated with alterations in the level of cyclic AMP-dependent protein kinase activity (Ref. 61,62). The detailed mechanism by which the steroid hormones affect the autophosphorylation of the regulatory subunit of cyclic AMP-dependent protein kinase is not yet understood. However, the effect of the steroids is indirect and requires de novo protein synthesis, in contrast to the direct effect of cyclic AMP on protein kinases. The effect of steroid hormones and of cyclic AMP on the same cyclic AMP-dependent protein kinase (Fig. 9) may provide a molecular basis for the ability of the steroid hormones to act synergistically (permissive action of the steroid hormones) or antagonistically with those hormones and neurotransmitters whose effects are mediated through cyclic AMP (Refs. 61,62).

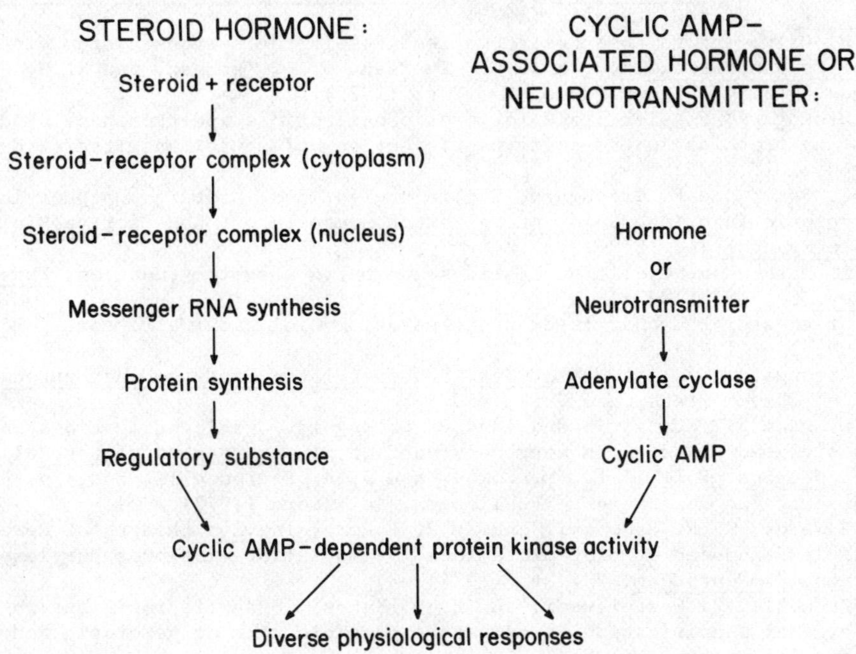

Fig. 9. A possible molecular basis for the biological interactions of steroid hormones with those hormones and neurotransmitters whose effects are mediated through cyclic AMP. The steroid hormones, through a mechanism involving *de novo* protein synthesis but not involving cyclic AMP, regulate cyclic AMP-dependent protein kinases in their target tissues. The cyclic AMP-associated hormones and neurotransmitters, through a mechanism not involving *de novo* protein synthesis, cause an increase in cyclic AMP which directly affects protein kinase activity. For simplicity, mechanisms of steroid hormone action independent of the cyclic AMP-dependent protein kinase system are not shown. (Taken from Ref. 7.)

In conclusion, a variety of experimental data, summarized in this article, and in greater detail elsewhere (Refs. 7,8), indicate that protein phosphorylation represents a final common pathway in biological regulation. Several regulatory agents, including cyclic AMP, cyclic GMP, calcium and the steroid hormones have been shown to affect the state of phosphorylation of specific proteins in the nervous system and elsewhere. Clarification of the detailed manner in which these phosphoproteins may regulate function is an exciting area for future research.

REFERENCES

(1) G. A. Robison, R. W. Butcher, and E. W. Sutherland, *Cyclic AMP*, Academic Press,

New York (1971).
(2) F. E. Bloom, Reviews of Physiology, Biochemistry, and Experimental Pharmacology, Springer, Berlin (1975).
(3) J. W. Daly, The role of cyclic nucleotides in the nervous system, in Handbook of Psychopharmacology, 5 (L. L. Iversen, S. H. Iversen, and S. H. Snyder, eds.), p. 47 Plenum Press, New York (1975).
(4) P. Greengard, Possible role for cyclic nucleotides and phosphorylated membrane proteins in the postsynaptic actions of neurotransmitters, Nature 260, 101 (1976).
(5) K. G. Beam, and P. Greengard, Cyclic nucleotides, protein phosphorylation and synaptic function, Cold Spring Harbor Symposia on Quantitative Biology, The Synapse 40, 157 (1976).
(6) J. A. Nathanson, Cyclic nucleotides and nervous system function, Physiol. Rev. 57, 157 (1977).
(7) P. Greengard, Phosphorylated proteins as physiological effectors. Science 199, 146 (1978).
(8) P. Greengard, Cyclic Nucleotides, Phosphorylated Proteins and Neuronal Function, Raven Press, New York, (1978).
(9) P. D. Kanof, T. Ueda, I. Uno, and P. Greengard, Cyclic nucleotides and phosphorylated proteins in neuronal function, in Approaches to the Cell Biology of Neurons, Vol. II (W. M. Cowan, and J. A. Ferrendelli, eds.) p. 399, Society for Neuroscience Publishers, Baltimore (1977).
(10) C. Edwards, S. R. Nahorski, and K. J. Rogers, In vivo changes of cerebral cyclic AMP induced by biogenic amines: association with phosphorylase activation, J. Neurochem. 22, 565 (1974).
(11) M. Brunelli, V. Castellucci, and E. R. Kandel, Synaptic facilitation and behavioral sensitization in Aplysia: possible role of serotonin and cyclic AMP, Science 194, 1178 (1976).
(12) J. F. Kuo, and P. Greengard, Cyclic nucleotide-dependent protein kinases. IV. Widespread occurrence of adenosine 3',5'-monophosphate-dependent protein kinase in various tissues and phyla of the animal kingdom. Proc. Nat. Acad. Sci. U.S.A. 64, 1349 (1969).
(13) E. G. Krebs, Protein kinases, Curr. Top. Cell Regul. 5, 99 (1972).
(14) T. A. Langan, Protein kinases and protein kinase substrates, Adv. Cyclic Nucl. Res. 3, 99 (1973).
(15) C. S. Rubin, and O. M. Rosen, Protein phosphorylation, Annu. Rev. Biochem. 44, 831 (1975).
(16) H. G. Nimmo, and P. Cohen, Hormonal control of protein phosphorylation, Adv. Cyclic Nucl. Res. 8, 145 (1977).
(17) T. Ueda, H. Maeno, and P. Greengard, Regulation of endogenous phosphorylation of specific proteins in synaptic membrane fractions from rat brain by adenosine 3':5'-monophosphate, J. Biol. Chem. 248, 8295 (1973).
(18) T. Ueda, and P. Greengard, Adenosine 3':5'-monophosphate-regulated phosphoprotein system of neuronal membranes. I. Solubilization, purification and some properties of an endogenous phosphoprotein, J. Biol. Chem. 252, 5155 (1977).
(19) I. Uno, T. Ueda, and P. Greengard, Enzyme systems involved in phosphorylation and dephosphorylation of two endogenous phosphoproteins in neuronal membranes, Arch. Biochem. Biophys. 183, 480 (1977).
(20) S. M. Lohmann, U. Walter, W. Sieghart, and P. Greengard, manuscript in preparation.
(21) W. Sieghart, J. Forn, R. Schwarcz, J. T. Coyle, and P. Greengard, manuscript in preparation.
(22) J. T. Coyle, and R. Schwarcz, Lesion of striatal neurones with kainic acid provides a model for Huntington's chorea, Nature 263, 244 (1976).
(23) E. G. McGeer, and P. L. McGeer, Duplication of biochemical changes of Huntington's chorea by intrastriatal injections of glutamic and kainic acids, Nature 263, 517 (1976).

(24) T. Hattori, and E. G. McGeer, Fine structural changes in the rat striatum after local injections of kainic acid, Brain Res. 129, 174 (1977).
(25) R. S. Cohen, F. Blumberg, K. Berzins, P. Siekevitz, T. Ueda, and P. Greengard, manuscript in preparation.
(26) F. E. Bloom, E. Battenberg, T. Ueda, and P. Greengard, manuscript in preparation.
(27) S. M. Lohmann, T. Ueda, and P. Greengard, Ontogeny of a synaptic phosphoprotein in brain, Proc. Nat. Acad. Sci. U.S.A. in press (1978).
(28) I. Uno, T. Ueda, and P. Greengard, Adenosine 3':5'-monophosphate-regulated phosphoprotein system of neuronal membranes. II. Solubilization, purification and some properties of an endogenous cyclic AMP-dependent protein kinase, J. Biol. Chem. 252, 5164 (1977).
(29) J. Forn, and P. Greengard, manuscript in preparation.
(30) U. Strombom, J. Forn, and P. Greengard, manuscript in preparation.
(31) B. Katz, The Release of Neural Transmitter Substances, Thomas, Springfield (1969).
(32) P. F. Baker, A. L. Hodgkin, and E. B. Ridgway, Depolarization and calcium entry in squid giant axons, J. Physiol., 218, 709 (1971).
(33) B. Katz, and R. Miledi, Study of synaptic transmission in the absence of nerve impulses, J. Physiol. 192, 407 (1967).
(34) W. W. Douglas, Stimulus-secretion coupling: the concept and clues from chromaffin and other cells, Br. J. Pharmacol. 34, 451 (1968).
(35) R. L. Patrick, and J. D. Barchas, Stimulation of synaptosomal dopamine synthesis by veratridine, Nature 250, 737 (1974).
(36) V. H. Morgenroth, III, M. C. Boadle-Biber, and R. H. Roth, Activation of tyrosine hydroxylase from central noradrenergic neurons by calcium, Mol. Pharmacol. 11, 427 (1975).
(37) B. K. Krueger, J. Forn, and P. Greengard, Depolarization-induced phosphorylation of specific proteins, mediated by calcium ion influx, in rat brain synaptosomes, J. Biol. Chem. 252, 2764 (1977).
(38) H. Schulman, and P. Greengard, Stimulation of brain membrane protein phosphorylation by calcium and an endogenous heat-stable protein, Nature 271, 478 (1978).
(39) W. Y. Cheung, Cyclic 3':5'-nucleotide phosphodiesterase: demonstration of an activator, Biochem. Biophys. Res. Commun. 38, 533 (1970).
(40) S. Kakiuchi, and R. Yamazaki, Calcium-dependent phosphodiesterase activity and its activating factor isolated from brain: studies on cyclic 3':5'-nucleotide phosphodiesterase, Biochem. Biophys. Res. Commun. 41, 1104 (1970).
(41) C. O. Brostrom, Y.-C. Huang, B. McL. Breckenridge, and D. J. Wolff, Identification of a calcium-binding protein as a calcium-dependent regulator of brain adenylate cyclase, Proc. Nat. Acad. Sci. U.S.A. 72, 64 (1975).
(42) T. S. Teo, H. Wang, and J. H. Wang, Purification and properties of the protein activator of bovine heart cyclic adenosine 3',5'-monophosphate phosphodiesterase, J. Biol. Chem. 248, 588 (1973).
(43) H. Schulman, and P. Greengard, Activator of the calcium-dependent protein kinase from brain is identical to the "calcium-dependent regulator" of phosphodiesterase, Proc. Nat. Acad. Sci. U.S.A. in press (1978).
(44) N. D. Goldberg, R. G. O'Dea, and M. K. Haddox, Cyclic GMP, Adv. Cyclic Nucl. Res. 3, 155 (1973).
(45) J. F. Kuo and P. Greengard, Cyclic nucleotide-dependent protein kinases. VI. Isolation and partial purification of a protein kinase activated by guanosine 3',5'-monophosphate, J. Biol. Chem. 245, 2493 (1970).
(46) P. Greengard, and J. F. Kuo, On the mechanism of action of cyclic AMP, in Role of Cyclic AMP in Cell Function, Volume 3 of Advances in Biochemical Psychopharmacology, (E. Costa and P. Greengard, eds.), p. 287, Raven Press, New York (1970).
(47) J. F. Kuo, G. R. Wyatt, and P. Greengard, Cyclic nucleotide-dependent protein

kinases. IX. Partial purification and some properties of guanosine 3',5'-monophosphate-dependent and adenosine 3',5'-monophosphate-dependent protein kinases from various tissues and species of arthropoda, J. Biol. Chem. 246, 7159 (1971).

(48) F. Hofmann, and G. Sold, A protein kinase activity from rat cerebellum stimulated by guanosine 3',5'-monophosphate, Biochem. Biophys. Res. Commun. 49, 1100 (1972).

(49) J. F. Kuo, Changes in relative levels of guanosine 3',5'-monophosphate-dependent and adenosine 3',5'-monophosphate-dependent protein kinases in lung, heart, and brain of developing guinea pigs, Proc. Nat. Acad. Sci. U.S.A. 72, 2256 (1975).

(50) Y. Takai, K. Nishiyama, H. Yamamura, and Y. Nishizuka, Guanosine 3':5'-monophosphate-dependent protein kinase from bovine cerebellum, J. Biol. Chem. 250, 4690 (1975).

(51) K. Nakazawa, and M. Sano, Partial purification and properties of guanosine 3':5'-monophosphate-dependent protein kinase from pig lung, J. Biol. Chem. 250, 7415 (1975).

(52) J. E. Casnellie, and P. Greengard, Guanosine 3',5'-cyclic monophosphate-dependent phosphorylation of endogenous substrate proteins in membranes of mammalian smooth muscle, Proc. Nat. Acad. Sci. U.S.A. 71, 1891 (1974).

(53) H. R. DeJonge, Cyclic nucleotide-dependent phosphorylation of intestinal epithelium proteins, Nature 262, 591 (1976).

(54) D. J. Schlichter, J. E. Casnellie, and P. Greengard, An endogenous substrate for cGMP-dependent protein kinase in mammalian cerebellum, Nature 273, 61-62 (1978).

(55) B. Weiss, and J. Crayton, Gonadol hormones as regulators of pineal adenyl cyclase activity, Endocrinology 87, 527 (1970).

(56) M. G. Rosenfeld, and B. W. O'Malley, Steroid hormones: effects on adenyl cyclase activity and adenosine 3',5'-monophosphate in target tissues, Science 168, 253 (1970).

(57) P. S. Ross, V. C. Manganiello, and M. Vaughan, Regulation of cyclic nucleotide phosphodiesterases in cultured hepatoma cells by dexamethasone and $N^6,O^{2'}$-dibutyryl adenosine 3':5'-monophosphate, J. Biol. Chem. 252, 1448 (1977).

(58) G. Senft, G. Schultz, K. Munske, and M. Hoffmann, Effects of glucocorticoids and insulin on 3',5'-AMP phosphodiesterase activity in adrenalectomized rats, Diabetologia 4, 330 (1968).

(59) E. B. Thompson, and M. E. Lippman, Mechanism of action of glucocorticoids, Metabolism 23, 159 (1974).

(60) A. Y.-C. Liu, and P. Greengard, Aldosterone-induced increase in protein phosphatase activity of toad bladder, Proc. Nat. Acad. Sci. U.S.A. 71, 3869 (1974).

(61) A. Y.-C. Liu, and P. Greengard, Regulation by steroid hormones of phosphorylation of specific protein common to several target organs, Proc. Nat. Acad. Sci. U.S.A. 73, 568 (1976).

(62) A. Y.-C. Liu, I. Uno, U. Walter, and P. Greengard, manuscript in preparation.

Cholinergy

The Use of Affinity Constants for Studying Interactions Between Pharmacodynamic Groups and the Muscarinic Receptor

R.B. Barlow

Department of Pharmacology, Medical School, University of Bristol, Bristol BS8 1TD, U.K.

MEASUREMENT OF AFFINITY

Competitive Antagonists

The affinity of a drug for a receptor is the most fundamental measure of its biological activity. If the receptors can be obtained in an isolated form, for instance as membrane fragments, it should be possible to measure the affinity directly by studying the adsorption of labelled drug. Alternatively the affinity of an unlabelled drug can be studied by competition with that of a labelled drug which binds to the same receptors. Correction must be made for non-specific binding of the drug (at sites which are not receptors) and it is essential that there is equilibrium between the drug and the receptor material. Usually the law of mass action is applied to the adsorption process and the results are fitted to the Lanmuir adsorption isotherm (Ref. 1) by Lineweaver-Burk plots (Ref. 2) or Scatchard plots (Ref. 3), or Hill plots (Ref. 4) are made in order to test for cooperativity. It is also possible to fit the results directly to a logistic expression by the method of least-squares (Ref. 5). The biggest problem with this kind of experiment, however, is to know whether the isolated receptor material is structurally identical with the functional receptors present on intact cells which lead to the pharmacological response.

It has been known for a long time how to measure the affinity of compounds which are competitive antagonists for these functional receptors in intact tissues. The methods were developed largely by Schild (Ref. 6), based on ideas of Clark (Ref. 7) and Gaddum (Ref. 8,9) and involve the measurement of the dose-ratio (DR) produced by a concentration (B) of the antagonist in equilibrium with the preparation and the application of the Gaddum-Schild equation: $DR = 1 + BK_B$, where K_B is the affinity constant of the antagonist. The dose-ratio should depend on the concentration of the antagonist (B) and on its affinity (K_B). For a competitive antagonist the graph of log (DR-1) against log (B) should be a straight line with a slope of 1; $\log B = -\log K_B$ when $\log (DR-1) = 0$ (Ref. 10). Schild defined pA_x as log (1/B) when DR = X so pA_2 is the same as $\log K_B$. Usually, however, the dose-ratio is greater than 2, because there is less error in (DR-1) with bigger dose-ratios, and K can be measured more accurately (Ref. 11). With these experiments, as with experiments with receptor preparations made from membrane fragments, it is necessary to allow time for the antagonist to come into equilibrium and with some potent compounds this can take almost an hour at 37°C, though with weaker compounds equilibration occurs within a minute.

The competitive nature of the antagonism can be checked by making measurements over a wide range of concentrations or by testing the compound in the presence of a known competitive antagonist (Ref. 12). The dose-ratio should be independent of the agonist used to activate the receptors; for example, the value of log K_B for a compound at the muscarinic receptors in the guinea-pig ileum should be the same whether the agonist is acetylcholine, carbachol, or n-pentyltrimethylammonium (Ref. 12). For values of log K up to 10 the error in the mean of a group of 5 or more estimates is usually less than 0.1 log units (Ref. 13). It is difficult to obtain accurate values, however, with extremely potent compounds (Ref. 14).

Receptor Identification

A test to see whether receptors in "isolated" preparations made from membrane fragments are identical with receptors on intact cells can therefore be made by comparing the affinities of competitive antagonists in the two systems. If measurements of affinity are made at the same temperature, and in the same physiological salt solution, a drug should have the same affinity in both tests. If several compounds are tested, a graph of log K_B in one system against log K_B in the other should be a straight line with a slope of 1, passing through the origin. This method has been used to try to establish the identity of receptors in different tissues (Ref. 15) but although it is often easy to see similarities in binding, positive identification is virtually impossible; experiments can only show that within the limits of error there is no significant difference. The situation is similar to trying to establish the identity of a substance by bio-assay. It is, nevertheless, the only available way of comparing the identities of isolated and functional receptors.

Agonists and Partial Agonists

For substances which are agonists and produce a response from the tissue the problem is more complicated. Activity depends not only on affinity for receptors but on ability to activate them and drugs differ in their ability to activate receptors, just as substrates differ in their rate-constants for breakdown by an enzyme. The pharmacologist can compare the concentrations of two drugs which produce comparable biological responses but the ratio of the concentrations, which expresses the ratio of activities, depends on ability to activate receptors as well as on affinity and it does not follow that the more active drug is necess-arily more strongly bound to the receptors. Sometimes the concentration or dose producing a half-maximal response from the tissue (EC_{50} or ED_{50}) is used to indicate the activity of a compound and log ($1/EC_{50}$) is referred to as pD_2 (Ref. 16, 17). It is well established, however, that with many systems, including the isolated guinea-pig ileum, the maximum response to many drugs is tissue-limited, rather than drug-limited, and active compounds produce a maximal response when only a very small proportion of receptors has been activated (Ref. 18,19). In these circumstances pD_2 is not a measure of affinity. The situation is not comparable with the interaction between substrate and enzyme where the maximum rate is obtained with saturation of the active sites. To measure the affinity of an agonist for the receptors in a tissue it is necessary to be able to calculate the proportion of receptors occupied by the drug. It is sometimes possible to attempt this with a suitable irreversible blocking agent (Ref. 20,21) but the interpretation of the results is complicated by the possibility that the receptor may behave allosterically (Ref. 22).

With most tissues there are substances which are weak and do not produce the maximum response of which the tissue is capable, even in very high concentration.

These are termed partial agonists (or competitive duallists) and log ($1/EC_{50}$), where EC_{50} produces half the maximum response of which the drug is capable, is also referred to as pD_2 for these compounds (Ref. 17). With a preparation such as the guinea-pig ileum, where active agonists produce a maximum response with only a small proportion of the receptors activated, pD_2 is no more a measure of the affinity of a partial agonist than it is of a full agonist. The affinity of a partial agonist may, however, be measured by experiments in which it is allowed to compete with an active agonist (Ref. 18).

In experiments with receptor fragments the affinity of an agonist, measured from the adsorption of labelled material or by competition with some other labelled ligand, need bear no relation to the concentration producing the biological effect (pD_2), because this depends also on ability to activate receptors. If the receptor behaves allosterically there will be two binding sites for agonists, but even if the binding characteristics of an agonist for both are known it is still not yet possible to relate these to properties measured from the log.dose-response curve by the pharmacologist. It is only with partial agonists or competitive antagonists that it is possible to measure affinity for receptors satisfactorily; if the receptors behave allosterically, this is the affinity for the inactive form (Ref. 22).

CONTRIBUTIONS OF GROUPS TO AFFINITY

Forces between Drug and Receptor

Present ideas about the forces between drug and receptor which appear in text books (Ref. 23,24,25) are based largely on calculations made by Pauling (Ref. 26) but these may not allow adequately for the effects of water in the receptor environment (Ref. 27). It is therefore desirable to see whether they are supported by experimental evidence. The problem can be studied by observing the effects of chemical groups on affinity and it is particularly easy to do this with drugs acting at the muscarinic acetylcholine receptors of the isolated guinea-pig ileum because results have been obtained (Ref. 12,13,14,28,29) with over 200 closely related compounds which have the general structure shown in Fig. 1.

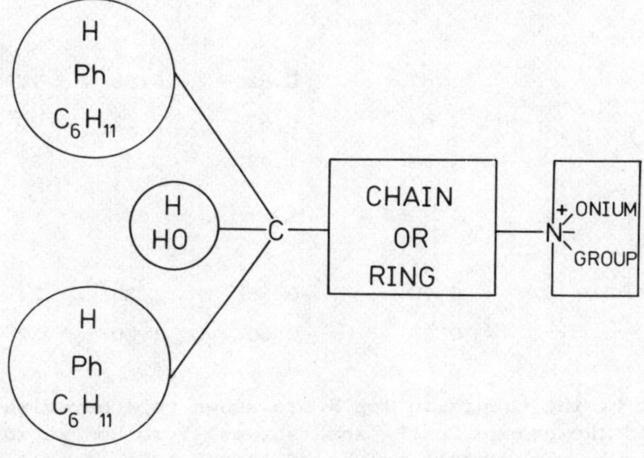

Fig. 1

For example, the interaction between whole units of charge, such as a positively charged group in a drug with a negatively charged group in a receptor, is expected to contribute 4-6 kcal (17-25 kJ)/mole to the free energy of adsorption, ΔG. The effect on log K can be calculated from the Van't Hoff relation, -ΔG = RTlnK, and at 37°C the presence of the charge should increase log K by 2.8-4.2 log units. The contribution made by the charge can be observed experimentally by comparing the affinities of antagonists which are esters of choline with those of the corresponding esters of 3,3-dimethylbutan-1-ol ($HOCH_2CH_2CMe_3$). In 4 such comparisons (Ref. 28), however, the biggest increase in log K (observed with the phenylcyclohexylglycollic esters) was only 1.8 log units corresponding to an increase in ΔG of 2.5 kcal (10.6 kJ)/mole. With the mandelic esters the presence of the charge actually reduced log K by 0.2 units.

The contribution from a hydrogen bond between drug and receptor is expected to contribute between 1.5 and 7.5 kcal (6 and 31 kJ)/mole to the free energy of adsorption and the increase in log K should be between 2.1 and 10.6 log units. In 59 comparisons of compounds with -H replaced by -OH the biggest increase in log K was 1.9 (Table 1).

Table 1 Effects of Changes in Structure on log K for Muscarinic Receptors of the Guinea-pig Ileum at 37°C

Replacement of $-\overset{+}{N}Me_3$ by:	Δlog K max	min	range	n	$\Delta\phi^o_v$
CMe_3	0.20	-1.78	1.98	4	> 10
$\overset{+}{N}Me_2Et$	0.72	-0.19	0.91	23	15.6
methylpyrrolidinium	0.61	-0.23	0.84	23	19.9
$\overset{+}{N}MeEt_2$	0.89	-0.34	1.23	23	30.0
methylpiperidinium	1.08	-0.48	1.56	23	33.6
ethylpyrrolidinium	1.04	-0.19	1.23	23	34.6
$\overset{+}{N}Et_3$	1.25	-0.30	1.55	23	44.7
ethylpiperidinium	0.99	-0.65	1.65	23	48.0
Replacement of -H by:					
Ph	3.51	0.62	2.89	77	63
C_6H_{11}	4.04	0.87	3.17	60	79
OH	1.94	-1.49	3.43	59	1
Replacement of $-CH_2-CH_2-$ by:					
-CO-O-	0.70	-0.65	1.35	24	-13.8
$-CH_2-O-$	0.34	-0.66	1.00	32	-10.9

The extreme values of the change in log K are shown together with the number of comparisons (n) and the change in the apparent molal volume at infinite dilution ($\Delta\phi^o_v$; cm^3/mole, 25°C). To convert values of Δlog K into changes in free energy, multiply by 1.42 (kcal/mole) or 5.93 (kJ/mole).

We have recently compared atropine with (±)hydratropyltropine, which lacks the hydroxyl group present in atropine, and found a difference in log K of 2.1 units for the hydrochlorides and 2.2 units for the quaternary metho-salts (Ref. 30). These results indicate that hydrogen bonds between these drugs and receptors are relatively weak. With many compounds the hydroxyl group markedly lowers affinity.

In contrast, the effects of hydrocarbon groups on affinity (Table 1) are much bigger than would be expected from Van der Waal's forces and indicate the hydrophobic nature of the interaction. This explains the adverse effects of a hydroxyl group in a position where it cannot form a hydrogen bond to the receptor but merely disrupts hydrophobic interactions. The replacement of hydrogen by hydroxyl in the 5-, 4-, 3- positions of n-pentyltriethylammonium (n-pentyl$\overset{+}{N}Et_3$), for instance, reduces log K for muscarinic receptors by 1.2, 1.0 and 0.7 units respectively (Ref. 31). As expected, the maximum contributions made by hydrocarbon groups can be related to their size in solution, as assessed from the increment in apparent molal volume at infinite dilution ($\Delta\phi_v^o$).

Variation in Effects of Groups on Affinity

Table 1 shows that the effects of a group on affinity are not constant (compare the maximum and minimum values) but vary from one drug to another, even among closely related compounds. This variation can be explained by supposing that the contribution which a group could make to affinity may be offset by the disturbance which its introduction causes to existing binding (Ref. 12). The contribution which a hydroxyl group can make by hydrogen bonding must be offset by the reduction it causes in hydrophobic bonding. Likewise a large hydrocarbon group may make a large contribution to hydrophobic bonding but its bulk may sterically impede binding and with these groups the range of effects in Table 1 (as well as the maximum effect), is related to the size in solution ($\Delta\phi_v^o$). With groups containing polar oxygen atoms the effects are unrelated to size but with any group the range of effects is related to the maximum effects (Fig. 2).

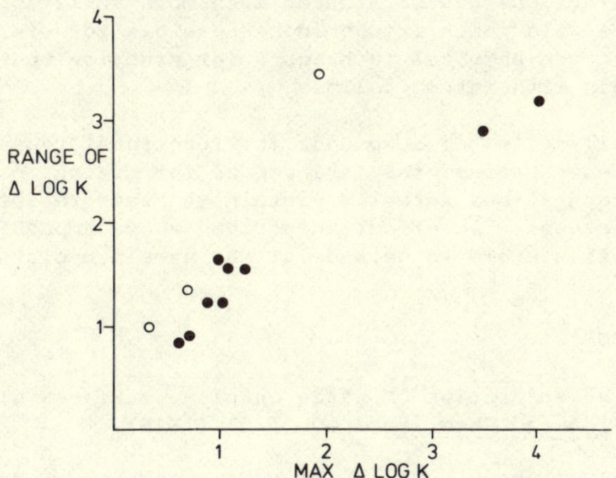

Fig. 2
The difficulty of predicting affinity. Groups which can make big contributions are associated with a big range. Closed circles, hydrocarbon groups; open circles, groups containing oxygen.

This illustrates the extreme difficulty of attempting any useful predictions of affinity. Changes in structure which might be expected to produce big increases in affinity are also those where the range of effects is big. The results make it difficult to see how affinity could be correlated quantitatively with chemical properties by an equation of the Hansch type (Ref. 32,33).

Effects on the Enthalpy and Entropy of Adsorption

Because a group does not produce a constant increment in the free energy of adsorption, it must have different effects on the enthalpy of adsorption (ΔH) and entropy of adsorption (ΔS) in different molecules. It should be possible to obtain some idea of the effects on enthalpy by measuring log K at two different temperatures because

$$\frac{\Delta \log K}{\Delta \frac{1}{T}} = -\frac{\Delta H}{R}, \quad \text{where R is the gas constant.}$$

Changes in entropy can then be calculated ($T\Delta S = \Delta H - \Delta G$). It is only possible to lower the temperature of the guinea-pig ileum from 37°C to about 29°C, so estimates of ΔH calculated in this way are very inaccurate but it has been found that there are significant differences between the values for some compounds (Ref. 34). Usually lowering the temperature increases the affinity and adsorption is associated with an increase in entropy, but results obtained with (-)-hyoscine, (-)-hyoscine methiodide and (-)-hyoscyamine methiodide (but not the (+)-enantiomer) all indicate a decrease in entropy (Ref. 35).

Effects on entropy are to be expected in the interactions between drug and receptor, which is a highly organized structure at least partly in an aqueous environment. It has been suggested that effects on entropy and on water may be involved in agonist activity (Ref. 36,37) but such ideas cannot be tested with the limited information which can be obtained from relationships between chemical structure and the affinity of antagonists or partial agonists for functional receptors. The interactions can be studied much more satisfactorily with isolated receptor preparations with which it should be possible to work with a wider range of temperature and to use physical techniques for studying the binding process which are not possible with intact cells.

Measurement of the affinities of compounds for functional receptors remain important, nevertheless, because they are needed for trying to establish that isolated receptor preparations actually contain the same receptors as those present in intact tissues. The effects described above emphasize the need for measurements with both systems to be made at the same temperature.

REFERENCES

1. I. Langmuir, The adsorption of gases on plane surfaces of glass, mica and platinum, J. Amer. Chem. Soc. 40, 1361 (1918).

2. H. Lineweaver and D. Burk, The determination of enzyme dissociation constants J. Amer. Chem. Soc. 56, 658 (1934).

3. G. Scatchard, The attractions of proteins for small molecules and ions, Ann. N.Y. Acad. Sci. 51, 660 (1949).

4. W.E.L. Brown and A.V. Hill, The oxygen-dissociation curve of blood and its thermodynamical basis, Proc. Roy. Soc. B, 94, 297 (1923).

5. R.B. Parker and D.R. Waud, Pharmacological estimation of drug receptor dissociation constants. Statistical evaluation I. Agonists, J. Pharmacol. 177, 1 (1971).

6. H.O. Schild, pA, a new scale for the measurement of drug antagonism, Br. J. Pharmac. 2, 189 (1947); pA_x and competitive drug antagonism, ibid., 4, 277 (1949).

7. A.J. Clark and J. Raventos, The antagonism of acetylcholine and of quaternary ammonium salts, Quart. J. Exp. Physiol. 26, 375 (1937); A.J. Clark, (1937) General Pharmacology, Handbuch der Experimentellen Pharmakologie IV, Springer, Berlin.

8. J.H. Gaddum, The quantitative effects of antagonistic drugs, J. Physiol. 89, 7P (1937).

9. J.H. Gaddum, Theories of drug antagonism, Pharmacol. Rev. 9, 211 (1957).

10. O. Arunlakshana and H.O. Schild, Some quantitative uses of drug antagonists, Br. J. Pharmac. 14, 48 (1959).

11. Edinburgh Staff, (1974), Pharmacological Experiments on Isolated Preparations, 2nd edition, Churchill-Livingstone, Edinburgh, p 28.

12. F.B. Abramson, R.B. Barlow, M.G. Mustafa and R.P. Stephenson, Relationships between chemical structure and affinity for acetylcholine receptors, Br. J. Pharmac. 37, 207 (1969).

13. R.B. Barlow, F.M. Franks and J.D.M. Pearson, Studies on the stereospecificity of closely related compounds which block postganglionic acetylcholine receptors in the guinea-pig ileum, J. Med. Chem. 16, 439 (1973).

14. F.B. Abramson, R.B. Barlow, F.M. Franks and J.D.M. Pearson, Relationships between chemical structure and affinity for postganglionic acetylcholine receptors of the guinea-pig ileum, Br. J. Pharmac. 51, 81 (1974).

15. R.B. Barlow, F.M. Franks and J.D.M. Pearson, A comparison of the affinities of antagonists for acetylcholine receptors in the ileum, bronchial muscle and iris of the guinea-pig, Br. J. Pharmac. 46, 300 (1972).

16. L.C. Miller, T.J. Becker and M.L. Tainter, The quantitative evaluation of spasmolytic drugs in vitro, J. Pharmac. 92, 260 (1948).

17. E.J. Ariëns and J.M. Van Rossum, pD_x, pA_x and pD'_x values in the analysis of pharmacodynamics, Arch. int. Pharmacodyn. 110, 275 (1957).

18. R.P. Stephenson, A modification of receptor theory, Br. J. Pharmac. 11, 379 (1956).

19. M. Nickerson, Receptor occupancy and tissue response, Nature 178, 697 (1956).

20. R.F. Furchgott (1966), The use of β-haloalkylamines in the differentiation of receptors and in the determination of dissociation constants of receptor-agonist complexes, Advances in Drug Research 3, Academic Press, London, p 21.

21. D. Mackay (1966), A new method for the analysis of drug-receptor interactions Advances in Drug Research 3, ed. Harper and Simmonds, Academic Press, London, p 1.

22. D. Colquhoun (1973), The relation between classical and cooperative models for drug action, Drug Receptors, ed. Rang, Macmillan London, p 149.

23. A. Albert (1973), Selective Toxicity, 5th Edn., Chapman and Hall, London, p 222.

24. J.M. Van Rossum (1968), Drug-receptor theories, Recent Advances in Pharmacology, 4th Edn. ed. Robson and Stacey, Churchill London, p 107.

25. A. Goldstein, L. Aronow and S.M. Kalman (1974), Principles of Drug Action, 2nd edn., Wiley New York, p 2.

26. L. Pauling (1939), The Nature of the Chemical Bond, Cornell University Press, (1960) ibid., 3rd edition.

27. E.W. Gill (1965), Drug receptor interactions, Progress in Medicinal Chemistry 4, ed. Ellis and West, Butterworth London, p 39.

28. R.B. Barlow and J.H. Tubby, Actions of some esters of 3,3-dimethylbutan-1-ol (the carbon analogue of choline) on the guinea-pig ileum, Br. J. Pharmac. 51, 95 (1974).

29. R.B. Barlow, J.B. Bremner and K.S. Soh, The effects of replacing ester by amide on the biological properties of compounds related to acetylcholine, Br. J. Pharmac. 62, 39 (1978).

30. R.B. Barlow, K.N. Burston and S. Ramtoola, unpublished.

31. R.B. Barlow, The effects of a hydroxyl group on some chemical and biological properties of n-pentylammonium salts, J. Pharm. Pharmac. in press.

32. C. Hansch, P.P. Maloney, T. Fujita and R.M. Muir, Correlation of biological activity of phenoxyacetic acids with Hammett substituent constants and partition coefficients, Nature, 194, 178; C. Hansch and T. Fujita, ρ-σ-π Analysis. A method for the correlation of biological activity and chemical structure, J. Amer. Chem. Soc. 86, 1616 (1964).

33. C. Hansch (1977), On the predictive value of QSAR, Biological Activity and Chemical Structure, ed. Keverling Buisman, Elsevier Amsterdam, p 47.

34. R.B. Barlow, K.J. Berry, P.A.M. Glenton, N.M. Nikolaou and K.S. Soh, A comparison of affinity constants for muscarine-sensitive acetylcholine receptors in guinea-pig atrial pacemaker cells at 29°C and in ileum at 29° and 37°C, Br. J. Pharmac. 58, 613 (1976).

35. R.B. Barlow and K.N. Burston, unpublished.

36. B. Belleau, A molecular theory of drug action based on induced conformational perturbations of receptors, J. Med. Chem. 7, 776 (1964).

37. B. Belleau, Water as the determinant of thermodynamic transitions in the interaction of aliphatic chains with acetylcholinesterase and the cholinergic receptors, Ann. N.Y. Acad. Sci. 144, 705 (1967); (1968) Patterns of ligand-induced changes on a receptor surface: the water extrusion hypothesis, Physico-chemical Aspects of Drug Action, ed. Ariëns, Pergamon Press, Oxford, p 207.

Structure and Activity of Some New Muscarine Derivatives and Acetylcholine Analogues

P.G. Waser and W. Hopff

Institute of Pharmacology, University of Zürich, 8006 Zürich, Switzerland

For many years we have been investigating the pharmacological properties of different derivatives of muscarine in order to find relations between their structure and their cholinergic activity (Waser, 1961). The cyclic muscarine compounds have distinct advantages compared to acetylcholine or other aliphatic cholinomimetic compounds:

a) The side chains have a fixed steric position on the tetrahydrofurane ring.

b) Only the cationic head is flexible with the restriction that its position is limited by steric hindrance arising from the proximity of its methyl groups to the hydrogen atoms on the tetrahydrofurane ring.

c) The two oxygen functions are separated into two isolated groups of the ether oxygen and the hydroxy group in muscarine, or the carbonyl group in muscarone. In muscarine the methylammonium group represents the choline moiety of the acetylcholine model.

We want to discuss here the activity of some new compounds. Desether-muscarine, desoxy-muscarine and desether-muscarone were synthesized by R.S. Givens, University of Kansas in Lawrence, USA (Fig. 1). They represent the racemates and were compared with DL-muscarine and DL-muscarone on isolated guinea pig ileum, and isolated frog rectus muscle in a tyrode bath at 37°C with oxygenation. We obtained complete dose-response curves for these compounds and compared their activity further in four-point assays. Acetylcholine was used as reference. Complementary experiments concerned the action on the blood pressure of anaesthetized rats.

Results

On the isolated guinea-pig ileum acetylcholine and DL-muscarine had an identical activity, but DL-desether-muscarine and DL-desoxy-muscarine were 10 times less active (Table 1). In the four-point assay this factor was even slightly bigger. Muscarone, interesting enough, was 10 times more active than acetylcholine, which was equal with DL-desether-muscarone in the ED 50 but with a slightly different inclination of the dose-response curve. Again DL-desether-muscarone was 10 times less active than DL-muscarone. On the frog rectus muscle preparation DL-muscarine and DL-desether-muscarine were completely inactive in concentrations of 10^{-5}-10^{-3}m and produced no contractions. Only DL-desoxy-muscarine produced small contractions at very high

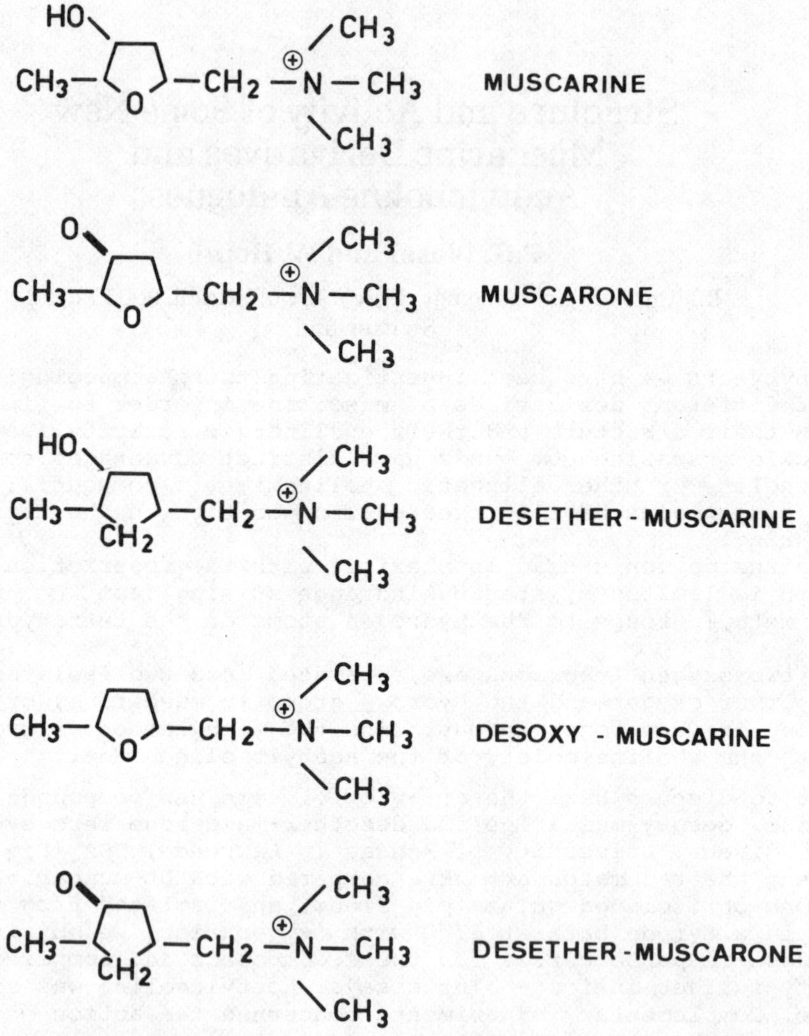

MUSCARINE

MUSCARONE

DESETHER - MUSCARINE

DESOXY - MUSCARINE

DESETHER - MUSCARONE

Fig. 1. Derivatives of Muscarine

concentrations (10^{-3}-10^{-2}m). Acetylcholine was twice as active as DL-muscarone, DL-desether-muscarone showing a similar intermediate activity.

The blood pressure of anaesthetized rats was lowered by DL-muscarine and DL-desether-muscarone in a similar degree as by acetylcholine (10^{-5}m/kg), DL-muscarone being more active, DL-desether-muscarine and DL-desoxy-muscarine being less depressive (Table 1). After atropine blood pressure was slightly increased by some derivatives as with DL-muscarine.

TABLE 1

Racemates of	Molar concentrations for ED 50 Contraction of		Equal decrease of Rat blood pressure (mol / kg)
	Guinea pig ileum (muscarinic action)	Frog rectus muscle (nicotinic action)	
Acetylcholine	6×10^{-8}	2×10^{-5}	10^{-5}
Muscarine	7×10^{-8}	$> 10^{-3}$	10^{-5}
Desethermuscarine	10^{-6}	$> 10^{-3}$	10^{-4}
Desoxymuscarine	7×10^{-7}	10^{-3} - 10^{-2}	10^{-4}
Muscarone	9×10^{-9}	4×10^{-5}	10^{-6}
Desethermuscarone	10^{-7}	$3{,}5 \times 10^{-5}$	10^{-5}
Thiomuscarine	$1{,}3 \times 10^{-5}$	$> 6 \times 10^{-3}$ (all four thiomuscarines)	
Allo-thiomuscarine	2×10^{-4}		
Epi -thiomuscarine	$3{,}5 \times 10^{-4}$		
Epiallo-thiomuscarine	$6{,}5 \times 10^{-5}$		

Thiomuscarine, another derivative of muscarine in which the furane ring is replaced by a thiophane ring, was investigated together with the racemates of its 4 stereoisomeric forms (synthetized by C.H. Eugster, K. Allner, P. Rüedi and M.Giannella, University of Zürich) (Fig. 2). DL-thiomuscarine was 530 times less active contracting the guinea pig ileum than DL-muscarine, similar to what we had already described for its muscarinic activity in cat and frog (Waser, 1961). The stereoisomers were even less active (1500-7500 times) with only little difference between them. The allo transposition of the methyl and methylammonium groups on the ring seems to be more favorable for cholinergic activity than the epi position of the hydroxy group. On frog rectus muscle all 4 racemates were inactive to a concentration of 6×10^{-3}m, equal to the nicotinic inactivity of other isomers of muscarine.

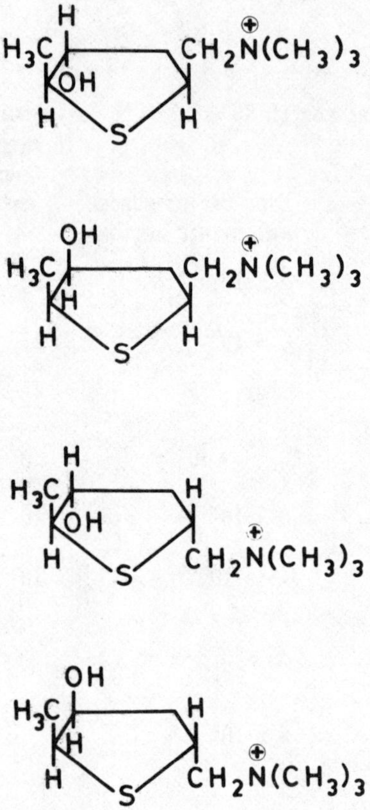

THIOMUSCARINE

EPI-THIOMUSCARINE

ALLO-THIOMUSCARINE

EPI-ALLO-THIOMUSCARINE

Fig. 2. Stereoisomers of Thiomuscarine

Pure acetylcholinesterase isolated from the electric organs of Torpedo marmorata was inactivated by all muscarine and muscarone compounds and thiomuscarine isomers only at relatively high concentrations $>10^{-5}$m (Table 2). There exists a parallel inhibition especially for the investigated stereoisomers of thiomuscarine at high concentrations (10^{-4}-10^{-2}m). This leads to the assumption that the enzymatic center engaged in hydrolizing acetylcholine is not directly and specifically blocked, but indirectly changed over an allosteric mechanism, as the muscarine derivatives, and the thiomuscarines are no substrate of the enzyme. DL-muscarone acts in the same manner,

TABLE 2

INHIBITION OR STIMULATION OF ACETYLCHOLINESTERASE (EC 3. 1. 1. 7)

RACEMATES OF	10^{-5}	10^{-4}	$2x10^{-4}$	10^{-3}	$5x10^{-2}$	10^{-2} M
MUSCARINE		105 %	108 %	116 %	100 %	
MUSCARONE		82 %	59 %			
DESOXYMUSCARINE	96 %			81 %		
DESETHERMUSCARINE				76 %		
THIOMUSCARINE		79 %	64 %			27 %
EPI - THIOMUSCARINE		73 %	57 %			23 %
ALLO - THIOMUSCARINE		66 %	52 %			20 %
EPIALLO - THIOMUSCARINE		74 %	44 %			17 %

but DL-muscarine produces at the high concentration of 10^{-3}m a surprising increase (+ 15%, $p < 0.01$) of the hydrolytic activity. This is at the moment hard to explain, but as there cannot be any interference with the active centre there might be another indirect allosteric interaction, as the enzyme is in solution without stabilizing phospholipids.

Beside the musccarine derivatives we have been investigating some <u>other acetylcholine analogues</u> (Fig. 3) which have a rigid structure in the choline part of the molecule (synthetized by J. Büchi and M.S. Malik, ETH, Zürich). Either the α or the β-C-atom or both C-atoms of the choline are part of a phenyl ring and therefore the substituants are in a sterically fixed position. The degree of rigiditiy of the molecule is different for these 3 different structure types including at most the complete choline molecule or one of its atoms. The kationic ammonium heads are formed either by alkyl groups or N-methyl pyrimidine.

None of these compounds showed any muscarinic activity on the guinea pig ileum or nicotinic contraction of frog rectus muscle up to concentrations of 10^{-2}m. But they show an atropine-like antagonism against contractions induced by acetylcholine in both preparations.

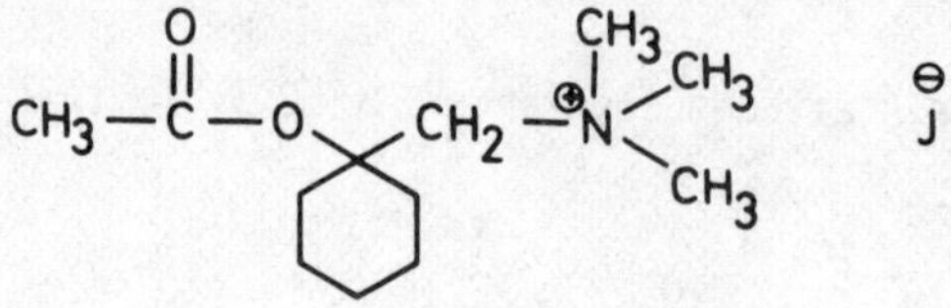

W 1

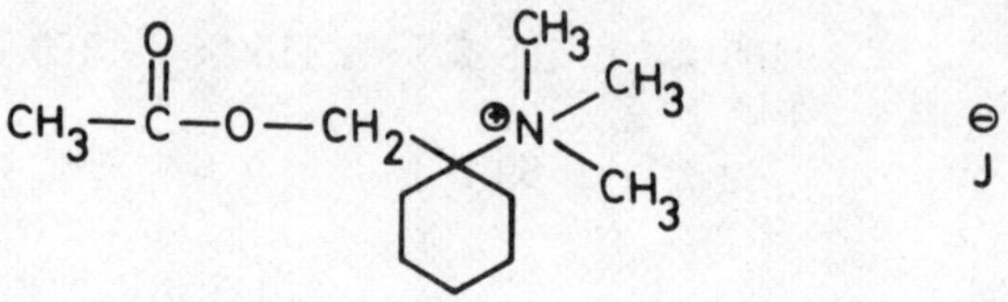

W 2

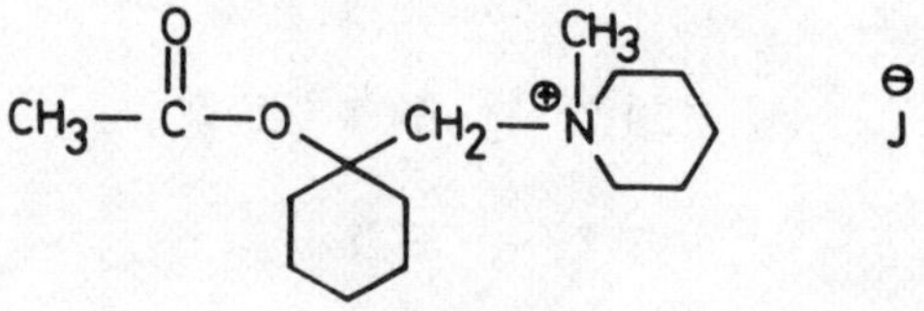

W 3

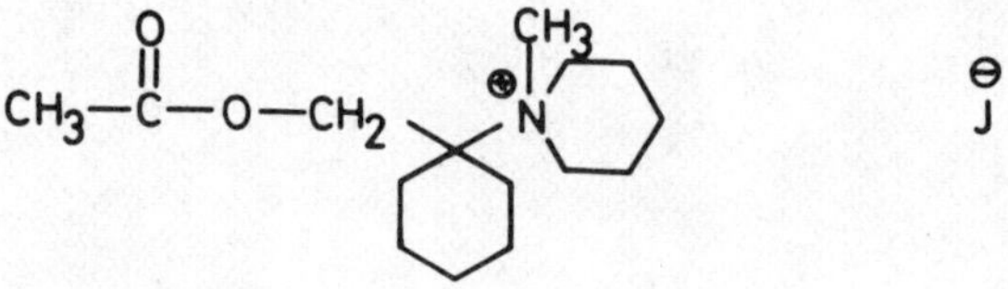

W 4

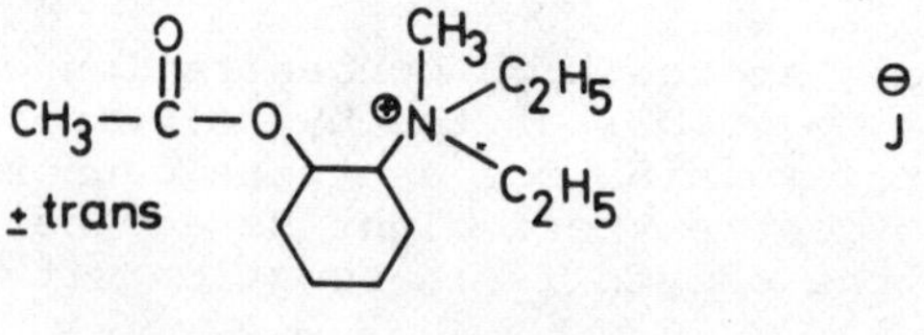

W 5

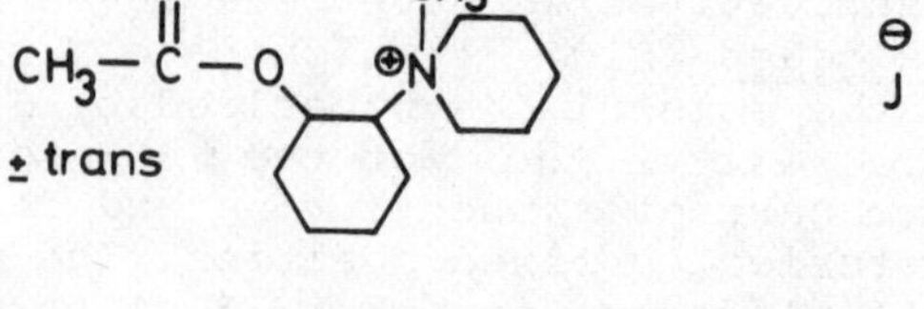

W 6

Fig. 3. Cyclohexyl analogs of Acetylcholine

This action is not prominent (50% antagonism with 10^{-5}-10^{-4}m compared to atropine 10^{-8}m!) and limited mainly to 3 compounds (W3, W4, W6) with large pyridinium heads. The blood pressure of rats reacted only at very high concentrations (10^{-2}m/kg) with a decrease comparable to 10^{-5}m/kg acetylcholine which got smaller after atropine (0,5 mg) blocking all vascular cholinergic action sites. This cardiovascular effect probably is unspecific and caused by direct smooth muscle spasmolysis but not a muscarinic action. Pure acetylcholinesterase from Torpedo marmorata and unspecific serum-cholinesterase were blocked by high concentration (4.10^{-5}-4.10^{-4}m) of these derivatives, especially when the β-C-atom of the choline was part of a substituting phenylring (W1, W3). Selfhydrolysis of some compounds was very small and unimportant.

Discussion

The first derivatives of muscarine, in which the ether oxygen function was exchanged, are D,L-thio-muscarine and D,L-thio-muscarone (Waser, 1960). The tetrahydrofurane ring of these compounds was replaced by tetrahydrothiophene. The drastically reduced biological activity (in cats and frogs) of thio-muscarine (1/5000) and thio-muscarone (1/40) was explained by the inability of the less electronegative sulfur atom to form hydrogen bonds to the cholinergic receptor, and the increase in ring size influencing the fit of all functional groups on the receptor area. The cyclo-pentane ring is comparable in size to that of tetrohydrofurane, but the desether-compounds are unable to form electrostatic or hydrogen bonds to the CH_2-groups (Givens and Rademacher, 1974; Cingolani, Gamba, Pigini, Re, Rossini, Giannella,Gualtieri, Melchiorre and Pigini, 1974). Nevertheless their muscarinic activity is remarkable and at most only 10 times lower compared to the parent compounds and the nicotinic activity of desether-muscarone is not changed.

One might conclude from this investigation, that the ether-oxygen is of minor importance for cholinergic action, and that the hydroxy or carbonyl groups and the quaternary nitrogen are mainly responsible for the binding. However it is conceivable that the electrostatic interactions of the ether-oxygen may be replaced by Van der Waals forces on the binding of the desether-compounds, as long as these compounds fit the receptor site well. This might especially explain the small difference in nicotinic activity of the muscarone compounds, in which the planar carbonyl group forms hydrogen bonds to the receptor instead of the ether-oxygen.

The importance of the sterical position of the side chains can be discussed only in regard to the racemates which were investigated. All stereoisomers of thiomuscarine were considerably less active on the guinea pig ileum than muscarine, but of a similar activity as thiomuscarine. The relative order of activity was decreasing from the natural to the epiallo, allo- and epi-stereoisomer, exactly as in the stereoisomer series of muscarine. The epi-form is always the most unfavorable, the epiallo-form the nearest to the natural muscarine activity.

Probably only a few conformations have to be considered for the molecule in solution or in a biological system, in addition to the well defined crystal structure of Jellinek (1957). As no indication exists that the biologically active form of a molecule acting on the receptor is the same as the rigid form adopted in the crystal state, we think it is more elucidating to investigate and discuss a restricted number of conformations, which seem to us nearer to reality.

This point is strengthened by many attempts of different research groups, as ours, to freeze the rest of the cholinergic molecule by fixing the cationic head to the furane ring, or by using other ring systems, which contain the functional side groups of the cholinergic chain in a rigid form. The result of all our investigations was a dramatic loss of cholinergic activity and in many cases even a change to weak antagonistic activity, as demonstrated by our cyclo-hexane-derivatives in which the choline part of the molecule is involved. In comparison to this the muscarine derivatives have mobile cationic ammonium groups, but the acetic ester part is incorporated and frozen in the furane ring. This points to the importance of a freely moving methylammonium group for cholinergic activity.

Obviously what we observe in case of a completely rigid structure, as in a crystal, is not what happens really on the living membrane. The change in a living membrane is induced by the interaction of two structures, agonist and biopolymer, probably both of them mobile, causing the highest state of activation. Muscarine or muscarone (Fig. 1), both molecules being only partly rigid, but flexible in the choline-like side chain, produce highest muscarinic and nicotinic activity. For this effect mainly the three pharmophore groups, especially in muscarine (cationic nitrogen head, ether oxygen, hydroxy group) are responsible. In addition, since of all possible stereoisomers only L (+) muscarine is fully active, roughly 1000 times more then its antipode D (-) muscarine, the steric position of these is essential. Opposed to this, the stereospecifity of muscarone is quite low, only three-fold, and the D(-) isomer is more potent (Waser, 1958). This intriguing and paradoxical finding, which was reproduced by other investigators (Gyermek and Unna, 1958), is not yet explained.

Conclusions

The exchange of an active ether-oxygen in muscarine and muscarone to an inactive CH_2-group has influenced their biological activity less than expected. Obviously the desether-compounds are able to adjust themselves on the receptor area in such a way, that the final stimulus is only little influenced. The exchange of the ether-oxygen in muscarine to a sulfur atom diminishes the muscarinic activity of all stereoisomers markedly. Some new cyclohexyl derivatives of acetylcholine, with immobilized choline chain, have no cholinergic but anticholinergic properties. The free moving methylammonium group permitting allosteric changes of the binding receptor protein seems to be essential for cholinergic activity. A rigid structure is not able to perform this way. The hydrolytic action of acetylcholinesterase is blocked only in high concentrations of all derivatives, and muscarine even produces a stimulation of the enzyme. The compounds are not associated to the active center of the enzyme, but interfere indirectly.

Acknowledgments

This investigation was sponsered by the Swiss National Foundation, project No. 3.086-0.76. We thank also Mr. CH. Spiess and Mr. G. Engler for their excellent technical help.

<u>References</u>

Cingolani M., Gamba G., Pigini P., Re L., Rossini L., Giannella M., Gualtieri F., Melchiorre C. and Pigini M. Biological characterization of some cyclopentane analogues of muscarone and muscarine. <u>Brit.J. of Pharmacol</u>. 469P-470P (1974).

Givens R.S. and Rademacher D.R. Further studies on carbocyclic analogs of muscarine. Oxidation of desether-muscarine to desether-muscarone. <u>J. Medicinal Chem</u>. 17, 457-459 (1974).

Gyermek C. and Unna K.R. Relation of structure of synthetic muscarines and muscarones to their pharmacological action. <u>Proc. Soc. exptl. Biol.</u>, New York 98, 882-885 (1958).

Jellinek F. The structure of muscarine. <u>Acta crystallographica</u>, Cambridge 10, 277-280 (1957).

Waser P.G. Struktur und Wirkung des Muscarins, des Muscarons und ihrer Stereoisomeren. <u>Experientia</u> 14, 356-358 (1958).

Waser P.G. Struktur und Wirkung von muscarinähnlichen Verbindungen. <u>Experientia</u> 16, 347-349 (1960).

Waser P.G. Chemistry and pharmacology of muscarine, muscarone and some related compounds. <u>Pharmacological Reviews</u>, 13, 465-515 (1961).

The Conformation and Flexibility of Cholinergic Molecules

N.V. Khromov-Borisov
Institute for Experimental Medicine, Leningrad, USSR

In an attempt to estimate the "active conformation" of various cholinergic molecules attention was drawn to "rigid conformers" with fixed key fragments. The transformation of a flexible molecule into a rigid form usually requires the introduction of an additional group in its structure. However, it must be taken into account that this group itself may prevent or promote the interaction of the molecule with a cholinoreceptor (ChR). For example, cis and trans isomers of cyclopropane derivatives I and II (Fig. 1) were considered in several papers (Ref. 1, 2) as rigid models of acetylcholine (ACh).

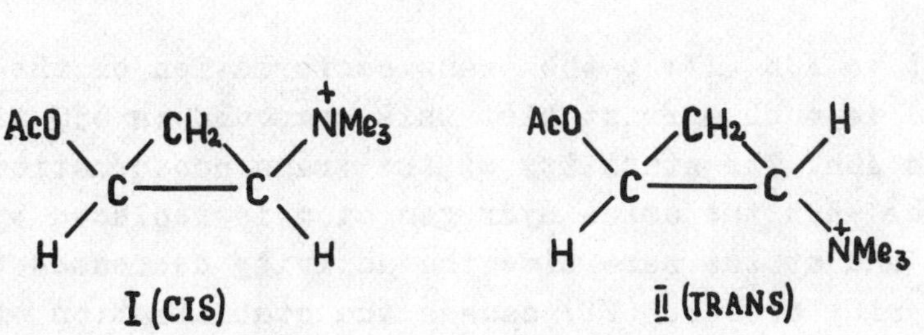

AcO, CH_2, $\overset{+}{N}Me_3$, C, C, H, H — I (CIS)

AcO, CH_2, H, C, C, H, $\overset{+}{N}Me_3$ — II (TRANS)

Fig. 1. Cyclopropane derivatives of ACh

These models can serve as muscarinomimetic but not nicotinomimetic agents because the CH_2-group in cyclopropane ring replaces two hydrogen atoms in α- and β-position of ACh, and it is known that acetyl-ß-methylcholine is almost devoid of nicotinomimetic activity.

In order to estimate the conformation of cholinergic ligands responsible for nicotinic activity we synthesized and tested a series of nitrogen derivatives of ACh (see Table 1).

It is known that ACh itself is able to adopt either the gauche or the trans conformation of the key fragment. Calculations showed that the gauche conformation of ACh is slightly more preferable.

TABLE 1. Nitrogen derivatives of ACh. Conformation and nicotinic activity (Frog, rectus abdominis m.)

			α	EC50 (M/l)
III	$CH_3CO{-}O{-}CH_2{-}CH_2{-}\overset{+}{N}(CH_3)_3$	GAUCHE	0.98	1.3×10^{-7}
IV	$CH_3CO{-}N(H){-}CH_2{-}CH_2{-}\overset{+}{N}(CH_3)_3$	TRANS	0.98	1.3×10^{-4}
V	$CH_3CO{-}N(CH_3){-}CH_2{-}CH_2{-}\overset{+}{N}(CH_3)_3$	TRANS	0.97	4.1×10^{-4}
				×368
VI	$CH_3CO{-}N(CH_2{-}CH_2)_2\overset{+}{N}(CH_3)_2$	GAUCHE	0.96	1.1×10^{-6}

In contrast to ACh (III), the trans conformation of the amide derivative (IV) is much more stable. This compound is 650 times less active than ACh. The stability of the trans conformation becomes still higher when the amide hydrogen atom is replaced by a methyl group (V), and at the same time the activity decreases threefold; piperazine ring closure (VI) causes the stabilisation of the gauche conformer (piperazine ring is virtually fixed in the chair form) and simultaneously leads to a very significant increase in the nicotinic activity (368 times).

It should be noted that the transformation of the methylamide derivative (V) into piperazine (VI) is carried out merely by the introduction of a chemical bond in the molecule; the α- and ß-positions remain unaffected. Therefore we can suppose that the increase in the activity is due only to the conformational change. These results suggest that gauche is the active conformation in the interaction of ACh and its analogues with the ChR of skeletal muscles.

To verify this suggestion, a comparative investigation of the corresponding di-cations with the interonium distance of 20 Å was carried out. The results are shown in Tables 2 and 3.

TABLE 2. Piperazine ring closure in di-cations (Frog, rectus abdominis m.)

		EC_{50} (M/l)
$Me_3\overset{+}{N}-CH_2-CH_2-O-CO-(CH_2)_8-CO-O-CH_2-CH_2-\overset{+}{N}Me_3$	GAUCHE	5.3×10^{-8}
$Me_3\overset{+}{N}-CH_2-CH_2-NH-CO-(CH_2)_8-CO-NH-CH_2-CH_2-\overset{+}{N}Me_3$	TRANS	8.5×10^{-5}
$Me_2\overset{+}{N}(CH_2-CH_2)_2N-CO-(CH_2)_8-CO-N(CH_2-CH_2)_2\overset{+}{N}Me_2$	GAUCHE	3.5×10^{-8}

TABLE 3. Piperazine ring closure in di-cations (Cat, m. tibialis)

		EC_{50} (mcM/kg)
$Me_3\overset{+}{N}-CH_2-CH_2-NH-CO-(CH_2)_8-CO-NH-CH_2-CH_2\overset{+}{N}Me_3$	TRANS	0.15
$Me_2\overset{+}{N}(CH_2-CH_2)_2N-CO-(CH_2)_8-CO-N(CH_2-CH_2)_2\overset{+}{N}Me_2$	GAUCHE	0.006
$Me_3\overset{+}{N}-CH_2-CH_2-NH-SO_2-C_6H_4-C_6H_4-SO_2-NH-CH_2-CH_2-\overset{+}{N}Me_3$	TRANS	0.05
$Me_2\overset{+}{N}(CH_2-CH_2)_2N-SO_2-C_6H_4-C_6H_4-SO_2-N(CH_2-CH_2)_2\overset{+}{N}Me_2$	GAUCHE	0.04

In accordance with our assumptions the results indicate that the closure of piperazine ring in di-cations leads to a considerable increase in the cholinomimetic activity. These findings confirm the better steric fit of the gauche conformation of the key fragment of nicotinomimetic agents with respect to ChR.

According to our hypothesis the above mentioned di-cations react with two anionic sites of the tetrameric ChR, along the diagonal of a square (Fig. 2). (Ref. 3).

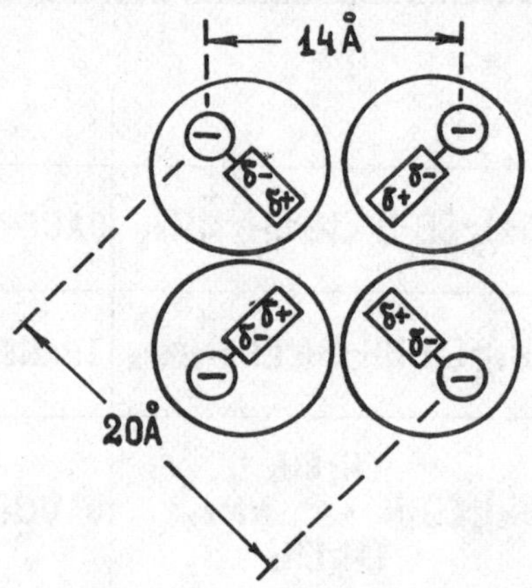

Fig. 2. Scheme of tetrameric ChR.

The tetrameric scheme of ChR of skeletal muscles was developed on the basis of pharmaco-chemical investigations. Later De Robertis (4) has proposed a tetrameric model for the cholinergic receptor which gives a logical explanation for the results obtained in the study of artificial membranes.

The mutual arrangement of anionic sites in Fig. 2 refers to the conformation of free ChR. At present there is no doubt that the primary effect of the interaction of ChR with cholinomimetic agents is the conformational change in the receptor macromolecule.

As a result of the investigation of the type of action of various di-cations (depolarizing and non-depolarizing) we came to the conclusion that when a complex of ChR with ACh or cholinomimetics is formed, the distances between the anionic sites in the activated ChR become shorter (Ref. 5).

It is interesting that Karlin and co-workers (6) have come to the same conclusion on the basis of the investigation of cholinergic alkylating agents. These authors write: "These results suggest that in the active state the acetylcholine-binding-site is in a shortened conformation relative to its conformation in the inactive state".

We may say that ACh and cholinomimetics act as effectors of the conformational change in ChR and hence, they stimulate the depolarization of the postsynaptic membrane.

The investigation of the type of action of various di-cations led us to the conclusion that when the activation of the tetrameric ChR takes place, the distance between the anionic centres decreases as is shown on Fig. 3: the sides of the square become 8.5 Å long and, hence, the diagonales become 12 Å long. Therefore, the degree

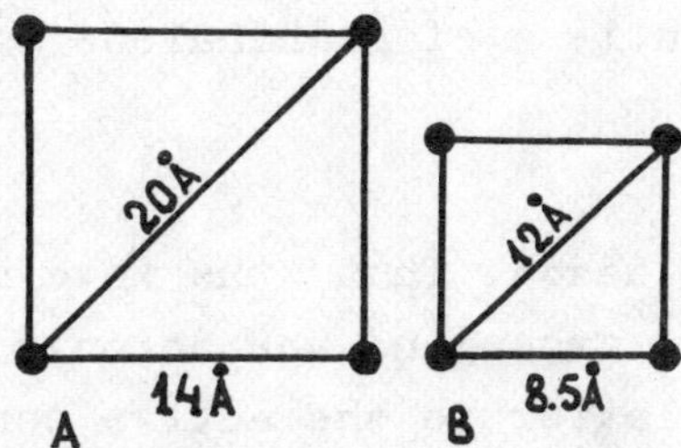

Fig. 3. Free (A) and activated (B) ChR

of flexibility of di-cations should affect their type of action. Depolarizing di-cations with the maximum N-N distance of 14 Å should be able to adopt the conformation with the N-N distance of 8.5 Å. For di-cations with the N-N distance of 20 Å the conformation with the N-N distance of 12 Å should be possible. Rigid structure is characteristic of di-cations of the non-depolarizing type of action (Ref. 7). This criterion was tested by considering several examples. One of them is shown in Fig. 4.

The di-cation with an absolutely rigid structure is a myorelaxant of the non-depolarizing type of action.

When benzene rings are replaced by flexible tetramethylene chains (lower formula), the action of the myorelaxant becomes depolarizing.

It has been shown previously that the pharmacological properties of cholinergic cations depend to a considerable extent on the struc-

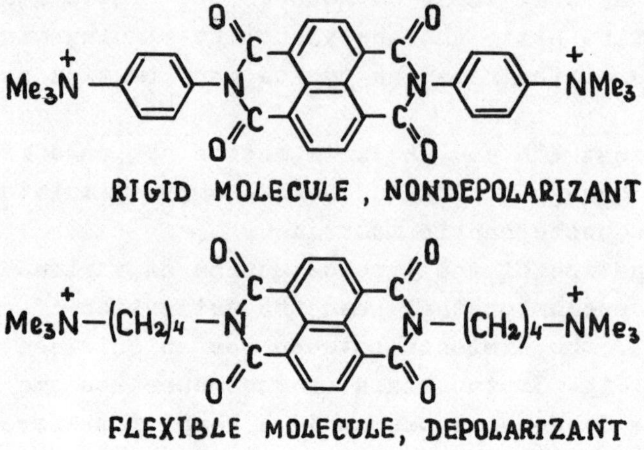

$$Me_3\overset{+}{N}-C_6H_4-N(CO)_2C_{10}H_4(CO)_2N-C_6H_4-\overset{+}{N}Me_3$$

RIGID MOLECULE, NONDEPOLARIZANT

$$Me_3\overset{+}{N}-(CH_2)_4-N(CO)_2C_{10}H_4(CO)_2N-(CH_2)_4-\overset{+}{N}Me_3$$

FLEXIBLE MOLECULE, DEPOLARIZANT

Fig. 4. Derivatives of naphthalene tetracarboxylic acid.

ture of their cationic heads. Thus, the intermolecular contacts formed between the $\overset{+}{N}Me_3$ group and the anionic site of ChR stimulate the transition of the latter to the active conformation. Such contacts may be named stimulating or shortly S-contacts. A cationic head like $\overset{+}{N}Et_3$ is able to form intermolecular bonds (cross-links) which can inhibit the conformational transformation of the drug-receptor complex. We name such contacts inhibiting or I-contacts.

The above suggestion of the geometrical parameters of free and activated ChR permits the formulation of the following requirements for non-depolarizing di-cations: 1. Optimum N-N distance (14 Å or 20 Å), 2. Rigid structure of the molecule, and 3. The ability of cationic heads to form I-contacts.

This criterion was confirmed by various examples. The results obtained support our suggestion. On the other hand, we make use of this working hypothesis for the design and synthesis of curariform drugs. Taking into account the above requirements we synthesized a myorelaxant which represent a bis-triethylammonium derivative of p-terphenyl which was named Tercuronium (see Table 4).

The structure of Tercuronium is rigid, the N-N distance is 14 Å and the cationic heads are capable of forming I-contacts.

Tercuronium was tested as a muscle relaxant in surgical operations. It is a non-depolarizing muscle relaxant of high potency,

TABLE 4. Curare-like activity of rigid di-cations (Rabbit, HDD).

	N-N DISTANS	HDD (mcM/kg)
$Et_3\overset{+}{N}$–C₆H₁₀–C₆H₄–C₆H₄–$\overset{+}{N}Et_3$ TERCURONIUM	14.2Å	0.022
$Me_3\overset{+}{N}$–C₆H₄–C₆H₄–C₆H₄–$\overset{+}{N}Me_3$	14.2Å	0.25
$Et_3\overset{+}{N}$–C₆H₄–C₆H₄–$\overset{+}{N}Et_3$	10.0Å	6.0
$Me_3\overset{+}{N}$–C₆H₄–C₆H₄–$\overset{+}{N}Me_3$	10.0Å	16.0

free from the disadvantages inherent in the known relaxants. It is completely safe in clinical use. It exhibits highly selective and specific activity without affecting other physiological systems. It is possible to use Tercuronium for intubation of the trachea. The blocking effect of Tercuronium is 8 times higher than that of d-Tubocurarine and only slightly lower than that of Pancuronium. However, Pancuronium causes a stable rise in the arterial pressure, whereas with a 10-fold dose of Tercuronium the blood pressure was only slightly reduced. Recently it was patented in several countries.

For checking up, a corresponding di-cation with trimethylammonium cationic heads (absence of I-contacts) and two analogous di-cations with the N-N distance of 10 Å were also synthesized. These examples confirmed the expected relationships: the bis-trimethylammonium analogue of Tercuronium was found to be 10 times less active. The corresponding biphenyl derivatives were found to be even less active (up to hundred-fold less).

For further pharmaco-chemical checking of our suggestions on the molecular mechanism of the myoparalytic action of di-cations, we synthesized rigid di-anions exhibiting geometrical and charge complementarity with respect to rigid di-cationic myorelaxants.

Figure 5 shows a highly active myorelaxant Ritetronium and its di-anionic analogue with sulpho groups replacing the cationic heads.

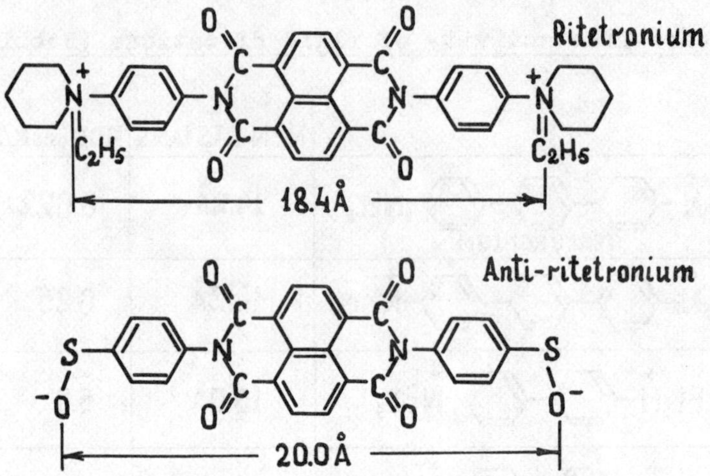

Fig. 5. Di-cation Ritetronium and di-anion Anti-ritetronium.

Investigations in vitro showed that when the sodium salt of this di-anion interacts with Ritetronium, a stable salt-like complex is formed without any myoparalytic action. Experiments on cats showed that the disulpho sodium salt of very low toxicity exhibits the anti-curare action: it rapidly stops the muscle block caused by Ritetronium

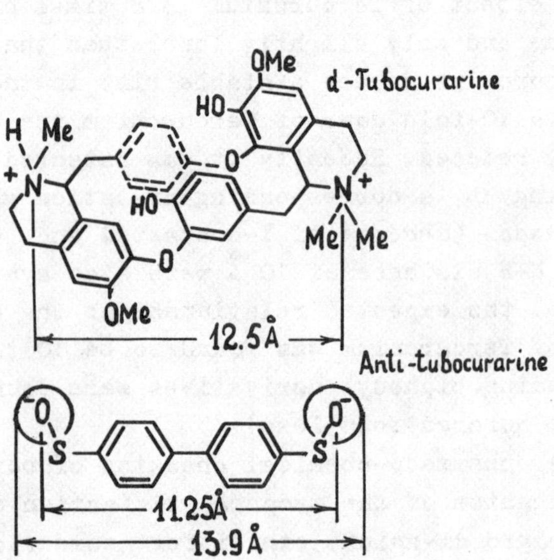

Fig. 6. Di-cation Tubocurarine and di-anion Anti-tubocurarine.

Fugure 6 shows another anti-curariform di-anion complementary with respect to d-Tubocurarine. It rapidly stops the muscle relaxation caused by d-Tubocurarine or other rigid di-cations for which the N-N distance is close to the inter-anion distance of this disulpho derivative. The Figure also shows the permissible limits of geometrical complementarity calculated with regard for the ionic radii of oxygen atoms carrying negative charges.

In experiments on cats d-Tubocurarine in a dose of 0.25 mg/kg causes complete muscle block which continues for 23 $\pm$ 2 min.

If anti-tubocurarine (120 mg/kg) is introduced, the first muscle contractions are observed after 30-40 sec. and after 1.3 $\pm$0.4 min. the effect of d-Tubocurarine completely disappears.

Further investigations in this field are in progress.

REFERENCES

(1) P.D.Armstrong, J.G.Cannon, Small ring analogs of acetylcholine. Synthesis and absolute configurations of cyclopropane derivatives, Journ. Med. Chem. 13, 1037 (1970)

(2) C.Chothia, P.Pauling, Absolute configuration of cholinergic molecules; the christal structure of (+)-trans-2-acetoxy cyclopropyl trimethylammonium jodide, Nature 226, 65 (1970)

(3) N.V.Khromov-Borisov, M.J.Michelson, The mutual disposition of cholinoreceptors of locomotor muscles, and the change in their disposition in the course of evolution, Pharmacol. Reviews, 18, 1051 (1966)

(4) E.De Robertis, Molecular biology of synaptic receptors, Science (Wash. D. C.) 171, 963 (1971)

(5) N.V.Khromov-Borisov, In: Ergebnisse der experimentellen Medizin, Bd.17, T. 1, Berlin, 1974, Pharmacochemical investigations on the neuromuscular synapse, pp.117-125.

(6) A.Karlin, D.A.Cowburn, M.J.Reiter, In: Drug receptors, a Symposium ed. H.P.Rang, University Park Press, Baltimore, 1973, Molecular properties of the acetylcholine receptors, pp.193-209.

(7) N.V.Khromov-Borisov, The significance of conformations in the interaction of biological active molecules with receptors, Chim.-Pharmaceut. Journ. (in Russ.) 10, No.11, 5 (1976).

An NMR Study of the Conformations of Atropine and Scopolamine Cations in Aqueous Solution

J. Feeney, E.A. Piper and R. Foster

National Institute for Medical Research, Mill Hill, London NW7 1AA

ABSTRACT

Nmr data on atropine and scopolamine cations in aqueous solution have been used to deduce the conformations of these species in solution and to provide information about multiple conformations and molecular flexibility.

INTRODUCTION

During the last few years there have been several attempts to correlate the activities of muscarinic agonists of acetylcholine with features of their three dimensional structure (1,2,3). Following on the early work of Pfeiffer (4) and Beckett (5) who had recognised the structural homologies between drugs active at parasympathetic sites, Beers and Reich (1) pointed out that most acetylcholine muscarinic agonists contain a charged quaternary nitrogen atom and a group containing a lone pair which can take up equivalent positions to those of the charged nitrogen and the ester oxygen in acetylcholine. They envisaged the lone pair on the oxygen being involved in the formation of a hydrogen bond approximately 4.4 Å from the centre of the positive charge. It was also noted that the presence of an alkyl group corresponding in position to the acetoxy methyl group in acetyl choline also increases the agonist interaction with the binding sites. Baker and coworkers (2) using detailed conformational information from X-ray and NMR studies have made similar correlations. Although they could find no single conformation which described all the known muscarinic agonists (2,6) they were able to put limits on the range of torsion angles which are found in the potent agonists. They noted that certain agonists would fit the general rules outlined above if their conformations on binding to the receptor are somewhat modified from those observed in the crystal structure. They also pointed out that not only must the essential binding groups of the agonist be able to take up their correct relative positions for optimum binding but the remaining parts of the molecule must not sterically interfere with the agonist-receptor interactions (2,3). The powerful muscarinic antagonists atropine (I) and scopolamine (II) also have a charged nitrogen and an ester oxygen approximately in the same positions as those in the agonists. While competitive antagonists would need to retain some of the structural features required for strong binding to the muscarinic sites they must possess additional structural features which lead to optimal antagonist activity. To understand fully the factors controlling antagonist activity one would need to have information about the conformations of agonists and antagonists in their free state and when bound to the receptor and also information about the conformational changes induced in the receptor by the binding of agonists and antagonists. At the present time it is not possible to measure the conformations of agonists and antagonists in the bound state but X-ray and NMR methods can be used to give such information about the free species. This is of only limited value in the absence of information about the bound species but it will be required eventually to determine the energetics of the interaction with the receptor.

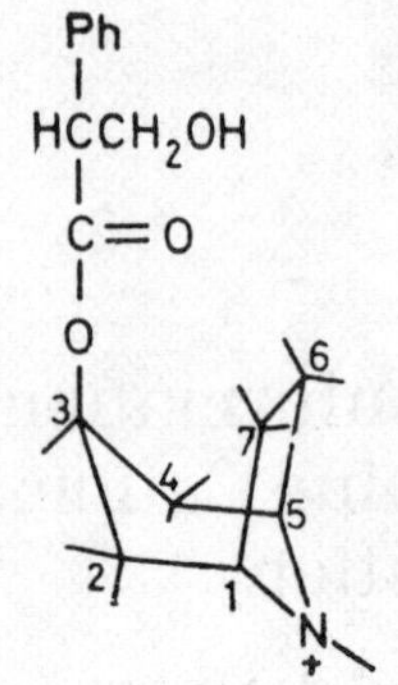

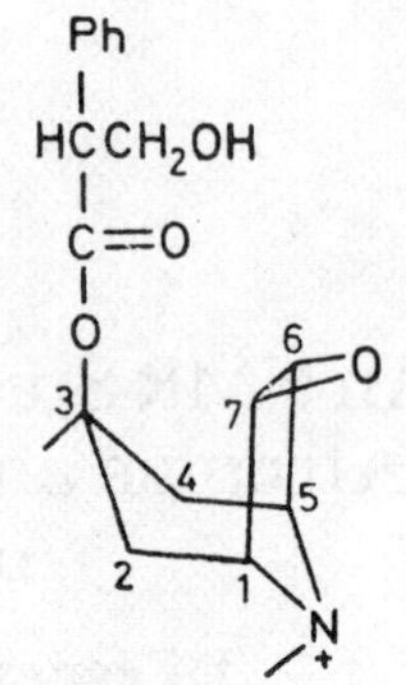

The X-ray structures of atropine and scopolamine salts have already been determined by Pauling and Petcher (6). In this paper we describe the use of NMR measurements to establish their conformations in solution. The NMR method usually does not give the detailed overall conformational information obtainable from X-ray studies on crystals but it does have the advantages that it provides the conformation in the solution state and can sometimes give additional information about the presence of molecular flexibility and multiple conformations. Such information might be useful in helping us to understand how the antagonists bind to receptors.

MATERIALS AND METHODS

Scopolamine hydrobromide (BDH Ltd.) and atropine sulphate (Sigma Chemicals Ltd.) were kindly provided by Dr. P.J. Pauling and were used without further purification. The ^{1}H nmr spectra were obtained at 100, 220 and 270 MHz using Varian (HA 100, HR 220), Perkin Elmer (R34) and Bruker (WH 270) spectrometers, and the ^{13}C spectra at 25.2 MHz using a Varian XL-100 equipped with proton noise decoupling facilities. The ^{13}C spectra and some of the proton spectra were recorded using the Fourier transform mode of operation.

The compounds were examined as 10% w/w solutions in D_2O using DSS (sodium 4,4-dimethyl-4-silapentane-1-sulphonate) as an internal reference for the ^{1}H spectra and dioxane as a reference material for the ^{13}C studies.

Some ^{1}H nmr measurements were also made at lower concentrations (5 mM and 50 mM) to assess the effects of intermolecular interactions on the chemical shifts. The effects on the shielding differences for symmetrical pairs of protons in the tropane ring were negligible.

RESULTS

^{1}H and ^{13}C NMR Studies

In the 270 MHz ^{1}H nmr spectra of the cations of atropine and scopolamine all the tropane ring protons have different chemical shifts and are thus magnetically non-equivalent. This is illustrated in the ^{1}H spectrum of scopolamine hydrobromide shown in Fig.1. The protons on the (C1, C2, C3) and (C3, C4, C5) fragments give separate ABMX type spectra from which we have calculated the chemical shifts and coupling constants shown in Tables 1 and 2. As noted previously (7-9), the non-equivalent 6 and 7 protons in the epoxide ring of the scopolamine cation give rise to two AB doublets (J = 3.6 Hz) broadened by a small coupling to the adjacent protons $J_{56} \approx 1$ Hz).

The ^{1}H spectrum of the atropine cation is much more complicated and it was not possible to carry out a detailed analysis of all of the tropane ring proton signals; however, the two ABMX spin systems for protons on (C1, C2, C3) and (C3, C4, C5) could be analysed (see Table 2). The shifts of the signals from the remaining ring protons (ABCD system) were estimated from the centres of their multiplets. The coupling constants measured for the atropine cation are tabulated in Table 2 and are seen to be similar to the corresponding coupling constants in scopolamine hydrobromide.

The spectra of the CH_2CH fragments in both molecules were analysed as ABC systems and gave chemical shifts and coupling constants in good agreement with each other and also with those observed previously for tropic acid (10).

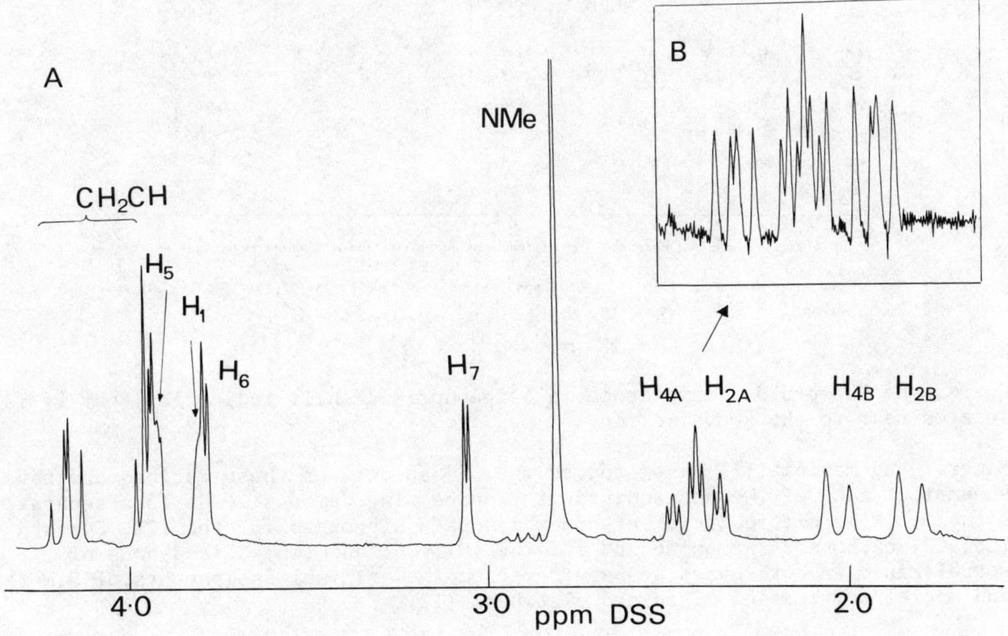

Fig.1. Part of the 270 MHz ^{1}H spectrum of scopolamine hydrobromide in D_2O solution

TABLE 1 The ^{1}H and ^{13}C Chemical Shifts in the tropane rings of the atropine and scopolamine cations

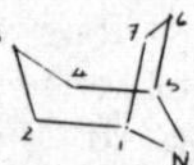

	ATROPINE				SCOPOLAMINE	
	N-Me equatorial		N-Me axial		N-Me axial	
	^{1}H chemical shift	Differences	^{1}H chemical shift	Differences	^{1}H chemical shift	Difference
H1	3.71	0.12			3.82	0.12
H5	3.83				3.94	
H2	2.28	0.06			2.40	0.07
	1.88				1.82	
H4	2.34	0.21			2.47	0.22
	2.09				2.04	
H3	5.04				5.05	
H7	1.54	0.53			3.07	0.73
	1.95					
H6	2.07	0.13			3.80	
	2.08					
	^{13}C chemical shift	Differences	^{13}C chemical shift	Differences	^{13}C chemical shift	Differences
C1	62.94	~ 0.10	59.37	~ 0	58.13	0.14
C5	62.94		59.37		58.27	
C2	35.04	0.14	28.56	~ 0.1	25.20	0.15
C4	35.18		28.56		25.20	
C6	24.01	0.25	26.02	0.24	53.92	0.35
C7	23.76		25.78		53.57	
C3	66.33				64.68	
NCH_3	39.33		32.4		25.72	

^{1}H shifts referenced to DSS. ^{13}C shifts were measured from a dioxane reference and transferred to a TMS reference scale.

Errors in ^{1}H chemical shifts $\pm$ 0.01 ppm except for the H6 and H7 of atropine ($\pm$ 0.05 ppm)

Errors in ^{13}C chemical shifts $\pm$ 0.03 ppm.

One of the CH_2 protons experiences a large upfield shift indicating that it is located near to the aromatic ring.

Simeral and Maciel (11) have studied the ^{13}C spectra of these cations and have shown that each of the non-equivalent tropane ring carbons leads to a separate ^{13}C signal. It is difficult to make unambiguous assignments for the (C2, C4) and (C6, C7) carbons of atropine and for the (C1, C5) and (C6, C7) carbons of scopolamine: in both cases we have reversed the original assignments of Simeral and Maciel for reasons discussed elsewhere (12).

The (C2', C6') and (C3', C5') pairs of carbon nuclei each give rise to only a single carbon resonance in both species: rapid flipping of the aromatic ring sbout the $\emptyset_2$ torsion angle would lead to equivalence within these pairs of nuclei.

TABLE 2 The ^1H-^1H spin coupling constants for the tropane ring nuclei in the atropine and scopolamine cations

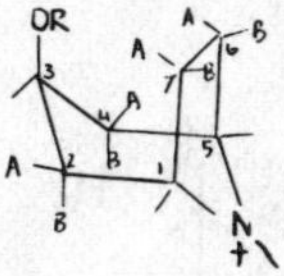

Coupling Constant Hz	Scopolamine	Atropine
J_{32A}	1.0	< 2
J_{32B}	3.8	4.4
J_{2A2B}	17.3	16.8
J_{12B}	4.8	4.2
J_{54B}	4.8	4.7
J_{12A}	1.3	-
J_{56A}	< 2.0	-
J_{6A7A}	3.6	-
J_{17A}	< 2.0	-

Errors: Scopolamine ± 0.1 Hz Atropine ± 0.2 Hz

Conformation of the (Ph) (COOR)CH-CH_2OH Moiety

We have measured the J_{AC} and J_{BC} vicinal coupling constants for the side-chain fragment (III) of both molecules in order to determine the conformation about the $C_\alpha C_\beta$ bond. The coupling constants are found to be the same for both molecules (J_{AC} = 6.5, J_{BC} = 8.2 Hz) and also very similar to the values reported previously for tropic acid (10) (5.55, 8.88 Hz) indicating that all the molecules have similar side-chain conformations. Consideration of a Karplus relationship between the coupling constants and torsion angles indicates that these observed

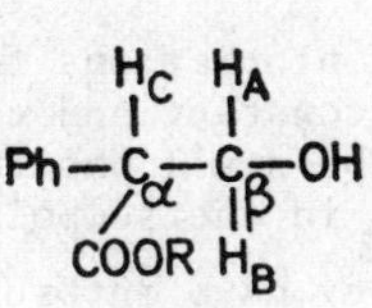

(III)

coupling constants cannot arise from a single rigid conformation about the C-C bond. It is usual to consider substituted ethanes of this type as rapidly interconverting mixtures of the three staggered rotamers IV to VI

with fractional populations P_{IV}, P_V and P_{VI} such that

$$P_{IV} + P_V + P_{VI} = 1 \tag{1}$$

From consideration of coupling constants in model compounds and the effects of substituent electronegativity on their values we have estimated the component coupling constants in the rotamers to be $J_{HH}^{gauche} = 2.3$ and $J_{HH}^{trans} = 12.1$ Hz*
Thus the observed averaged vicinal coupling constants are given by:

$$J_{AC} = (P_{IV} + P_{VI})\, J_{HH}^{gauche} + P_V\, J_{HH}^{trans} \tag{2}$$

$$J_{BC} = (P_V + P_{VI})\, J_{HH}^{gauche} + P_{IV}\, J_{HH}^{trans} \tag{3}$$

Equations (1) to (3) can be used to estimate the fractional populations from the measured values of J_{AC} and J_{BC}. For these molecules it was found that only rotamers IV and V are significantly populated ($P_{IV} = 0.60$, $P_V = 0.40$, $P_{VI} = 0.0$). These results are essentially the same as those found previously for tropic acid (10) i.e. conformations with phenyl and hydroxyl (or acetyl) groups in trans positions predominate. The presence of a single averaged spectrum for the mixture of conformers means that there is rapid interconversion between the different forms.

Conformation of the tropane ring

The tropane ring configurations and conformations for scopolamine (VII) has been determined previously using coupling constant and chemical shift data. Mandava and Fodor (5) have pointed out that the small observed coupling constants between the H_2 and H_3 protons (1.0 and 3.8 Hz in the scopolamine cation) show that H_3 occupies the equatorial position. They have interpreted the remaining ring coupling constants in terms of a distorted chair conformation. In their 60 MHz

---

* This analysis assumes that the J_{HH}^{gauche} and J_{HH}^{trans} values are the same in all three rotamers and will lead to some errors in the estimated population($\pm$ 10%)

studies they did not observe the non-equivalence of the pairs of protons at the (2,4) and (1,5) positions seen in this present work. Fortunately there are no differences in coupling constants between the non-equivalent nuclei and their neighbours. We have estimated the torsion angles in the tropane ring of the scopolamine cation from the crystal structure and found that the values (χ_{C1-C2} = 60 ± 5, 60 ± 5°; χ_{C1-C7} = 70 ± 5°; χ_{C2-C3} = 40 ± 5, 80 ± 5°) are consistent with the small vicinal coupling constants observed in the NMR studies. Thus the tropane ring appears to have a similar conformation in the solution and in the crystal state. For atropine only limited information about the coupling constants is available (see Table 2) but the values which could be measured are similar to those in scopolamine.

Crystal studies have shown that the configuration of the N-Me group is different in the atropine (I) and scopolamine (II) cations (1,2). NMR studies have confirmed that the predominant configurations found in solution are the same as those in the crystal state (5,7).

Overall molecular conformation using ring current shifts

Three-bond coupling constants can only provide local conformational information about the particular torsion angle characterised by the coupling constants. For non-rigid acyclic molecules containing several bonds it is usually impossible to construct the overall molecular conformation from such measurements. Sometimes the relative positions of non-bonded groups in a molecule can be determined by considering the shielding effects of anisotropic shielding groups (such as aromatic rings) on nuclei in neighbouring groups. For example, in the structures of both atropine and scopolamine there is an aromatic residue which will cause appreciable ring current shifts of neighbouring nuclei in a manner which depends on the positions of the nuclei with respect to the ring. In principle this method of conformational analysis can lead to the overall shape of the molecule. However, in practice the method is usually difficult to apply because of the problem of isolating the shielding contributions arising solely from ring current effects. Ideally one requires a model compound which lacks the aromatic ring but retains the same conformation. For the molecules under consideration here this problem can be circumvented by considering the shielding differences between corresponding nuclei in the tropane ring rather than the individual shifts. These shielding differences are given in Table 1 and arise from the unequal anisotropic shielding effects of the aromatic ring in the presence of hindered rotation about the bonds in the -O-C(=O)-C-C fragment. By comparing these shielding differences with the values calculated for an extensive set of conformations it has proved possible to determine the position of the aromatic ring with respect to the tropane ring.

Johnson and Bovey (9) have estimated the ring current chemical shift contributions at various coordinate positions near an aromatic ring using the equation

$$\delta_{ppm} = \frac{ne^2}{6\pi mc^2 a} \cdot \frac{1}{[(1+p)^2+z^2]} \left[K + \frac{1-p^2-z^2}{(1-p)^2+z^2} E\right] \qquad (4)$$

where n is the number of circulating electrons in a loop of radius a, p and z are cylindrical coordinates expressed in units of a, and K and E are elliptical integrals. Thus, starting from a set of Cartesian coordinates for an aromatic derivative it is a relatively straightforward matter to calculate the aromatic ring current shifts at any nucleus in the molecule (12). We have calculated the ring current shift differences for the 7 pairs of corresponding nuclei in the tropane ring of scopolamine for a wide range of defined conformations (obtained by systematically varying the torsion angles ϕ_2, ϕ_3 and ϕ_5 in Structure VII). The calculated shift differences were then compared with experimental values to find the conformations which gave the best fit to the observed data, as measured by the smallest value of the sum of the squares on the residuals Σr_i^2. In these calculations the tropane ring is considered to be rigid and in the same conformation as was found in the crystal structure.

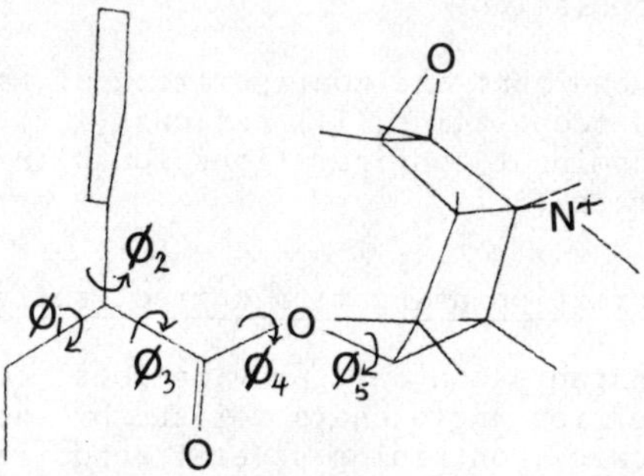

(VII)

Some additional information can be obtained by considering the ^{1}H shifts of scopine (VIIIa) and scopolamine (VIIIb) in $CDCl_3$ (9): these data indicate that the H_6 and H_7 protons of scopolamine have upfield shift contributions of 0.23 and 0.76 ppm respectively which can be assumed to arise mainly from aromatic ring current effects. Although there will be some error in assuming these values are the same for the scopolamine cation in aqueous solution we have ensured that the solutions formed using the shielding differences in Table 1 are also generally consistent with the approximate values of the total ring current shift contributions estimated above.

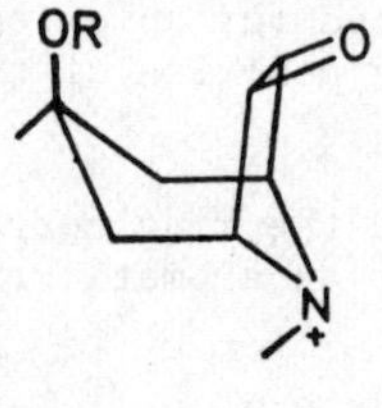

(VIII)

(a) Scopine R = H

(b) Scopolamine R = $COCH(CH_2OH)C_6H_5$

δ_{H6}, δ_{H7} = 2.84, 3.37 ppm

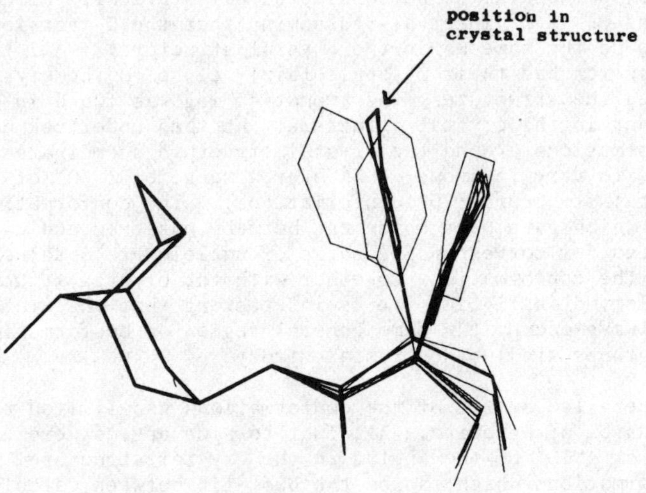

Scopolamine hydrobromide

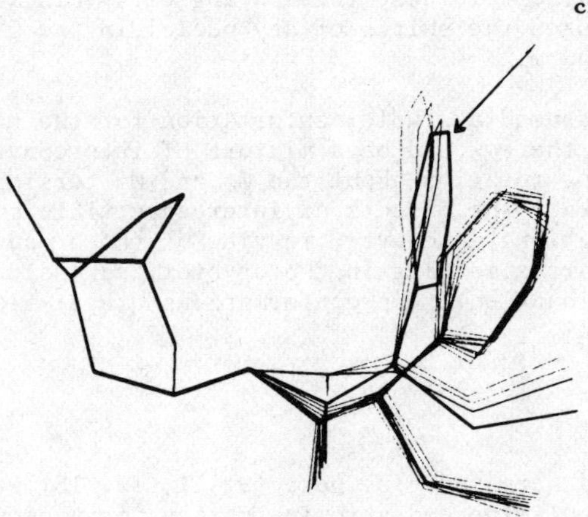

Atropine sulphate

Fig.2. The family of conformations which show good agreement between observed and calculated ring current shielding contributions for the tropane ring nuclei in scopolamine and atropine. The positions of the aromatic rings in the crystal structures are also shown.

In the calculations on scopolamine we considered all sterically allowed torsion angles of $\emptyset_2$, $\emptyset_3$ and $\emptyset_5$ at 6° intervals, assuming that the $\emptyset_4$ torsion angle (the ester CO-O bond) to be the same as in the crystal structure*. All the allowed solutions of this search had their $\emptyset_5$ angle fairly close to the crystal values and, when we plotted the structures, the aromatic ring was found in conformations quite similar to that in the crystal structure. We then undertook a more detailed search of the conformations around the crystal structure coordinates allowing all four torsion angles to vary in steps of 3° over a range of ± 30° of the torsion angles in the crystal structure (20^4 conformations). The conformations of the scopolamine cation which gave best agreement between observed and calculated shielding differences for corresponding pairs of nuclei were obtained from this search and some of the conformations together with the crystal structure conformation are plotted in fig.2a. It is interesting that the aromatic ring and the tropane ring always occupy the same general region of conformational space near to where the groups are in the crystal structure.

For atropine, the detailed search of the conformations was limited to those around the crystal coordinates of atropine. All four torsion angles were varied in steps of 3° over a range of ± 30° of the angles in the crystal structure. Figure 2b indicates the conformations which showed the best fit between calculated and observed shielding differences. Again the aromatic ring is generally in the same region of conformational space in both the crystal and solution conformations.

For both atropine and scopolamine there are several combinations of torsion angles which result in the aromatic and the tropane rings being in the same positions relative to each other. Thus while this method can give us information about the positions of the rings it cannot give much useful information about the exact values of these torsion angles. This is not surprising since we have not used the shifts of any nuclei in the O-C(=O)-CH-C fragment in our fitting procedure.

In the calculations we have assumed a single conformation for the molecules although it seems likely that there would be a mixture of interconverting conformers especially involving rotation about the $\emptyset_2$ and $\emptyset_3$ torsion angles. It is not possible to incorporate the effects of interconvertible multiple conformations into the calculations. However, in view of the broad agreement between the positions of the aromatic ring in the crystal and 'solution' structure it does indicate that the dominant solution conformations are similar to the crystal structure conformation.

DISCUSSION

From the measurements reported here and elsewhere (9, 11, 12, 15) we can conclude that the conformations of scopolamine and atropine cations in aqueous solution have the following features:-

(i) The aromatic ring occupies approximately the same region of conformational space in both crystal and solution states (see Fig.2).

(ii) The CH_2OH side chain exists as a mixture of the rapidly interconverting rotamers IV and V with fractional populations 0.60 and 0.40 respectively.

(iii) The OCOR substituent on the tropane ring is in an axial position (9) and the tropane ring appears to have the same conformation as found in the crystal state.

* Scheiber and Nador (15) used dipole moment measurements to show that the atropine ester group has a cis planar conformation in solution

(iv) The ester linkage has a cis-planar conformation as shown by the dipole moment measurements of Scheiber and Nador (15).

(v) The aromatic ring is flipping rapidly about the $\emptyset_2$ torsion angle.

(vi) The N-methyl configurations in the atropine and scopolamine cations are different (as found in the crystal state) but the overall conformation of the two species is very similar.

Consideration of space-filling models indicates that one would expect considerable flexibility about the $\emptyset_1$, $\emptyset_2$ and $\emptyset_3$ torsion angle. The $\emptyset_4$ torsion angle within the planar ester group will remain constant because of the double bond character in the C-O bond. When atropine (or scopolamine) binds to the receptor it seems unlikely that there will be much deformation about the $\emptyset_4$ and $\emptyset_5$ torsion angles. If the tropane ring conformation remains unaltered in the interaction then the distance between the charged nitrogen and the ester oxygen (3.88 Å (6)) would not be too different from the N^+... O separations in rigid agonists known to bind strongly to the same receptor sites (e.g. in the agonist acetoxycyclopropyl-trimethylammonium the N^+... O distance is 3.69 Å (2)). If one considers the crystal structure of atropine (6) it is seen that the aromatic ring is fairly close to the ester oxygen atom: it is difficult to see how this oxygen could effectively interact with the receptor if the aromatic ring of the atropine does not change its position on binding. It seems likely that there would be a rotation about the $\emptyset_3$ on binding which would change the position of the aromatic ring and allow the oxygen atom to interact with the receptor more easily.

Associated with the 'flexible' torsion angles $\emptyset_1$, $\emptyset_2$ and $\emptyset_3$ we would expect to have multiple conformations. On binding to the receptor there would be a conformational selection process. This could involve the antagonist binding initially in any one of the solution conformations to form a nucleation complex with the receptor which has sufficient lifetime to allow the flexible fragments to rearrange into their final conformation (16).

We can conclude by noting that the NMR measurements give us a great deal of information about the conformations of the atropine and scopolamine cations in solution and that this information is of some limited usefulness in helping us to understand how the antagonists bind to their receptors.

REFERENCES

(1) W.H. Beers and E. Reich, Structure and activity of acetylcholine, Nature, 228, 917 (1970).

(2) R.W. Baker, C.H. Chothia, P.J. Pauling and T.J. Petcher, Structure and activity of muscarinic stimulants, Nature, 230, 439 (1971).

(3) C.H. Chothia, Interaction of acetylcholine with different cholinergic nerve receptors, Nature, 225, 36 (1970).

(4) C. Pfeiffer, Nature and spatial relationship of the prosthetic chemical groups required for maximal muscarinic action, Science, 107, 94, (1948).

(5) A.H. Beckett, Stereospecificity in the reactions of cholinesterase and the cholinergic receptor. Ann. N.Y. Acad. Sci., 144, Art.2, 675 (1967).

(6) P.J. Pauling and T.J. Petcher, The crystal structure of (-)-(S)-Hyoscine hydrobromide, Chem. Comm. 1001 (1969): Nature, 228, 673 (1970).

(7) S.R. Johns and J.A. Lamberton, Magnetic non-equivalence of the epoxide ring (C-6 and C-7) protons of scopolamine, Chem. Comm. 458 (1965).

(8) M. Ohashi, I. Morishima, K. Okada, T. Yonezawa and T. Nishida, ^{1}H nmr contact shifts of some tropanes, Chem. Comm. 34 (1971).

(9) N. Mandava and G. Fodor, Configuration of the ring nitrogen in N-oxides and the conformation of tropanes. Part XVIII, Can. J. Chem. 46, 2761, (1968).

(10) V.S. Dimitrov, S.L. Spassov, T.Zh. Radeva and J.A. Ladd, Nmr spectra of tropic acid and some derivatives. J. Mol. Struct., 27, 157 (1975).

(11) L. Simeral and G.E. Maciel, Carbon-13 chemical shifts of some cholinergic neural transmission agents. Organic Magnetic Resonance 6, 226 (1974).

(12) J. Feeney, R. Foster and E.A. Piper. Nmr study of the conformations of atropine and scopolamine cations in aqueous solution, J. Chem. Soc. Perk.II, 2016 (1977).

(13) J.B. Stothers, Carbon-13 NMR Spectroscopy, Academic Press (1972).

(14) C.E. Johnson and F.A. Bovey, Calculation of nmr spectra of aromatic hydrocarbons, J. Chem. Phys. 29, 1012 (1958).

(15) P. Scheiber and K. Nádor, Conformational analysis of atropine, Arzneim-Forsch 25, No.3, 375 (1975).

(16) A.S.V. Burgen, J. Feeney and G.C.K. Roberts, Binding of flexible ligands to macromolecules, Nature 253, 753 (1975).

The Conformation of Anticholinergic Substances

Peter Pauling

William Ramsay, Ralph Forster and Christopher Ingold Laboratories, University College London, Gower Street, London WC1

Dedicated to Hn. Prof. Dr. Peter Waser
in the thirtieth year of our friendship

An interactive computer graphic conformational analysis of observed and possible conformations of 24 antagonists of acetylcholine at the autonomic post-ganglionic parasympathetic nervous junction whose crystal structures are known and whose observed conformations vary widely has resulted in the determination of a consistent conformation for all 24 substances, and the determination of the chemical groups essential for specific pharmacological activity and their relative orientation. The possible conformations were monitored and refined using a simple, semi — empirical classical energy calculation. For 22 of the 24 substances studied, the energy of the consistent conformation is less than the energy of the conformation observed in crystals, and in the other two less than one kcal/mole higher. The results are best illustrated by smoothed histograms of analogous geometrical parameters, and by space filling molecular drawings. The observed crystal structures of (-)-S-hyoscyamine hydrobromide and (-)-S-hyoscine hydrobromide are that of the consistent conformation of the 24 substances.

KEYWORDS: Acetylcholine; Anticholinergic; Conformational Analysis; Antagonist; Muscarinic; Crystal Structure.

Abbreviations: Anticholinergics: antagonists of acetylcholine at the autonomic post-ganglionic parasympathetic (muscarinic) nervous junction; SAR: structure activity relationships.

Supported by the British Medical Research Council and the United States Public Health Service, National Institute of Neurological and Communicative Disorders and Stroke (5 RO1 NS 12021) and National Institute of Mental Health (5 RO1 MH 24284).

Structure activity relationships (SAR) are normally considered in terms of the two-dimensional chemical line diagram structures and the biological activities of substances. This approach, while leading to some advances in the understanding of the chemical requirements for a specific biological activity, has not lead to any great increase in, for example, the tailoring of drugs with specific major activity or without specific side effects nor in the efficiency of the generation of new or better drugs. Since the properties of molecules are determined by their three-dimensional structures in a total sense, what atoms in what chemical state are where, this limitation of conventional SAR is understandable, and it is unlikely that a real increase in the understanding and usefulness of SAR will occur only when it is the three-dimensional molecular structures that are considered.

The primary technique for the observation of the three-dimensional atomic structures of molecules is that of single-crystal x-ray diffraction analysis. The results of such analyses, while very precise, may not be relevant to the biologically important conformation of a molecule. Most substances of pharmacological interest are flexible in several ways, particularly rotation about single bonds. The conformation observed in the crystal is greatly influenced by the effects of intermolecular packing, effects far different to those between a single molecule and its biological receptor. These effects are particularly noticable in crystals of antagonists of acetylcholine at the autonomic post-ganglionic parasympathetic (muscarinic) nervous junction, (in this paper referred to as anticholinergics) typified by Atropine, (-)-S-hyoscyamine (Fig. 1.)

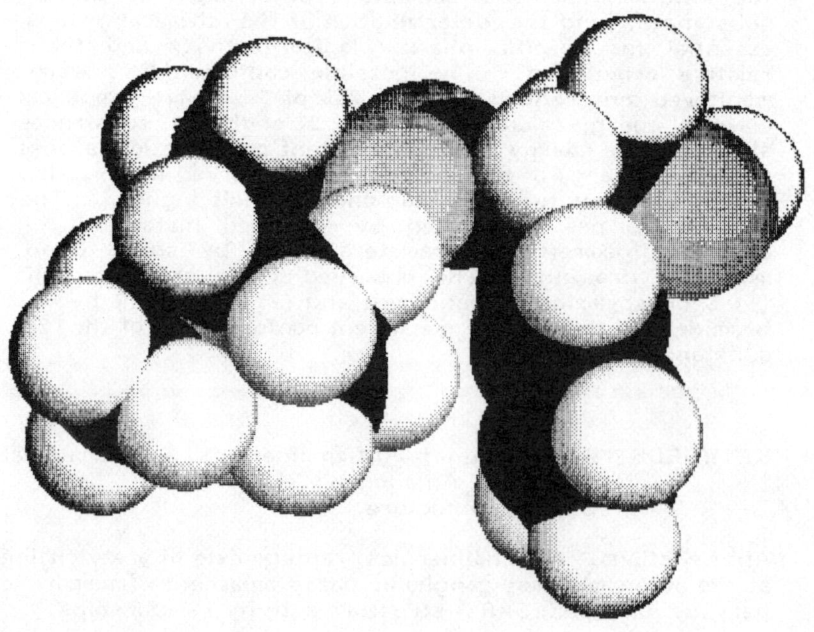

Fig. 1. The structure of Atropine, as observed in crystals of (-)-S-hyoscyamine gydrobromide (Petcher and Pauling, 1970).

Most anticholinergic substances are reasonably flexible, butterfly-shaped molecules with two rings (one almost universally a phenyl ring) as wings and a basic nitrogen chain tail, highly susceptible to variations in conformation from intermolecular packing forces. There is almost no consistency in the observed conformations of the 24 anticholinergics whose crystal structures are known, not using the anti-epileptic or anti-Parkinson's syndrome agents. Using a technique of human interaction with a computer graphics system, however, it has been possible by conformational manipulation of the observed structures, simultaneously monitoring the internal molecular energy of the individual molecules, to derive a single, consistent conformation that is energetically favourable for all 24 substances. This consistent conformation clearly indicates (except for the lack of pharmacological activities) the chemical groups and their relative orientation required for specific anticholinergic activity.

TECHNIQUE OF ANALYSIS

The Equipment

The method of conformational analysis used involves an inexpensive but very powerful interactive computer graphics system with programmes we have developed (Pauling and Richardson, unpublished.) The hardware consistes of a Digital Equipment Corporation (GT-44) PDP-11/40 processor, 28K core memory, two RK 05 disk cartridge drives (2.4 megabytes each,) VT11/VR17 display, LA30 DECwriter, card reader, paper tape reader, Versetec 1200 200 dot per inch printer/plotter, and slow and high speed communications systems via dial-up telephone to other computers. All the figures in this paper, except for Fig. 2, were produced in less than two minutes each by this system.

The programme used to analyse small molecules (up to 80 atoms; a derived programme for the study of proteins allows any number of atoms and has other special features) in addition to the usual facilities, such as rotation of molecular view and the calculation of geometrical parameters, has special features for this type of analysis. These include ability to invert an asymmetric molecule, invert a single asymmetric centre in a molecule with several asymmetric centres, delete and add atoms, particularly automatically adding hydrogen atoms in calculated positions, alter torsion angles, and calculate the internal energy of a molecule using a simple, two term, three parameter, semi-empirical, classical energy calculation (Giglio, 1969.) At all times, one of several types of molecular drawing appears on the screen and these and more sophisticated drawings can be transferred at any time to paper on the local matrix plotter or via telephone to the University of London Computer Centre microfilm plotter.

Proceedure of Analysis

Files containing the chemical, pharmacological, and crystallographic data of the 24 anticholinergics (and many other substances) are stored as individual, named disk files.. Each crystal structure was studied by rotating each variable torsion angle, which vary in number from 2 to 8 in the 24 substances, through 360°, plotting internal energy versus torsion angle, all other torsion angles being kept at their crystallographically observed values. After studying these curves, which clearly and reliably indicate impossible conformations but are not so reliable at small differences at minimum energy, the observed molecular conformations were altered by changing torsion angles in such a way as to approach a conformation that, based on continuing experience, would be consistent with possible conformations of the other substances. Since only one torsion angle and its resultant energy can be altered at a time, this technique leads to knowledge of a series of lines through an n-dimensional space where n is the number of variable torsion angles, and the process must be iterative: successively altering one torsion angle after another and repeating. While the internal energy of each molecule was monitored carefully during this process, the criterion of consistency of conformation was considered of greater importance, and we were rather surprised that the energy decreased in the consistent conformation. The low increase in energy for two substances in the consistent conformation (Glycopyrronium and Hexapyrronium) is consistent with the small change in pharmacological activity with change in absolute configuration at this end of the molecule for these substances.

THE ANALYSIS

The variable torsion angles parameters for most anticholinergic substances are defined in the artificial molecule shown in Fig. 2, but several important substances have no analogous torsion angle parameters.

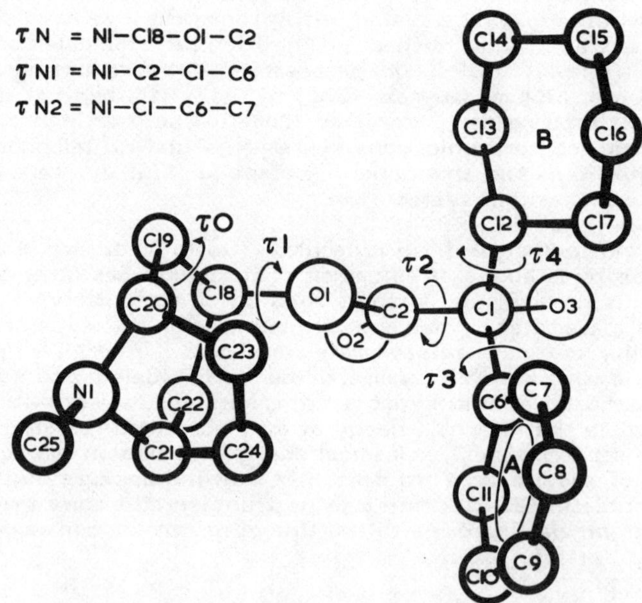

Fig. 2. A diagram of a typical but artificial anticholinergic substance defining the variable torsion angle parameters.
T2 = O1 - C2 - C1 - C6; T3 = C2 - C1 - C6 - C7; T4 = C2 - C1 - C12 - C13
(TN, T0, and T1 are not discussed in this paper.)
In Atropine and Scopolamine, there is no variable T4, but an additional one in that O3H is connected to C1 via an CH2 group.
This variable cannot be determined (except within limits of steric hindrance) without knowledge of the structure of the receptor.

Three artificial torsion angles have been defined (any four bonded or non-bonded atoms can define a "torsion angle." Of the 24 substances studied, the consistent, modified conformations of 22 have internal energy lower than that of the observed conformations by between -0.1 and -5.9 kcal/mole. The two substances with higher energy, Glycopyrronium and Hexapyrronium, have modified energies of +1.0 and +0.6 kcal/mole respectively. These values must be considered insignificant compared to the unknown energy available upon interaction with the receptor and the simplicity of the energy calculation. This result is also consistent with the lack of pharmacological sensitivity with change in absolute configuration at this end of these molecules, equivalent with interchange between the nitrogen atom and its carbon neighbour.

A convenient method of representing the differences between observed and modified conformations and the consistency of the modified conformations is Fig. 3. Each curve is a smoothed histogram of unit height and fixed, artificial standard deviation has been placed at each torsion angle and the result summed. It can be seen that the torsion angle population of the modified conformations is much more consistent than that of the observed conformations. It is difficult to define a set of universally applicable, analogous, quantitative set of geometrical conformational parameters for this class of molecules. Most anticholinergics are esters, for which an analogous set of conformational parameters can be defined and compared. Many important, such as the 1,3-dioxolan derivatives of Brimblecombe, Inch, Wetherall, and Williams (1971), and many unimportant anticholinergic substances cannot be defined in analogous terms. In Fig. 3, TN1 (a generalization of T2 and T1), TN2 (a generalization of T3 and T1), T2, T3 (which apply strictly only to esters) and the mean N1 - essential phenyl (C6 - C11) torsion angles and distance should be considered. TN1, defined as the artificial torsion angle N1 - C2 - C1 - C6, represents the "torsion angle" between the basic nitrogen atom N1 and the bonded axis of the phenyl ring C6 - C11 (the phenyl ring essential for cholinergic activity) looking along the backbone chain of the molecule (in esters,) a more generalised version of T2, Fig. 2. TN2, N1 - C2 - C6 - C7, represents a generalised version of the twist of the phenyl ring with the backbone; a generalised version of T3, Fig. 2..

In all cases where individual peaks occur in the modified histograms, these peaks can also be seen in the observed histograms. These peaks correspond to fixed or severely restricted torsion angles, such as the tricyclic methixene, and the compounds can be indivudually identified by reference to a table of numerical torsion angles. T4, while not important because anticholinergic activity is not sensitive to substitution in the ring B position, is interesting in that the type of ring, cyclopentyl or cyclohexyl, or phenyl, or some other group can be determined from the position of the peak in the modified histogram.

THE CONSISTENT CONFORMATION

The consistent conformation can be determined from a study of Fig. 3 and the chemical groups essential for anticholinergic activity determined by consistency of the groups in consistent conformation and a knowledge of pharmacological activity. The consistent geometrical parameters of the consistent conformation in terms of Fig. 2 are given in the Table.

The observed crystal structures of (-)-S-hyoscyamine hydrobromide (Kussather and Haase, 1972; Petcher and Pauling, 1970), (-)-S-hyoscine hydrobromide (Pauling and Petcher, 1969), and quinuclidinyl benzilate hydrobromide (Meyerhöffer and Carlström, 1969) are good observed examples of the consistent conformation, but that of Scopolamine N-Oxide (Hubor, Fodor, and Mandava, 1971) is not. The observed crystal structure of the 1,3-dioxolan derivative

The Conformational Parameters of the
Consistent Conformation of Anticholinergic Substances

TN1 = $-138°$ T2 = $55°$
TN2 = $38°$ T3 = $-100°$
T4 = $-27°$ for ring B = phenyl
T4 = $-58°$ for ring B = cyclopentyl or cyclohexyl
(C6 - C1 - C12 - C13 = 180°)

Mean distance N1 – essential phenyl (C6 - C11) = 5.9 A

VIIIH of Brimblecombe and colleagues (1971) 2S-(R-1-cyclohexyl-1-hydroxy-1-phenyl) methyl-4S-dimethylaminomethyl-1,3-dioxolan hydrochloride (Datta, Mondal, and Pauling, in press) also fits the consistent conformation well in terms of the overall shape of the molecule, although the chemical formula is completely different to that of most anticholinergics, which are esters. The observed conformation of Benactyzine hydrochloride (Petcher, 1974), on the other hand, does not, though not for obvious reasons.

The two chemical groups most important for anticholinergic activity are the phenyl group C6 - C11 and the basic nitrogen atom N1 surrounded by bulky hydrophobic groups. Although most anticholinergics are esters, oxygen atoms seem to be of much less importance in determining activity, although in the very high affinity 1,3-dioxolan substances of Brimblecombe and colleagues (1971), the oxygen atoms of the dioxolan ring are in nearly the same position as those of the ester oxygen atoms. The oxygen atom often bonded to C1 (Fig. 2) seems to add to affinity.

Except for the phenyl ring C6 - C11 and the bulky nitrogen group around N1, the vagueness of the last discussion illustrates the major fault in this analysis: I do not know the affinity binding constants to the receptor of this series of 24 substances. Such knowledge would greatly extend and increase the value of this form of analysis.

The artificial torsion angle N1 - C18 - O1 - C2 (in terms of Fig. 2) of acetylcholine in crystals of the chloride (Herdklotz and Sass, 1970) is observed at $-135°$. This value is very similar to that of TN1 of the consistent conformation of anticholinergics, although the torsion angles are not strictly analogous. N1 - O1 - C2 - C1 is 130° in acetylcholine chloride. The observed distance N1 - C1 in acetylcholine is 5.4 A and the mean distance N1 to the three hydrogen atoms of the acetyl methyl group is 5.7 A. This comparison was pointed out by Pauling and Petcher (1970). An analysis of muscarinic agonists was published by Baker and colleagues (1971), but this work will be repeated using interactive computer graphic techniques of analysis now available.

I am most greatful to Douglas Richardson for the interactive computer graphics programmes without which this analysis would not be possible, the late Narayandas Datta, Miss Panchali Mondal, and Trevor Petcher for crystallographic analyses, Miss Ann Steward and Miss Mary Lee for assistance, and Mr. John Cresswell for photography.

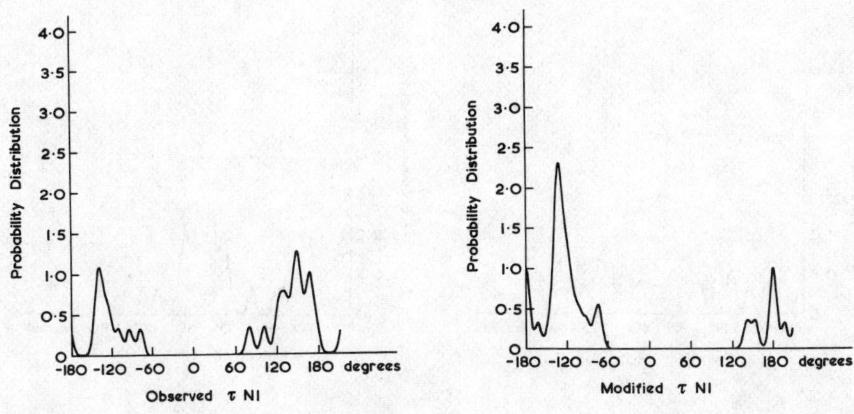

Fig. 3.1. TN1 = N1 - C2 - C1 - C6.

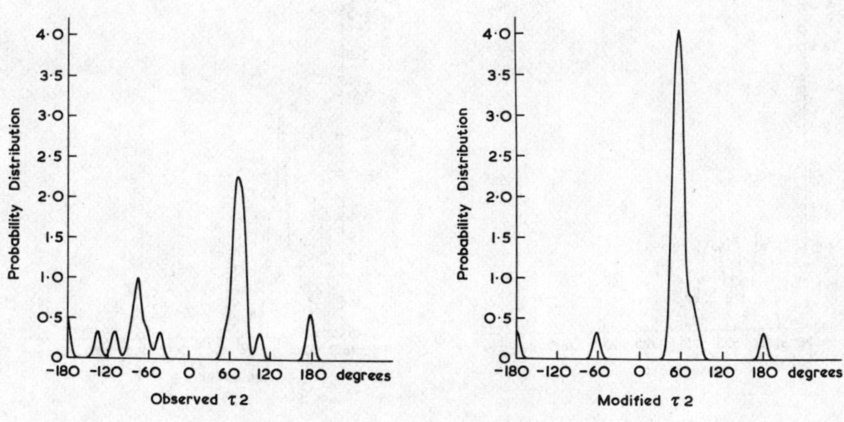

Fig. 3.2. T2 = O1 - C2 - C1 - C6.

Fig. 3. Smoothed histograms of torsion angle probability distributions in which a Guassian of unit height and unit area (*sigma* = 5°) has been placed at each torsion angle value of the observed and modified conformations of the 24 anticholinergic substances analysed and the result summed.

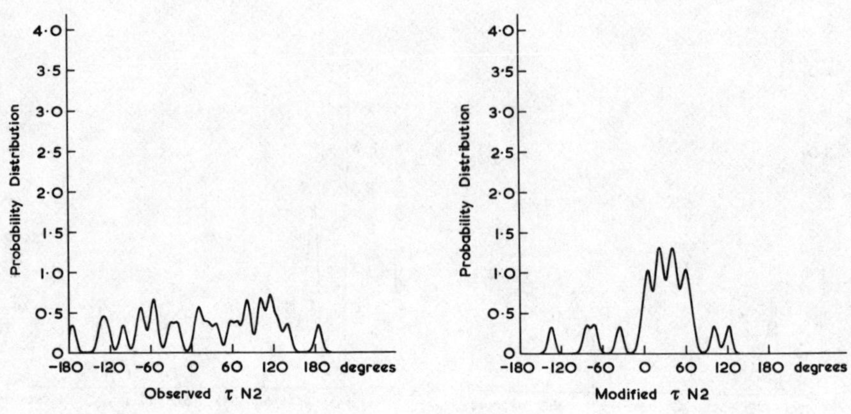

Fig. 3.3. TN2 = N1 - C1 - C6 - C7.

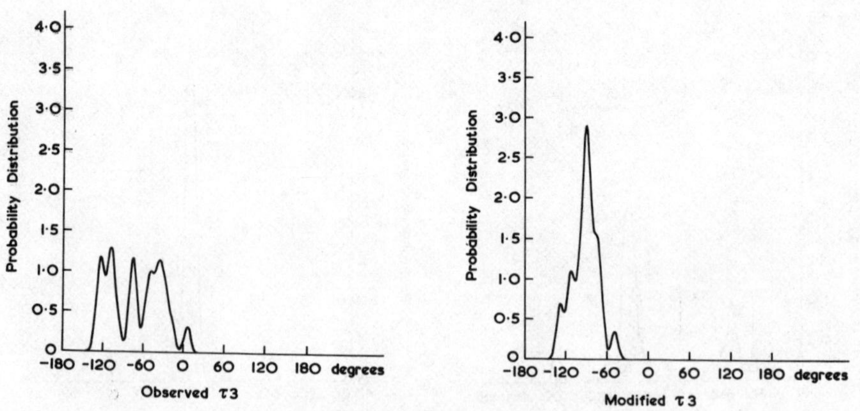

Fig. 3.4. T3 = C2 - C1 - C6 - C7.

Fig. 3. (cont.) It can be seen that the modified conformations are much more self-consistent than the observed. Individual peaks can be identified from a table of torsion angle values and generally correspond to substances of more or less rigid conformation whose chemical formulae differ significantly from that of Fig. 2.

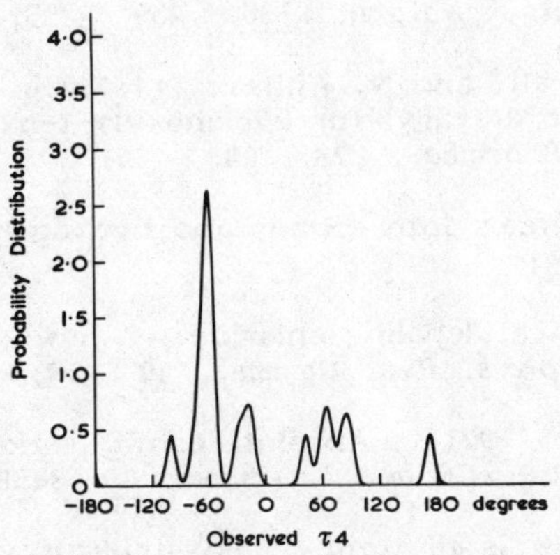

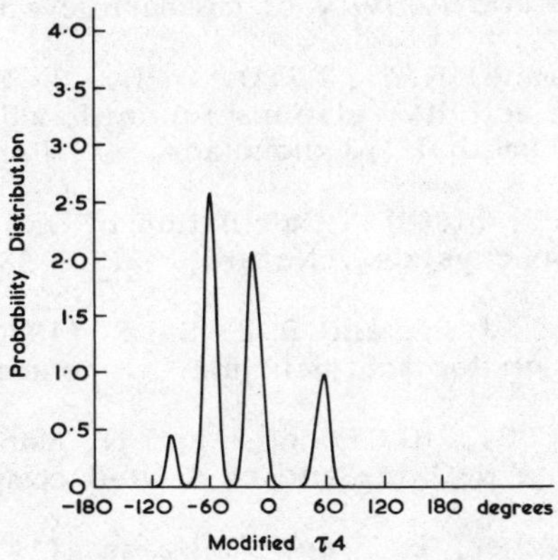

Fig. 3.5. T4 = C2 - C1 - C12 - C13.

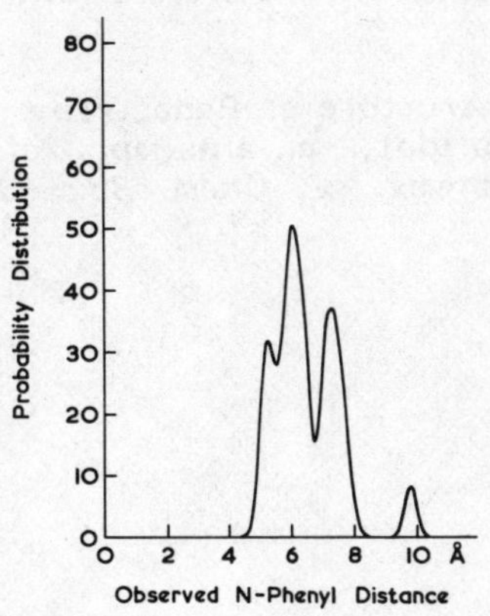

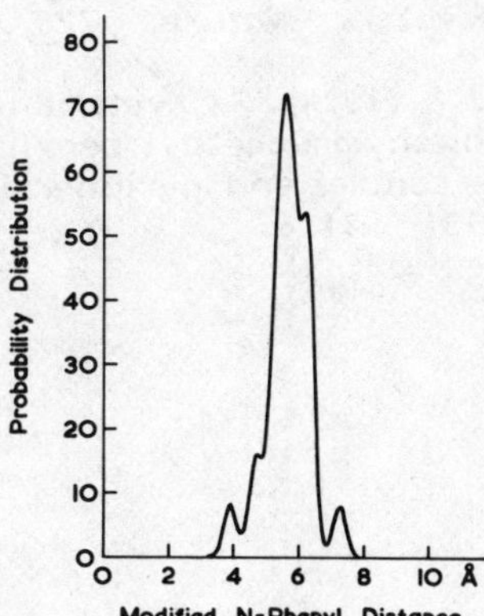

Fig. 3.6. The mean distance N1 to the essential phenyl ring A (C6 - C11) of Fig. 2. (*sigma* = 0.2 A).

References:

Baker, R.W., C.H. Chothia, P.J. Pauling, and T.J. Petcher (1971). Structure and Activity of muscarinic stimulants. *Nature*, 230, 439 - 435.

Brimblecombe, R.W., T.D. Inch, J. Wetherall, and N. Williams (1971). Structure activity relations for anticholinergic 2(1-aryl (or cyclohexyl)-1-hydroxy-1-phenyl) methyl-1,3-dioxolans. *J. Pharm. Pharmac.*, 23, 649 - 661.

Giglio, E. (1969). Calculation of van der Waals interactions and hydrogen bonding in crystals. *Nature*, 222, 339 - 341.

Herdklotz, J.K., and R.L. Sass (1970). Acetylcholine chloride: a new conformation for acetylcholine. *Biochem. Biophys. Res. Comms.*, 40, 583 - 588.

Huber, C.S., G. Fodor, and N. Mandava (1971). Absolute configuration of Scopolamine N-Oxide and of related compounds. *Canad. J. Chem.*, 49, 3258 - 3271.

von Kussäther, E., and J. Haase (1972). Kristall- und Molekülstruktur von 1-Hyoscyaminhydrobromid. *Acta Crystal.*, B28, 2896 - 2901.

Meyerhöffer, A., and D. Carlström (1969). The crystal and molecular structure of Quinuclidinyl Benzilate Hydrobromide. *Acta Crystal.*, B25, 1119 - 1126.

Pauling, P.J., and T.J. Petcher (1969). The crystal structure of (-)-S-Hyoscine Hydrobromide. *Chem. Comms.*, 1001 - 1002.

Petcher, T.J., and P.J. Pauling (1970). Interaction of Atropine with muscarinic receptor. *Nature*, 228, 673 - 674.

Petcher, T.J. (1974). Crystal and molecular structure of Benactyzine Hydrochloride (2-diethylaminoethyl benzilate hydrochloride), an antagonist of acetylcholine in the central and peripheral nervous systems. *J. Chem. Soc. (Lon), Perkin II*, 1151 - 1156.

Biodegradable Neuromuscular Blocking Agents

John B. Stenlake

Univeristy of Strathclyde, Glasgow, G1 1WX, Scotland

INTRODUCTION

For clinical acceptability the duration of action of neuromuscular blocking agents should be such that muscle paralysis will not be unduly prolonged beyond that required for completion of the operation or surgical procedure for which it is used. Savarese and Kitz (1) considered that three types of agent of (a) very short (b) medium and (c) long duration of action were required to provide adequate muscle relaxation for intubation and orthopoedic procedures, general surgery and open heart surgery respectively.

All neuromuscular blocking agents currently in clinical use are quaternary ammonium salts. They are, therefore, fully ionised irrespective of tissue pH, and do not diffuse across the phospholipid membrane boundaries of the gastro-intestinal tract, the blood-brain barrier, or the placenta. None-the-less, following intravenous injection they are rapidly distributed to motor endplates, in some cases within seconds (Ref. 2). The influence of cardiac output and muscle blood flow on onset time emphasises the key role of blood circulation in the distribution of neuromuscular blocking agents (Ref. 3).

Factors Affecting Duration of Action

Table 1 shows some typical mean recovery times for neuromuscular blocking agents in the cat (Ref. 4) and man (Ref. 5). The duration of action of neuromuscular blocking agents depends on the rate at which drug concentration in the extracellular fluid falls. This is dependent on several factors including dosage, the volume of distribution, metabolism and the neuromuscular blocking activity of metabolites, and the excretion kinetics of both drug and metabolites. The influence of some of these factors on the clinical pharmacokinetics of Tubocurarine, Suxamethonium, Pancuronium and Gallamine in patients with normal and impaired renal function has recently been described by Wingard and Cook (Ref. 5).

As quaternary ammonium salts of relatively high molecular complexity neuromuscular blocking agents are appreciably bound to plasma proteins and strongly acidic components of connective tissue, an important determinant of the volume of distribution. Likewise, as water-soluble salts, they are eliminated primarily and predominently via the kidney. In general, therefore, neuromuscular blocking agents, irrespective of whether they are depolarising or competitive blockers, fall into two groups, distinguishable by the extent of their metabolism. The first are essentially non-biodegradable, tissue bound, and hence of relatively long duration of action, especially in cases of renal insufficiency. In contrast, the second group, which may also have tissue depot sites, are biodegradable, and therefore exhibit durations of action which can be related to the rate and pathway of metabolism, but also in part to the neuromuscular blocking properties of their metabolites.

TABLE 1 Mean Recovery Times for Neuromuscular Blocking Agents

Compound	Mean recovery times (minutes) from full paralysing doses	
	Cat (100%) (Ref. 4)	Man (50%) (Ref. 5)
Tubocurarine	44	51
Dimethyltubocurarine	76	-
Gallamine	64	23(from 88% paralysis)
Alcuronium	71	-
Pancuronium	29	37
Fazadinium	20	-
Suxamethonium		10

Non-biodegradable Compounds

Gallamine, which is excreted unchanged 84% via the kidney in 24 hr with only 4% excretion in bile in 12 hr (Ref. 6), is bound to plasma proteins, albeit to an uncertain degree (Ref. 7) and is long acting in both cat (Ref. 4) and man (Ref. 5). It also shows elevated blood levels with ligation of the renal arteries in the dog (Ref. 6) and prolongation of neuromuscular blockade in patients with renal failure (Ref. 8,9).

Tubocurarine which is also long acting is similarly excreted substantially unmetabolised via the kidney. Some 36% of an i.v. dose appears in the urine of dogs in 3 hr and 75% in 24 hr (Ref. 10). Despite the molecular weight of tubocurarine chloride (696 anhydrous) being well above the normal cut off point for biliary excretion in the dog it is secreted in the bile to the extent of only 6% in 3 hr and 11% in 24 hr. This, however, increased to 39% in 24 hr with ligation of the renal arteries. In accord with these observations, prolonged paralysis and post-operative respiratory failure have been reported in a number of instances where large doses of Tubocurarine have been used in patients with renal failure (Ref. 11,12,13). A further study (Ref. 14), however, failed to show any significant differences in either the duration of paralysis or plasma concentration-time curves between patients with normal and impaired renal function; though patients with new kidney transplants showed markedly reduced ability to excrete Tubocurarine in the urine.

Binding to plasma proteins and other depot sites probably contributes to these effects. Muscle paralysis, which is only just apparent at plasma concentrations of 0.2μg/ml Tubocurarine, requires concentrations of 0.7μg/ml for complete neuromuscular blockade (Ref. 15), whereas protein binding only becomes appreciable (44%) at much higher concentrations (5μg/ml) (Ref. 16). At this concentration some 15% is bound to gamma globulin and only 24% to albumin, indicating a high degree of preferential binding (80 - 90%) (Ref. 17) to the former. This could well be significant at normal paralysing dose levels, in accord with observations of patient variation in the apparent volume of distribution (Ref. 14,18) and reports of correlations between gamma globulin levels and dosage required for full muscle paralysis (Ref. 19). These observations also correlate with reports that the closely related very long acting Dimethyltubocurarine (O,O,N-trimethyltubocurarine), which is excreted in the rat *via* the kidneys (Ref. 20), is not only some 40% bound to plasma proteins over a wide range of concentrations (0.006 - 88μg/ml), but is also appreciably bound in connective tissues to chondroitin sulphate and cartilage (Ref. 21).

Biodegradable Compounds

Pancuronium (1), which has a somewhat shorter duration of action, is excreted mainly via the kidneys, and although 58% is excreted unchanged after 8 hr in the cat, some 14.5% is eliminated as the 3-hydroxy-, 7% as the 17-hydroxy- and 4.5% as the 3,17-dihydroxy-derivative (Ref. 22). Similar results have been obtained in patients. Agostan *et al* (23) reported around 44% in the urine of which 20% was the 3-hydroxy-compound, together with 11% excretion in the bile over 30 hours whilst Burzello (24) found a total excretion of 50% in the urine in 12 hr, of which 10% consisted of metabolites, with about 10% excretion in bile over 24 hr. The precision of the assay methods used has, however, been criticised (Ref. 5) with some justification, and this may well explain the variation in the results obtained.

The metabolites of Pancuronium are still high molecular weight bis-quaternary steroids and the mono-hydroxy-compounds retain some muscle relaxant properties, though the 3,17-dihydroxy-compound is inactive (Ref. 24). Pancuronium is also strongly bound to plasma proteins, with some 87% bound at neuromuscular blocking levels in man (Ref. 25). Studies in mice have also shown substantial binding to chondroitin sulphate and mucopolysaccharides of connective tissue (Ref. 26). It is not surprising, therefore, that clearance rates of Pancuronium and its metabolites are reduced from 74ml/min to 20ml/min in renal failure (Ref. 27) and that the duration of neuromuscular blockade is substantially prolonged in patients with renal failure (Ref. 13,28,29,33).

PANCURONIUM BROMIDE ($2 Br^-$)

CAT METABOLISM AND URINARY EXCRETION (8 hr)

58% UNCHANGED

$CH_3.CO.OH$

METABOLITES

3-OH	(R^1 = H ; R^2 = Ac)	14.5%
17-OH	(R^1 = Ac; R^2 = H)	7%
3,17-DiOH	(R^1 = H ; R^2 = H)	4.5%

The relatively short action of Suxamethonium (2) is ascribed to its hydrolysis by plasma esterases first to monosuccinylcholine, which has only about 1/80th the neuromuscular blocking potency of Suxamethonium, and ultimately to choline and succinic acid which are completely inactive (Ref. 30).

$[Me_3\overset{+}{N}\cdot CH_2\cdot CH_2\cdot O\cdot CO\cdot CH_2\cdot CH_2\cdot CO\cdot O\cdot CH_2\cdot CH_2\cdot \overset{+}{N}Me_3]2X^-$ SUXAMETHONIUM (2)

PLASMA ESTERASE METABOLISM

$Me_3\overset{+}{N}\cdot CH_2\cdot CH_2\cdot OH\ X^-$

$[Me_3\overset{+}{N}\cdot CH_2\cdot CH_2\cdot O\cdot CO\cdot CH_2\cdot CH_2\cdot CO\cdot OH]\ X^-$ → $HOOC\cdot CH_2\cdot CH_2\cdot COOH$

MONOSUCCINYLCHOLINE SUCCINIC ACID

Evidence from human patient studies (Ref. 31,32) shows that hydrolysis *in vivo* is somewhat slower (3-7mg/litre/min) than *in vitro* (Ref. 33), which suggests that depot sites may have some, albeit limited, effect on its duration of action. Moreover, despite its short action, the major objection to Suxamethonium is its depolarising effect, and the consequent hazard to patients with plasma cholinesterase deficiency (Ref. 34,35).

Fazadinium Bromide (3) which in contrast has a competitive blocking action that is readily reversed by anticholinesterases, is similarly cleaved metabolically into fragments which are non-quaternary and inactive as neuromuscular blocking agents. Metabolism occurs rapidly in the liver by azoreductase in most laboratory animals. The principal products are non-quaternary, so that despite entero-hepatic circulation and excretion by both urinary and faecal routes (70% of the i.v. administered dose excreted in 24 hr as drug and metabolites in undisclosed proportions) Fazadinium is relatively short acting in the rat. It is, however, somewhat longer acting in the cat (Ref. 4) and man (Ref. 38,39) in which it is metabolised by the same pathway with cleavage of the molecule into inactive excretable fragments. Enterohepatic circulation is usually less pronounced in man. It would seem, therefore, that protein binding and deposition in connective tissue may influence its duration of action.

FAZADINIUM BROMIDE METABOLITE 1

pH - Dependent Biodegradable Compounds

Acceleration of molecular fragmentation to inactive products by chemical reaction with or without enzymic degradation offers an alternative method of achieving controlled neuromuscular block of limited duration. One example of this type of approach is seen in the compound AH 10407 and in its derivatives (4) (Ref. 40,41, 42). These ultra-short acting competitive blockers with onset times which can be measured in seconds, give complete block at doses of 2-4 mg/kg i.v. in the cat, dog, Cynamolgus monkey and cotton-eared marmoset with complete recovery 0.5 - 3.5 min. Compound AH 10407 is relatively stable under acid condition, but readily transformed in alkali and Krebs solution and in human blood 90% is degraded in 1.5 min. Degradation occurs as a result of nucleophilic attack by bicarbonate and other anions to non-quaternary and, hence, inactive compounds, half-life and potency increasing in parallel in the series AH 10407 < AH 11244 < AH 11056 in accord with increasing resistance to nucleophilic attack imposed by the respective phenyl ring substituents H < Me < MeO.

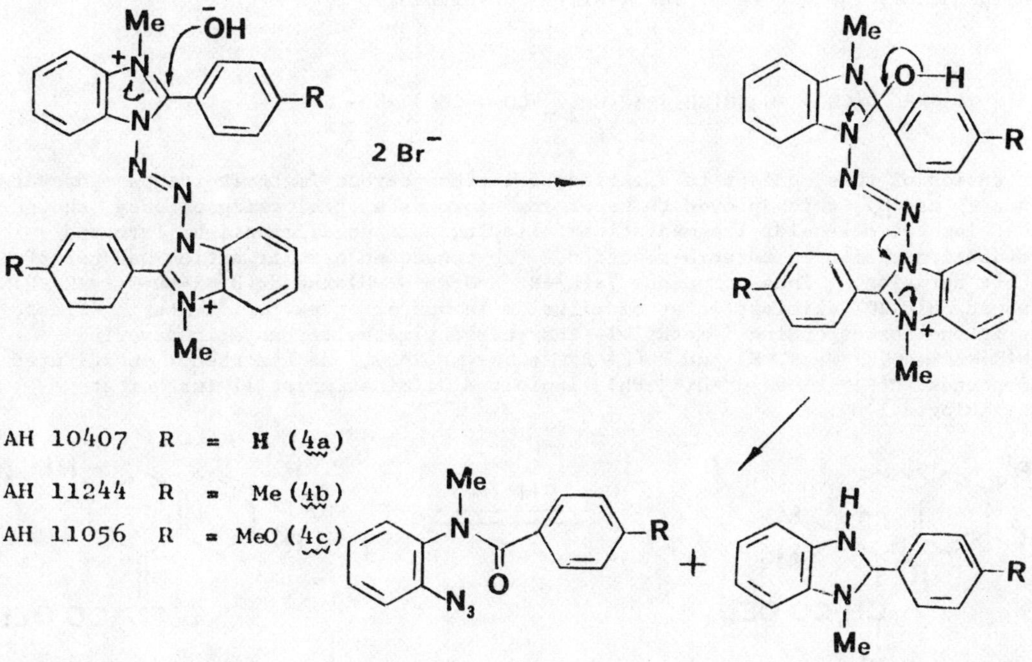

AH 10407 R = H (4a)

AH 11244 R = Me (4b)

AH 11056 R = MeO (4c)

Our own approach was founded on observations that the quaternary ammonium tetrahydropapaverinium alkaloid, petaline (5) readily underwent a typical but apparently facile Hofmann elimination on Amberlite IRA 400 (OH) (Ref. 43). The structural resemblance of petaline to the benzylisoquinolium units of Tubocurarine suggested the possibility of exploiting similar eliminations effective at physiological pH (7.4), as an alternative and novel means of terminating neuromuscular blocking effects more rapidly in an approach to the development of muscle relaxants of limited duration.

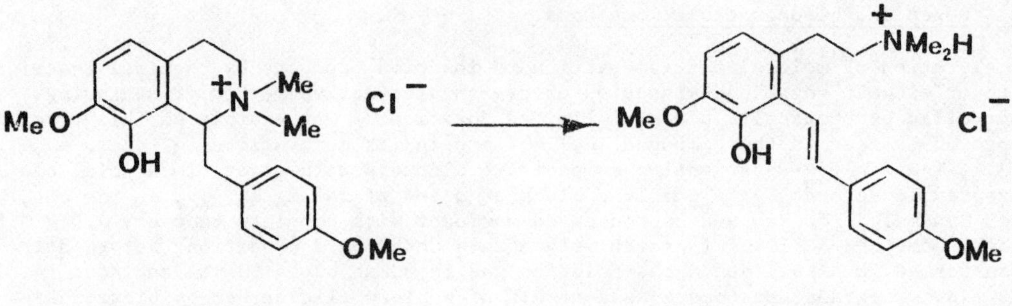

PETALINE CHLORIDE (5)

The potential feasibility of this approach had already been demonstrated in prior studies of a series of ester-based tris quaternary ammonium salts (6) (Ref. 44). These compounds, however, had low potency compared with Tubocurarine and Suxamethonium in laboratory animals, and reversibility by anticholinesterases was influenced by the nature of the N-alkyl substituents.

$$R_3\overset{+}{N}(CH_2)_2\cdot O\cdot CO(CH_2)_n\cdot \overset{+}{N}(Et)(R)(CH_2)_n\cdot CO\cdot O(CH_2)_n\cdot \overset{+}{N}R_3 \quad 3I^- \qquad (6)$$

Extension of this concept to a series of 1-ethoxycarbonylmethyltetrahydropapaverines (7 and 8), which proved to be of low neuromuscular blocking potency, showed that lack of molecular fragmentation following Hofmann elimination, favoured re-quaternisation in a retro-reaction. In consequence their action was not of short duration. Thus compounds 7a(R^1=R^2=MeO;R^3=R^4=H) and 8a(R^1=R^2=MeO;R^3=R^4=H) showed only 20% elimination at equilibrium in buffer at pH 7.4. Steric effects, as in the corresponding 1-methyl-1-ethoxycarbonylmethyltetrahydropapaverines, 7b(R^1=R^2=MeO;R^3=Me;R^4=H) and 8b(R^1=R^2=MeO;R^3=Me;R=H), and the phenyl substituted compounds 7c(R^1=R^2=MeO;R^3=H;R^4=Ph), inhibited elimination still further at physiological pH.

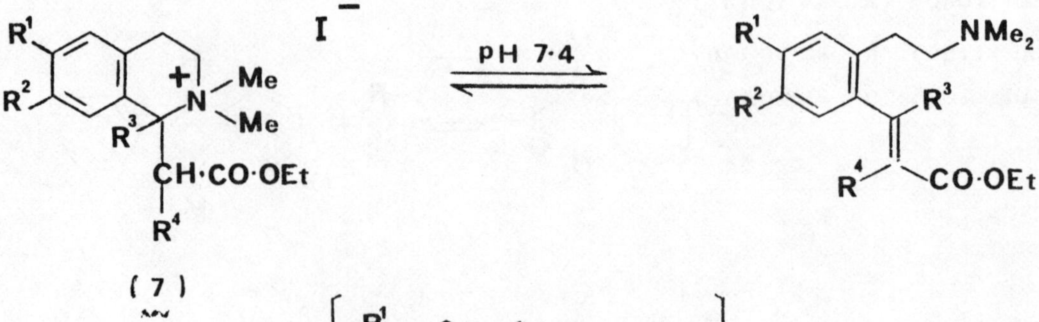

pH 7·4

(7)

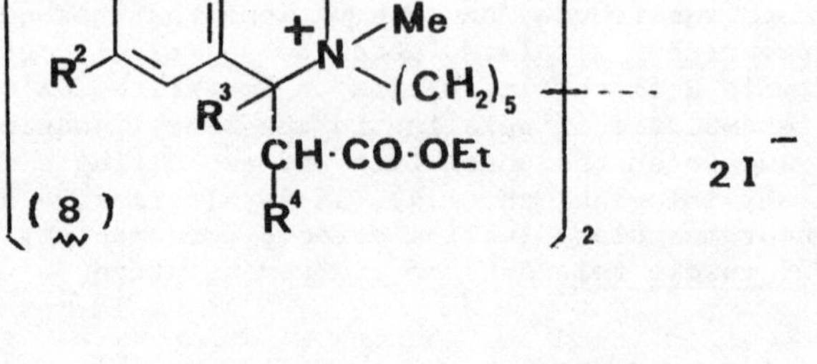

(8)

Similar studies on a second series of 1-phenacyl-1,2,3,4-tetrahydroisoquilinolinium salts (9 and 10) demonstrated that the rate of elimination was largely governed by electronic effects in the phenacyl ring. Thus, as anticipated, electron-donating groups (OMe) had a retarding effect, whereas electron-withdrawing effects (Cl) increased the rate of elimination to a point, where had the other requirements for potency been satisfied, block would have been very short. The extent of elimination, however, was dependent on steric factors, increasing with increasing bulk of the N-substituents in accord with the expected relief of strain in the open-chain structure (Table 2). Potencies, however, are extremely low, and the block was not reversed by Neostigmine.

TABLE 2 Elimination characteristics and neuromuscular blocking potencies of 1-phenacyl-1,2,3,4-tetrahydroquinolinium compounds

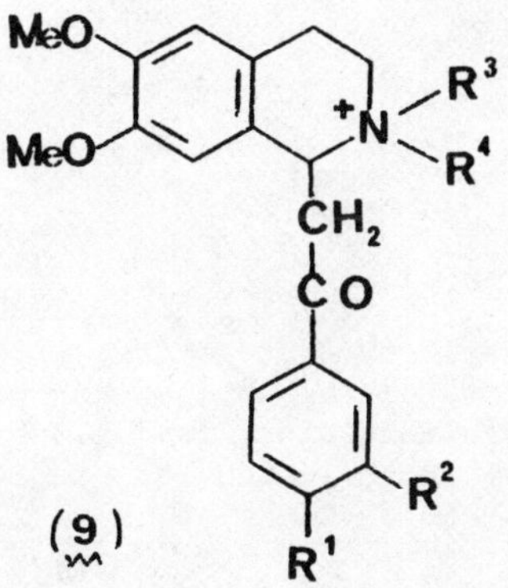

(9)

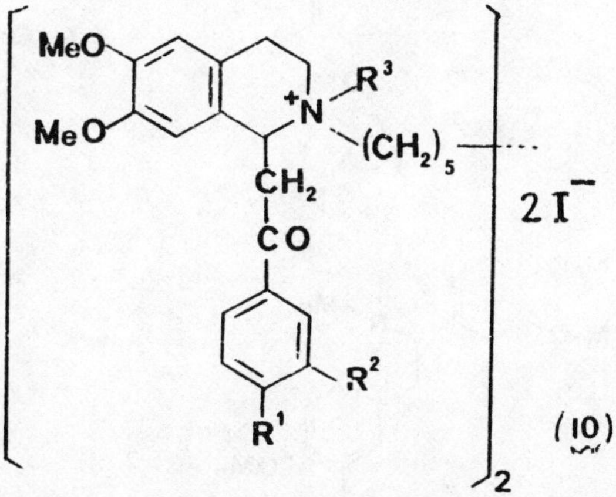

(10)

Compound	Substituents R^1	R^2	R^3	R^4	Hofmann elimination at pH 7.4 $t_{\frac{1}{2}}$ min to equilibrium	% Elimination at equilibrium	Neuromuscular Blocking Potency (TC=100)*
7a					30	20	1.7
9a	H	H	Me	Me	4	10	-
b	MeO	MeO	Me	Me	4	10	-
c	MeO	H	Me	Me	3.5	10	-
d	Cl	H	Me	Me	1.5	15	-
e	MeO	MeO	Et	Me	7.5	55	-
f	MeO	MeO	Et	Et	1.5	25	-
10a	H	H	Me	-	5	90	1.3
b	MeO	H	Me	-	7	90	0.7
c	MeO	MeO	Me	-	11	90	0.7
d	Cl	H	Me	-	2.5	90	0.5

* Rat phrenic nerve-diaphragm preparation

In extension of these findings, we have synthesised a series of non-depolarising neuromuscular blocking agents of structure (11) in which the requirements for potency and potential for pH/enzymic biodegradation have been optimised (Ref. 45). We are indebted to Dr. Roy Hughes (Ref. 46) of the Wellcome Foundation Ltd. for their pharmacological evaluation, which shows them to be of relatively high potency and that block is readily reversed by Neostigmine and potentiated by Halothane in the cat and rhesus monkey.

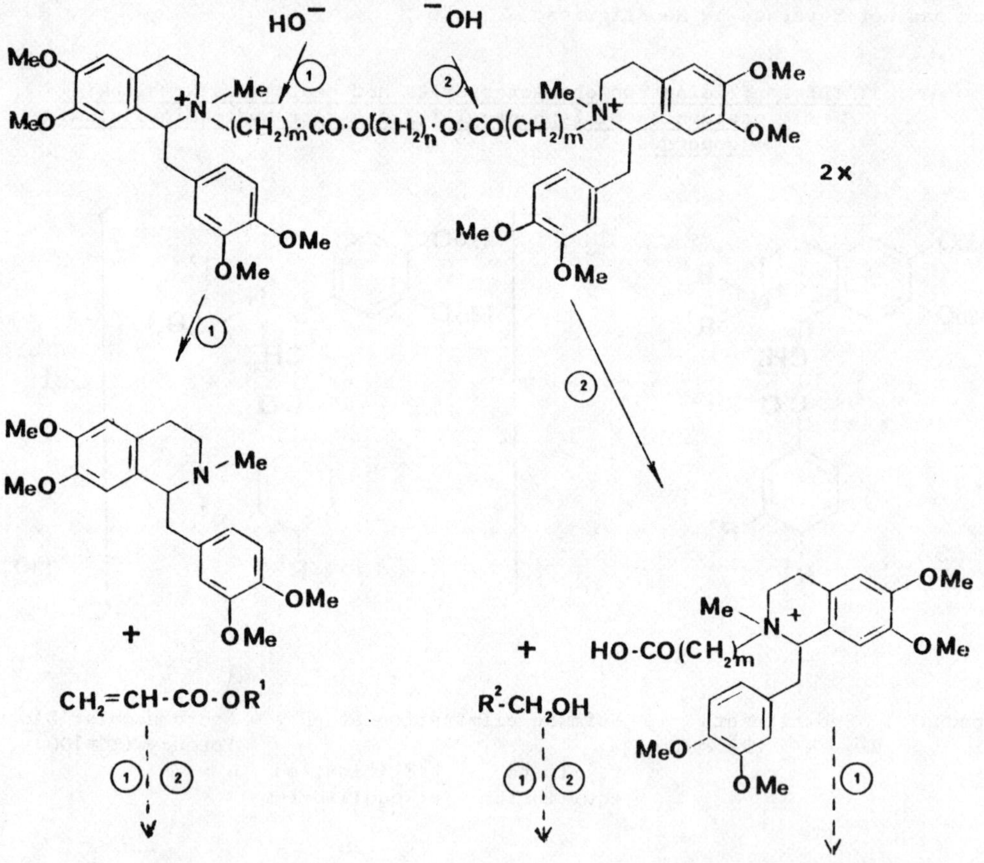

Some indication of their biodegradation stems from around 50% and 75% fall in potency seen in one of these compounds 11(m=2, n=5) when incubated for 30 min at 37° with pH 7.4 buffer and human plasma respectively. This is consistent with inactivation by a combination of esterase hydrolysis and pH-dependent Hofmann elimination. It is pertinent too, that compound 11(m=3, n=3) which has the same interquaternary separation as compound 11(m=2, n=5), but which is significantly less activated for Hofmann elimination since m=3 has potency of the same order but substantially longer duration of action.

REFERENCES

(1) J. J. Savarese and R. J. Kitz, Anaesthesiology 42, 236 (1975).
(2) D. Grob, R. J. Johns and H.S. McGehee, Johns Hopkins Bulletin 99, 195 (1956).
(3) V. A. Goat, M. L. Yeung, C. Blakeney and S.A. Feldman, Br. J. Anaesth., 48, 69 (1976).
(4) R. Hughes and D. J. Chapple, Br. J. Anaesth., 48, 59 (1976).
(5) L. B. Wingard and D. R. Cook, Clinical Pharmacokinetics 2, 330 (1977).
(6) S. A. Feldman, E. A. Cohen and R. C. Golling, Anaesthesiol. 30, 593 (1969).
(7) M. A. Skivington, Br. J. Anaesthes. 44, 1030 (1972).
(8) H. C. Churchill-Davidson, W.L. Way and R. H. deJong, Anaesthesiol. 28, 540 (1967).
(9) L. F. Prescott, Br. J. Anaesth. 44, 246 (1972).
(10) E. N. Cohen, W. H. Brewer and D. Smith, Anaesthesiol. 28, 309 (1967).
(11) J. Homi and E.R. Smith, 4th World Congress of Anaesthesiol. Excerpta Medica, Amsterdam. p 679 (1970).
(12) D. D. Riordan and A. A. Gilbertson, Br. J. Anaesth. 43, 506 (1971).
(13) R. D. Miller and D. J. Cullen, Br. J. Anaesth. 48, 253 (1976).
(14) R. D. Miller and R. S. Mattei, Amer. Soc. Anaesthiol. 181 (1976).
(15) R. S. Matteo, S. Spector and P. E. Harrowitz, Anaesthesiol. 41, 440 (1974).
(16) M. M. Ghoneim, S. E. Kramer, R. Barrow, H. Pandya and J. J. Routh Anaesthesiol. 39, 410 (1973).
(17) M. M. Ghoneim and H. Pandya, Br. J. Anaesth. 47, 853 (1975).
(18) P. E. Horowitz and S. Spector, J. Pharmacol. Exp. Therap.185, 94 (1973).
(19) J. Stovner, L. Theodorsen and Bjelke, Br. J. Anaesth. 43, 385 (1971).
(20) H. Shindo, E. Nakajima, N. Migakoshi and E. Shigehara, Chem. and Pharm. Bull. (Japan) 22, 2502 (1974).
(21) G. D. Olsen, E. M. Chan and W. K. Riker, J. Pharmac. Exp. Therap. 195, 242 (1975).
(22) S. Agoston, U. W. Kersten and D. K. F. Meijer, Acta Anaesth. Scand. 17, 129 (1973).
(23) S. Agoston, G. A. Vermeer, U. W. Kersten and D. K. F. Meijer, Acta Anaesth. Scand. 17, 267 (1973).
(24) W. Buzello, Anaesthetist 24, 13 (1975).
(25) J. M. Thompson, Anaesthesia 31, 219 (1976).
(26) P. G. Waser, Arch. Pharm. 279, 399 (1973).
(27) K. McLeod, M. J. Watson and M. D. Rawline, Br. J. Anaesth. 48, 341 (1976).
(28) E. Stojanov, Arzeimittel-Forschung 19, 1723 (1969).
(29) R. C. Miller, W. C. Stevens and W. I. Way, Anaesth. and Analg. 52, 661 (1973).
(30) H. W. Goedde, K. R. Held and K. Altland, Mol. Pharmac. 4, 274 (1968).
(31) E. I. Eger, Anaesthetic Uptake and Action, Williams and Wilkins. Baltimore p 315 (1974).
(32) H. Holst-Larson, Br. J. Anaesth. 48, 887 (1976).
(33) F. Hobbiger and A. W. Peck, Br. J. Pharmac. 37, 258 (1969).
(34) W. Kalow and W. Staron, Can. J. Biochem and Physiol. 35, 1305 (1957).
(35) H. Lehman and E. Ryan, Lancet, ii 124 (1956).
(36) E. E. Glover and M. Yorke, J. Chem. Soc. 3280 (1971).

(37) L. Bolgar, R. T. Brittain, D. Jack, M. R. Jackson, L. E. Martin, J. Mills, D. Pointer and M. B. Tyers, Nature (Lond.) 238, 354 (1972).
(38) R. T. Brittain and M. B. Tyers, Br. J. Anaesth. 45, 837 (1973).
(39) C. E. Blogg, T. M. Savage, J. C. Simpson, L. A. Ross and B. R. Simpson, Proc. Roy. Soc. Med., 66, 1023 (1973).
(40) E. E. Glover, R. T. Rowbottom and D. C. Bishop, J. Chem. Soc. Perkin Transactions, I. 842 (1973).
(41) C. E. Blogg, R. T. Brittain, B. R. Simpson and M. B. Tyers, Br. Pharmac. 53, 446p.
(42) R. T. Brittain, D. Jack and M. B. Tyers, Br. J. Pharmac., 61, 47 (1977).
(43) N. J. McCorkindale, D. S. Magrill, M. Martin-Smith, S. J. Smith and J. B. Stenlake, Tetrahedron Letters No. 51, 3841 (1964).
(44) F. M. Carey, J. J. Lewis, J. B. Stenlake and W. D. Williams, J. Pharm. Pharmac. 13 (Supplement) 103T (1961).
(45) J. B. Stenlake, R. D. Waigh, G. Dewar, J. Urwin and N. C. Dhar, UK Provisional Patent Application Nos. 50589/75 and 45028/76; Belgium Patent No. 76355 (1977).
(46) R. Hughes, unpublished.

Effects of Anticholinergic Drugs on the Actions of Anticholinesterases on Cat Skeletal Muscle

R.W. Brimblecombe*, Mary C. French and Sandra N. Webb +

Chemical Defence Establishment, Porton Down, Salisbury and Department of Physiology, University of Bristol, U.K.

Previous work by Brimblecombe and Everett (1, 2, 3 and 4) indicated that some anticholinergic drugs including N-ethyl-2-pyrrolidylmethyl-cyclopentylphenyl glycollate (PMCG), which was studied in most detail, caused augmentation of the indirectly or directly stimulated twitch of both slow-twitch and fast-twitch muscles of the cat hind limb. These same antagonists were capable of preventing or reversing twitch augmentation of the muscles produced by the organophosphate anticholinesterase agent, sarin. The present work was designed to investigate the latter phenomenon in more detail.

Experiments were carried out on cats weighing between 2 and 2.75 kg anaesthetised with intraperitoneal pentobarbitone sodium. The drugs studied were given intraperitoneally to either the soleus (a slow-twitch muscle) or flexor digitorum longus, FDL (a fast-twitch muscle), by way of a fine polythene cannula inserted retrogradely into the sural artery so that its tip reached the junction with the popliteal artery. Injections were made in a volume of 0.2 ml washed in with a further 0.2 ml of heparinised saline. The method of recording muscle contractions was by a transducer essentially as described by Buller and Lewis (5 and 6). The stimulus of 0.1 ms duration was delivered every 20 s to the cut proximal end of the motor nerve using a strength of 4x maximal.

In experiments with the relatively long-acting organophosphorus compounds the contralateral leg was used as a control. This procedure was validated by showing in control experiments that both sarin and soman produced similar responses when administered in equivalent doses to each leg - in each case a dose related twitch potentiation was followed by a twitch depression at higher doses (sarin 64 μg, soman 32 μg). Results are shown in Fig. 1.

---

*Present address: Department of Pharmacology, The Research Institute, Smith Kline and French Laboratories Limited, Welwyn Garden City, Hertfordshire, U.K.

+Present address: Department of Pharmacology and Therapeutics, The Medical School, University of Liverpool, Liverpool, U.K.

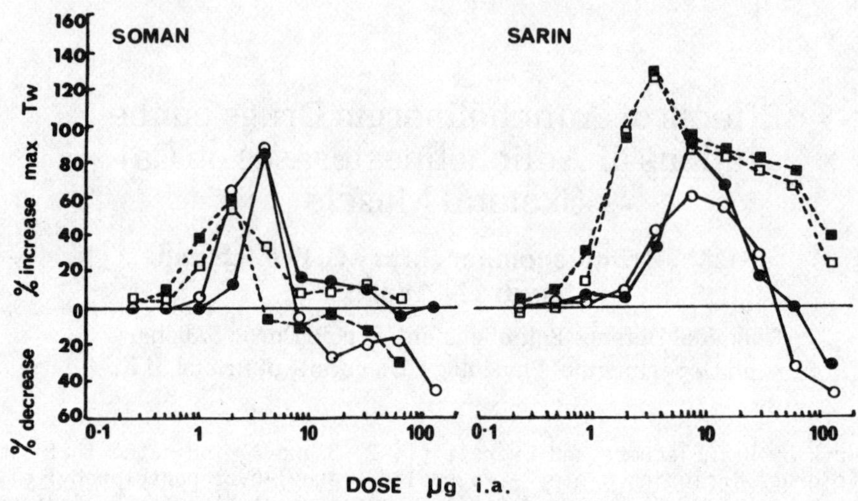

Fig. 1

Figure 1 shows the comparison of effects on maximal twitch response of soleus muscles from right and left legs when both legs were injected (i.a.) simultaneously with either sarin or soman. Results from two different experiments with each agent are shown. The filled symbols refer to the left leg in all cases. Note the similarity in the responses to sarin and soman of the muscles from the left and right leg of the same animal.

It was shown that PMCG protected the muscles from the effects of the organophosphate soman as well as from those of sarin and also from the carbamate anticholinesterase neostigmine.

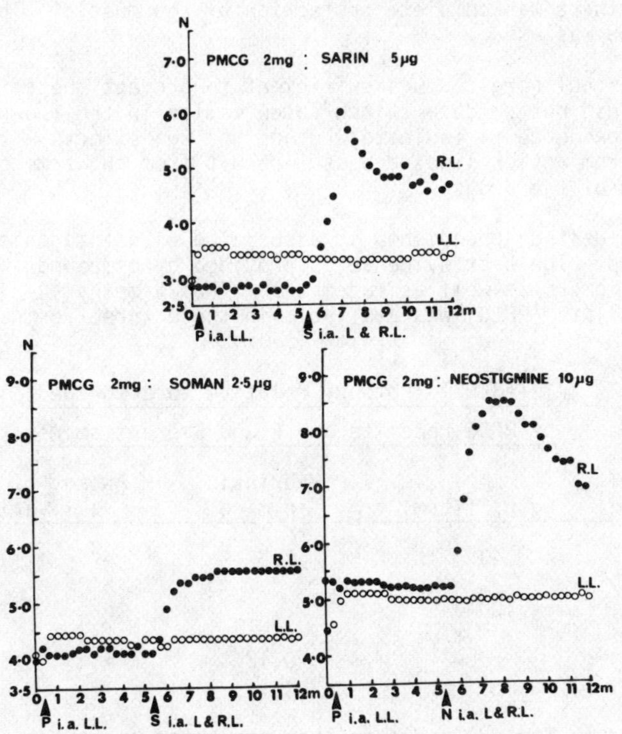

Fig. 2

Figure 2 - soleus muscle - shows results obtained from three different experiments in which PMCG was administered to one leg only 5 min prior to giving an organophosphate (sarin, soman) or carbamate (neostigmine) anticholinesterase agent to both legs. The pretreated muscles failed to respond in the characteristic manner to these agents. Open symbols, left leg (LL); filled symbols, right leg (RL).

Ordinate : maximal twitch tension in Newtons (N)

abscissa : time in minutes

The results shown in Fig. 2, indicate that when 2 mg PMCG was injected 5 min. prior to 5 μg sarin or 2.5 μg soman, doses producing 45 to 50% inhibition of muscle cholinesterase, there was complete protection of the muscle. This was also when 10 μg neostigmine was used.

Doses of the anticholinergic drugs sufficient to prevent the effects of the anti-cholinesterases did not produce twitch potentiation in their own right but there is insufficient evidence to indicate whether the two effects - twitch potentiation and protection from anticholinesterases - result from the same or different mechanisms of action of the drugs.

It is clear that neither phenomenon is associated with antagonist potency at muscarinic receptors. The best evidence is provided by compounds which were available as R and S enantiomers as well as racemates. Thus N-methyl-4-piperidyl-phenyl-cyclohexyl glycollate (PPCG) was available in these three forms:

TABLE 1 Potency relative to atropine of PPCG racemate and R and S enantiomers

PPCG	pA_2 (g.p. ileum)	Mydriasis in mice	Antag. of oxotremorine salivation in mice
Racemate	43	1.2	1.4
R	100	2.4	2.4
S	0.4	0.12	0.02

The results given in Table 1 indicate that the three compounds differ markedly in their potency as anticholinergic agents both *in vivo* and *in vitro*. Despite this all were approximately equipotent in reversing the effects of the anticholinesterases and also in producing twitch potentiation in untreated muscles.

The mechanism of action of the drugs in producing these effects at the neuromuscular junction is still not clear although various possibilities exist. Some of these have been explored by comparing the effects of the anticholinergics with other selected drugs. Firstly, it is known that curare-like compounds antagonise the effects of anticholinesterases on skeletal muscle. PMCG (Brimblecombe and Everett (4)), procaine (Harvey (7)) and quinine (Harvey (8)) all possess weak curare-like action and all behaved similarly in our experiments.

Local anaesthetics have also been shown by Harvey (7) to antagonise effects of anticholinesterases. We have shown that in addition to conventional local anaesthetics other membrane stabilizers, such as quinine and chlorpromazine, antagonise the effects of neostigmine on cat soleus muscle

TABLE 2 Local anaesthetic activity and activity in antagonising effects of neostigmine of some compounds studied

	Local anaesthetic activity relative to procaine (Frog sciatic nerve)	Effect on twitch response of cat soleus muscle to neostigmine (10 μg i.a.) Dose 1 mg i.a.
PMCG	0.4	+
Lignocaine	3	±
Procaine	1	0
Quinine	0.3	0
Chlorpromazine		±
Prilocaine		±
Triflupromazine		+

The anticholinergics studied also possess local anaesthetic properties (see Table 2) and although no obvious correlation is apparent this action cannot be completely ruled out as being responsible for the effects on muscle.

REFERENCES

(1) R.W. Brimblecombe and S.D. Everett, Actions of a cholinergic antagonist on mammalian skeletal muscle, Br. J. Pharmac. 36, 172-173P (1969a).

(2) R.W. Brimblecombe and S.D. Everett, Interactions between sarin and the cholinergic antagonist PMCG in fast and slow skeletal muscle of the cat, J. Physiol. Lond. 203, 19-20P (1969b)

(3) R.W. Brimblecombe and S.D. Everett, Actions of sarin on fast-twitch and slow-twitch skeletal muscles of the cat and the protective action by anticholinergic drugs. Br. J. Pharmac. 40, 57-67 (1970a).

(4) R.W. Brimblecombe and S.D. Everett, Actions of some cholinergic antagonists on fast-twitch and slow-twitch skeletal muscle of the cat. Br. J. Pharmac 40, 45-56 (1970b).

(5) A.J. Buller and D.M. Lewis, The rate of tension development in isometric tetanic contractions of mammalian fast and slow skeletal muscle. J. Physiol. Lond., 176, 337-354 (1965a).

(6) A.J. Buller and D.M. Lewis, Further observations on the differentiation of skeletal muscles in the kitten hind limb, J. Physiol. Lond., 176, 355-370 (1965b).

(7) A.M. Harvey, The actions of procaine on neuromuscular transmission, Johns Hopkins Hos. Bull., 65, 223-238 (1939b).

(8) A.M. Harvey, The actions of quinine on skeletal muscle, J. Physiol. Lond., 95, 45-67 (1939a).

Dual Mechanism of the Antidotal Action of Atropine-like Drugs in Poisoning by Organophosphorus Anticholinesterases

Thomas D. Inch and David M. Green

Chemical Defence Establishment, Porton Down, Salisbury, Wiltshire, England

ABSTRACT

Although more potent anti-acetylcholine drugs than atropine have been described, attempts to find atropine-like drugs that are significantly superior to atropine for the treatment of poisoning by anticholinesterases have been unsuccessful. Possible reasons for this have been investigated by comparing enantiomeric pairs of atropine-like drugs in a variety of tests for anti-acetylcholine activity (e.g. affinity constants on ileum, mydriasis, blockade of oxotremorine-induced tremors and salivation) and for their anticholinesterase antidotal properties. Large and consistent anti-acetylcholine enantiomeric potency ratios were observed, thereby providing an indication of a similar receptor structure in the PNS, CNS and in different species. However, the enantiomers of any compound studied showed equal antidotal properties. The conclusion reached was that atropine-like drugs have anti-convulsant as well as anti-acetylcholine properties.

INTRODUCTION

Although the biological properties of the (+) and (-) enantiomers of many naturally occurring and synthetic chemicals have long been known to often differ considerably, few in depth studies of these differences have been reported. Most studies have been concerned principally with establishing the absolute configuration and preferred conformation of the more active isomer and with relating that information directly to the requirements for an efficient drug - receptor interaction. Such studies in which investigations of cholinergic and anticholinergic drugs have had a prominent role have been a popular area for research over the last 20 years or so (Ref. 1). The purpose of this paper is to show, by reference to atropine-like drugs, that most previous work has not been taken far enough and that by careful comparisons of the biological properties of series of enantiomeric pairs it is possible to highlight key features of their biological reaction mechanisms which might otherwise go undetected.

The work to be described was initiated in 1968 following the development of a new general stereospecific method (Ref. 2) for preparing optically pure enantiomeric pairs of anticholinergic drugs. [For enantiomeric comparisons to be efficiently carried out it is essential that many pairs of optically pure enantiomers are readily available]. At that time a variety of tests for anticholinergic potency were being used routinely in our laboratories (e.g. measurement of affinity constants on the guinea pig ileum, antagonism of oxotremorine-induced salivation and tremors in mice, production of mydriasis in mice, elevation of **EEG** arousal threshold in cat encéphale isolé preparations) and several pairs of enantiomers were compared in these tests (Ref. 3).

One of the first lessons from these comparisons was that some of the test procedures were not entirely satisfactory. This became evident when considerable scatter was observed in the relation between the potency of the active enantiomer and the racemate. Where one enantiomer is much more potent than the other as was the case with many of the enantiomeric pairs examined the active isomer should be twice as active as the racemate unless there are good reasons to the contrary such as antagonism between the enantiomers. In the case of the anticholinergic drugs however the scatter in results could be attributed to the experimental protocol. Steps to modify the protocol were taken when it was shown that the time activity profiles of anticholinergic drugs depends critically on their activity, so much so that in any enantiomeric pair the active isomer has a slow time to onset of effects and a prolonged duration of action whereas the less active isomer has a rapid onset time and a short action time. By modifying the test procedures to allow measurements of potency at different times it was found that usually the active isomer was twice as potent as the racemate and that the ratio of the active isomer to the inactive isomer was constant (ca. 100) irrespective of the test or whether anticholinergic effects were measured on the central or the peripheral nervous system (Ref. 4).

The conclusion from these studies was that the anticholinergic receptors are the same in different areas of the body and in different species (Ref. 4, 5). The practical significance of this is that although structural modifications of anti-cholinergic drugs can cause variations in relative potencies in the CNS and PNS, no drug will be more potent in the CNS than the PNS. Further, apart from varying the PNS:CNS potency ratio it will not be possible to modify the structure of anti-cholinergic drugs to 'separate' effects that are anticholinergic in origin.

Against this background it was decided to re-evaluate the role of atropine-like drugs in providing protection against poisoning by organophosphorus anticholinesterases (Ref. 6). We now report results which show that only for protection against low doses of organophosphorus anticholinesterases does any relation exist between antimuscarinic and therapeutic activities of atropine-like drugs. Further, we show that some atropine-like drugs have anti-convulsant actions which may contribute to their ability to counteract the effects of poisoning by high doses of anticholinesterases.

METHODS AND RESULTS

The compounds studied were I, atropine sulphate; II, G3063 (N-methyl piperidin-4-yl-phenylcyclopentane-carboxylate hydrochloride), (Ref. 7, 8); III, PMCG (N-ethyl pyrollidine-2)methyl-2-cyclopentyl-2-hydroxy-2-phenylacetate hydrochloride, (Ref. 9); IV, hyoscine hydrobromide; R(-)V, N-methyl-piperidin-4-yl (R-)-2-cyclohexyl-2-hydroxy-2-phenylacetate hydrochloride (Ref. 3, 4); S(+)V, N-methyl-piperidin-4-yl(S+)-2-cyclohexyl-2-hydroxy-2-phenylacetate hydrochloride (Ref. 3, 4).

The results and methods used are given in the Table.

TABLE Protection, Anticonvulsant and Antimuscarinic Activity of Atropine-like Drugs

Drug	Protection against Sarin poisoning (a)		Antimuscarinic Activity (b)		Anticonvulsant Activity (c) (ED50 µmol/kg i.m.)		
	Maximal effective dose (µmol/kg i.m.)	LD50 protection ratio (24 h mortality)	Central: Antagonism of oxotremorine-induced tremors (ED50 µmol/kg i.p.)	Peripheral: Antagonism of oxotremorine-induced salivation (ED50 µmol/kg i.p.)	Maximal Electroshock (MES)	Nicotine	Metrazol
I	50	16 (11 - 23)	16.2 (10.0 - 26.6)	0.44 (0.3 - 0.66)	>800	(d)	>800
II	25	337 (130 - 873)	4.7 (1.9 - 11.3)	5.5 (3.4 - 8.6)	7.3 (2.2 - 14.1)	14.5 (7.8 - 29.4)	>400
III	25	158 (59 - 427)	3.5 (2.2 - 5.7)	1.7 (0.4 - 7.1)	4.3 (2.7 - 14.9)	27.7 (17.1 - 45.0)	>400
IV	25	44 (23 - 85)	1.1 (0.6 - 2.5)	.05 (0.02 - 0.08)	19.2 (19.3 - 34.8)	(d)	>400
R(-)V	50	51 (26 - 172)	0.56 (0.3 - 1.3)	0.18 (0.04 - 0.31)	10.8 (4.8 - 21.8)	17.2 (i.v.) (10.6 - 29.5)	>100
S(+)V	50	50 (23 - 190)	9.42 (5.2 - 16.6)	7.75 (3.8 - 10.2)	14.6 (5.7 - 50.0)	12.7 (i.v.) (7.7 - 21.4)	>100

a) Rats given atropine-like drugs and 30 mg/kg P2S (N-Methylpyridinium-2-aldoxime methane sulphonate) intramuscularly (i.m.) 15 min before subcutaneous (s.c.) injection of Sarin (Ref. 9). LD50 protection ratio = LD50 of Sarin in treated animals/LD50 of Sarin in untreated animals.

b) Previously published results (Refs. 3, 4, 9). ED50 values obtained in mice when oxotremorine (100 µg/kg) was given intravenously (i.v.) 15 min before intraperitoneal (i.p.) injection of atropine-like compound.

c) Rats given atropine-like drug i.m. 15 min before testing for antagonism of convulsions produced by:- (i) electric shock (100 mA strength, 0.2 s duration, frequency of 100 Hz with 1 ms pulse width) applied across the ears; (ii) 1 mg/kg nicotine hydrogen tartrate (i.v.); (iii) 70 mg/kg metrazol (i.m.).

d) The ED50 was not calculable due to an insignificant probit slope. The dose to produce antagonism of convulsions in 50% of animals was ca. 12.5 µmol/kg but a dose of 200 µmol/kg antagonised convulsions in < 80% animals.

DISCUSSION

The antimuscarinic activities of atropine and related drugs, together with a measure of their abilities, when used with an oxime, to protect against poisoning by isopropyl methylphosphonofluoridate (Sarin) in rats are given in the Table. Relations between protection and the administered dose of the atropine-like drugs are summarised in Fig. 1. All the drugs were superior to atropine in protecting

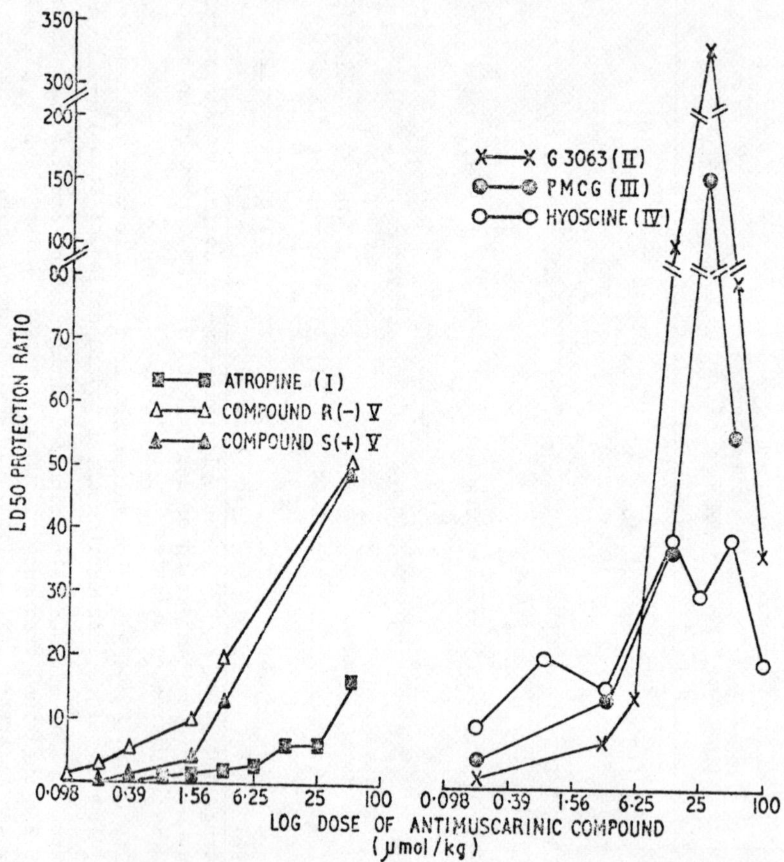

Fig. 1. Protection afforded in rats by various atropine-like compounds, in combination with 30 mg/kg P2S, given i.m. 15 minutes before poisoning with s.c. Sarin

against Sarin poisoning with II (G3063) and III (PMCG) being the most effective. Although R(−)V was a much more potent antimuscarinic drug than its enantiomer, S(+)V, in both the PNS and CNS, there was no difference in the protection that the enantiomers afforded against poisoning by Sarin, except perhaps at lower doses of Sarin (e.g. up to ca.5 LD50s) when for all six atropine-like drugs there does appear to be a relation between central antimuscarinic activity and protection.

Differences between the antimuscarinic and therapeutic actions of atropine-like drugs is further exemplified by a comparison of the time course of the antimuscarinic and therapeutic actions of the enantiomers of V. The highly stereoselective differences in the antimuscarinic time - activity profiles of R(-)V and S(+)V in mydriasis experiments (Fig. 2) have been described previously (Ref. 3, 4).

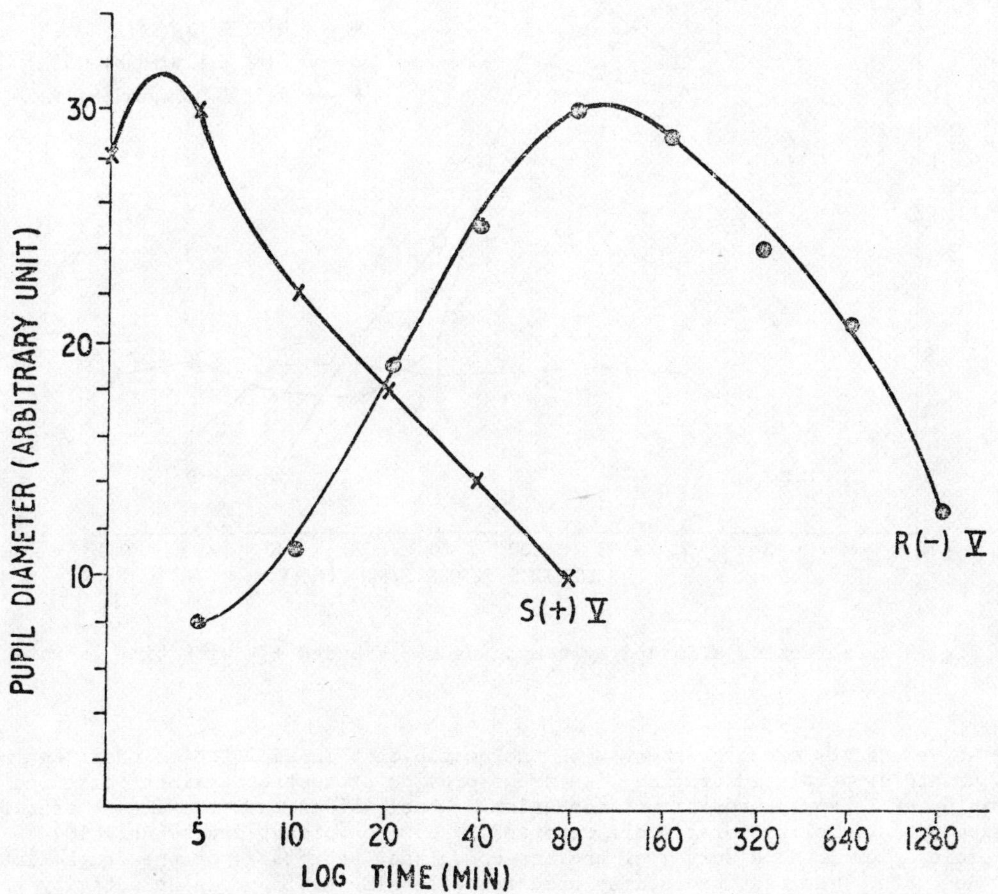

Fig. 2. Potency and duration of mydriatic action of compounds R(-)V and S(+)V in mice

The R(-)V, in addition to being more potent, has a far more protracted effect. For comparison the therapeutic time - activity profiles in Fig. 3 show that at equal doses the enantiomers of V show no overall stereoselectivity and have similar time - activity profiles which do not relate to the antimuscarinic results except perhaps after 240 minutes. After this time R(-)V still protects against 5 LD50s of Sarin, whereas S(+)V does not, the differences being statistically significant ($p < .05$). Also the protection afforded by 0.25 µmol/kg of R(-)V was more protracted than that afforded by 10 µmol/kg of S(+)V. These results again indicate that only for protection against low doses of Sarin there is a relation between protective action and antimuscarinic activity.

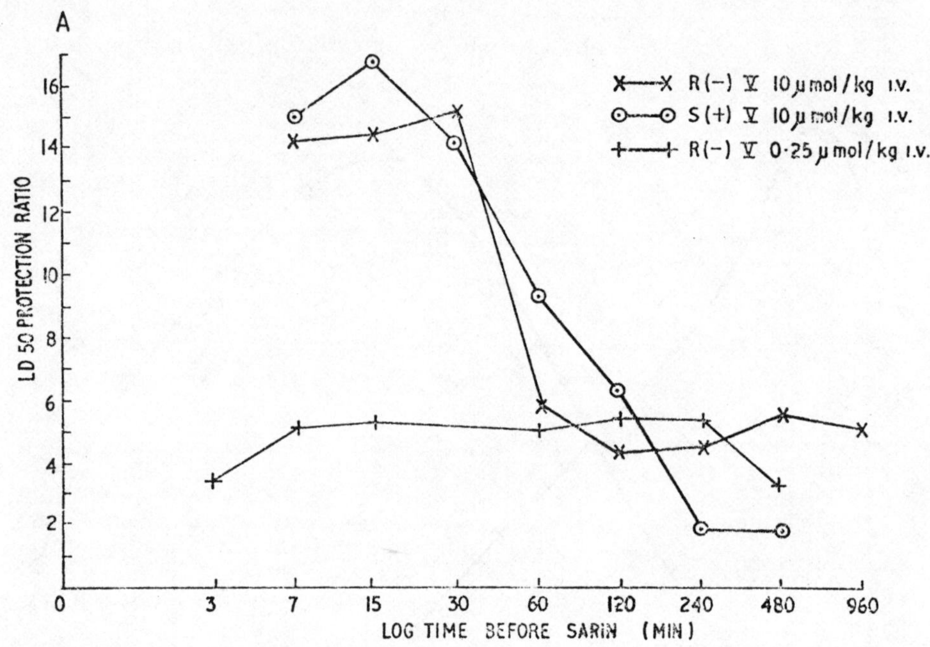

Fig. 3. Protection afforded by compounds R(-)V and S(+)V with time in rats

The above results are consistent with the notion that in addition to their antimuscarinic properties, atropine-like drugs provide protection against poisoning in Sarin by at least one additional mechanism. Observations that convulsant seizures caused by anticholinesterases are reversed by anticonvulsant drugs (Ref. 10) accordingly prompted a survey of the anticonvulsant activities of the drugs listed in the Table. The test procedures used for assessing anticonvulsant activity in rats were antagonism of convulsive activity produced by maximal electroshock (MES) and blockade of convulsions produced by nicotine and metrazol (pentylenetetrazol). The results are included in the Table.

None of the atropine-like drugs blocked the convulsions produced by metrazol. In the MES test atropine did not possess anticonvulsant activity whereas the other compounds that were superior to atropine in the protection experiments were all active. All the compounds were effective in blocking convulsions produced by nicotine. However neither for the MES test nor for the nicotine test was it possible to relate anticonvulsant activity quantitatively to protection.

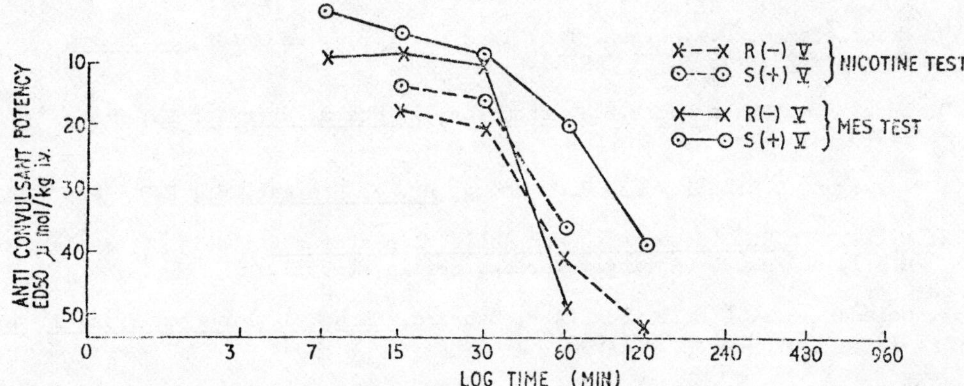

Fig. 4. Anticonvulsant activity of compounds R(-)V and S(+)V with time in rats

It was possible to relate time - activity profiles for protection and anticonvulsant time - activity profiles (Figs. 3, 4). The anticonvulsant time - activity profiles (Fig. 4) for R(-)V and S(+)V in the MES and nicotine tests showed no stereoselectivity and above the 5 LD50 protection level were similar to their therapeutic time - activity profiles. The dose of 0.25 μmol/kg R(-)V, which had insignificant anticonvulsant but significant antimuscarinic activity, protected at, but not above, the 5 LD50 level.

The above results are consistent with the hypothesis that the protection afforded by atropine-like drugs against low dosages of organophosphorus anticholinesterases is related to their antimuscarinic activity whereas the anticonvulsant properties of atropine-like drugs become important for protection against higher dosages of anticholinesterases.

In conclusion, use of enantiomeric comparisons as a research tool clearly highlighted a major difference between conventional antimuscarinic properties of atropine-like drugs and the therapeutic properties of such drugs against organophosphorus anticholinesterase poisoning. Further work is necessary to establish unequivocally the nature of this difference.

REFERENCES

(1) T.D. Inch and R.W. Brimblecombe, Int. Rev. Neurobiol. 16, 67 (1974).

(2) T.D. Inch, R.V. Ley and P. Rich, J. Chem. Soc. (C), 1683 - 1693 (1968).

(3) R.W. Brimblecombe, D.M. Green, T.D. Inch and P.B.J. Thompson, J. Pharm. Pharmac. 23, 745 - 757 (1971).

(4) T.D. Inch, D.M. Green and P.B.J. Thompson, J. Pharm. Pharmac. 25, 359 - 370 (1973).

(5) A.S.V. Burgen, C.R. Hiley and J.M. Young, Br. J. Pharmac. 51, 279 - 285 (1974).

(6) J.H. Wills. In Cholinesterases and Anticholinesterases (edit. by Koelle, G.B.), 883 - 920 (Springer-Verlag, Berlin) (1967).

(7) I.W. Coleman, P.E. Little and R.A.B. Bannard, Canad. J. Biochem. Physiol. 40, 815 - 826; 826 - 834 (1962).

(8) I.W. Coleman, P.E. Little and R.A.B. Bannard, Canad. J. Biochem. Physiol. 41, 2479 - 2491 (1963).

(9) R.W. Brimblecombe, D.M. Green, J.A. Stratton and P.B.J. Thompson, Brit. J. Pharmac. 39, 822 - 830 (1970).

(10) S. Rump, E. Grudzinska and Z. Edelwejn, Neuropharmacol. 12, 813 - 817 (1973).

Index

The page numbers refer to the first page of the contribution in which the index term appears

AMP, cyclic 123, 171
  in CNS 231
  and cardiac cell membranes 153
  and cardiac metabolism 143
  -dependent protein kinase 161, 171
  and drug mechanisms 193
  and hormone mechanisms 181
  immunocytochemical localization 207
  and muscle function 107
  neurotransmission 231
  rabbit heart 133
  radioimmunoassay 193
  theoretical system simulation 221
  urinary and vitamin D 25
  and vitamin D 37
ATPase 171
  in gizzard myosin 81
Acetylcholine 143, 293, 313
  activity of analogs 261
  antagonists 293
  cyclopropane derivatives 271
  and muscle tone 113
  nitrogen derivatives 271
  structure of analogs 261
Acetylcholinesterase 261
Actin 171
Acylation
  of cyclic nucleotides 193
Adenylate cyclase 181, 231
  agonists and antagonists 133
  desensitization 221
  and heart metabolism 143
  in heart and skeletal muscle 133
Adrenalin 153
β-adrenergic see also receptors
  desensitization 221
  response of atria 143
  stimulation 143, 153, 171
Affinity constants
  group contribution 253
Antibodies
  to cyclic nucleotides 193, 207
  to protein kinases 207
Anticholinergic see cholinergic antagonist
Anticholinesterases 313, 319
  antidotal properties 319
Anticonvulsant properties
  atropine-like drugs 319
Antimuscarinic activity
  atropine-like drugs 319
Arteriosclerosis 37
Atherosclerosis 37
Atropine
  -like drugs as antidotes 319
  NMR study of conformation 281
  structure 293
Autoradiography 15, 81, 231
Azide 123

Biodegradation 303
  pH dependence 303
Bone metabolism 3, 45
  mineralization 15
  resorbtion and diphosphonates 61

CMP, cyclic
  radioimmunoassay 181
CNS 231
Caesium 99
Calcification 61
Calcitonin 3, 15, 45, 61
  mode of action 45
  therapeutic uses 45
  and triglycerides 45
  ultimo-brachial fish gland 45
Calcium
  binding protein 15
  and cell motility 81
  channels 153
  contractile protein stimulation 171
  and cyclic GMP 113
  -dependent regulator 231
  distribution in body 3
  in extracellular fluid 3
  intracellular 3
  ions and muscle contraction 81
  permeability of cardiac cell membranes 153
  radiolabelled 25, 153
  -regulated protein modulator 181
  regulation of protein phosphorylation 231
  related diseases 25, 37, 45, 61
  store in cardiac muscle 81
  urinary and vitamin D 25
Calcium metabolism see also homeostasis
  and calcitonin 37, 45
  disorders 3
  and drugs 3
  and hormones 3
Catecholamines 133, 143, 153
Chapatti flour
  and calcium absorbtion 25
Cholecalciferol see vitamin D
Cholinergic
  antagonists 293, 313, 319

Cholinergic (cntd)
conformation of molecules 271
flexibility of molecules 271
stimulation 153
Cholinomimetics 261
Chromosomes
cyclic GMP fluorescence 207
Clinical pharmacology
of vitamin D 25
Conformation
of anticholinergics 293
of atropine and scopolamine 281
Crystal structure 293
Cyclic nucleotides see also AMP, GMP
analogs 153, 181
assay methods 193
CNS 231
diesterases 181
immunocytochemical localization 207
iodinated analogs 193
receptors 207

Dihydrocholesterol
reduced availability 25
Vitamin D biosynthesis 15, 25
Dihydroxyalprenolol 133
1, 25 dihydroxy vitamin D3 15
physiology and pharmacology 25
Diphosphonates 61
adverse reactions 61
and bone resorbtion 61
and calcification 61
clinical applications 61
and dental calculus 61
and osteoporosis 61
Displacement current 99
Drug action in CNS 231
Drugs
and cyclic nucleotides 113
receptor affinity 253
structural changes and receptor affinity 253
Ductus deferens
rat 113

Egg laying birds
and calcium metabolism 15
Enzymes 181
Ethyl 2 pyrrolidylmethyl cyclopentyl-phenyl glycollate (PMCG) 313
Excitation-contraction coupling 81

Fazadinium bromide 303
Free radicals 123

GMP, cyclic
bromo- 113, 153
in CNS 231
GMP, cyclic (cntd)
-dependent protein kinase 171
and drug mechanisms 181
and hormone mechanisms 181
immunocytochemical localization 207
and muscle function 107
radioimmunoassay 193
regulation of protein phosphorylation 231
and smooth muscle 113
smooth muscle relaxants and metabolism 123
Gallamine 303
Gating current 99
Gizzard 81
Glucagon 207
Glucocorticoids
and vitamin D metabolism 25
Glycogen metabolism 143
Glycogenolysis 143
Growth hormone and calcium 3
Guanine nucleotides see also GMP,
and adenylate cyclase 133
Guanylate cyclase 123
activation by nitro compounds 123
form and properties 123

Heart muscle 133
adenylate cyclase in microsomes 133
calcium permeability 153
contraction 107
contraction and cyclic nucleotides 143, 171
fractionated 133
protein kinase activity 143
Heart sarcolemma 153, 161
Homeostasis
calcium 3, 15, 25, 37, 61
phosphate 37
Hormones see also parathyroid
antagonist 45
and calcium metabolism 3
and cardiac protein kinase 143
and cyclic nucleotides 113
gill 45
mechanisms and cyclic nucleotides 181
thyroid and calcium 3
Hydroxyapatite 61
Hydroxylamine 113
25 hydroxyvitamin D3 15
reduced availability 25
Hyoscyamine hydrobromide 293
Hypocalcemia 3, 15
Hypoparathyroidism 25

Immunocytochemical
localization of cyclic nucleotides 207
Isoproterenol 153

induced cyclic AMP accumulation 143
Itai-itai
and cadmium 25

Kidney
and vitamin D 15
Kinetics
of adenylate cyclase activation 221
of smooth muscle kinases 171

Liver
protein kinase fluorescence 207
vitamin D accumulation 15

Membranes
cardiac cell 153
excitable and sodium channels 99
phosphoprotein 161
Methyl 4 piperidylphenylcyclohexyl-glycollate (PPCG) 313
Muscarine 261, 293
activity of derivatives 261
structure of derivatives 261
Muscarone 261
Muscle
relaxation of smooth 107
Muscle contraction
and calcium ions 81
and cyclic nucleotides 107
Mydriasis 313, 319
Myosin 81
light chain phosphorylation 81
phosphorylation 171
structure 171

NMR see Nuclear magnetic resonance
Neostigmine 313
Neuromuscular blocking agents
biodegragable 303
duration of action 303
recovery times 303
Neuronal function
and cyclic nucleotides 231
Neurotransmitters
antagonists 231
and cyclic AMP 231
Nitric oxide
and cyclic GMP metabolism 123
Nitro compounds 123
Nitroprusside 107
Noradrenalin
induced contraction 113
Norepinephrine
and calcium influx in gills 45
Nuclear magnetic resonance (NMR) 281

Oncorhyncus keta 45
Organophosphorous compounds 319
Osteomalacia 25
Osteoporosis 15, 45
calcitonin therapy 45
causes 3
effects of diphosphonates 61
and ovarian function 45
Oxytremorone tremors
blockade 319

PMCG see ethyl 2 pyrrolidylmethyl cyclopentylphenyl glycollate
PPCG see methyl 4 piperidylphenylcyclohexyl glycollate
Paget's disease 61
Pancuronium bromide 303
Parathyroid gland 15, 25
Parathyroid hormone 3, 15, 61
amino acid sequence 37
biosynthesis 37
metabolism 37
mode of action 37
Petaline chloride 303
Pharmacology
of calcium metabolism 3
Phenylephrine
and muscle tone 113
Phosphodiesterase 231
inhibitors 181
Phospholipase A 133
Phosphoproteins 161, 171, 231
Phosphorous metabolism 3, 15, 25
in gills 45
Phosphorylase
a 143
b kinase 143
Phosphorylation
in CNS 231
and contractile proteins 171
Plasma
calcium 3, 15, 61
membrane 133
Potassium channels 99
depolarization of mouse atria 143
Prostaglandin E1 143
Protein I 231
Protein kinases 81, 107, 143, 221
cyclic AMP dependent 161, 171, 231
immunocytochemistry 207
Protein modulator 81, 181
Protein phosphorylation 81, 161, 171, 231
regulation by steroids 231

Radioimmunoassay 45
of cyclic nucleotides 193
Receptors
adenylate cyclase coupled 133
β -adrenergic 221

Receptors (cntd)
  isolated 253
  muscarinic 253
  tetrameric cholinergic 271
Renal failure
  chronic 25
Rickets
  inherited 25
Ritetronium 271

Sarcolemma
  fractionation 161
  heart 161
  isolation from pigeons 161
Sarcoplasmic reticulum 81, 133
  fractionation from pigeon heart 161
Sarin 313
  protection by atropine-like drugs 319
Scopolamine
  NMR study of conformation 281
Simulation
  of AMP cyclic system 221
Skeleton
  and calcium balance 3
Smooth muscle
  nitro compound relaxants 123
  relaxation 113
Sodium channels
  in excitable membranes 99
Soman 313
Steroids 231
  regulation of protein phosphorylation 231
  sex and calcitonin 45
  sex and vitamin D 3, 15
Structural flexibility
  of cholinergic molecules 271
Structure-activity relations 253
  of muscarone derivitives 261
Superoxide dismutase 123
Suxamethonium 303
Synaptic transmission
  modulation 231
Synaptosome 231

Taenia caecum
  guinea pig 81
Tercuronium 271
Tetraethylammonium
  and potassium channels 99
Tetrodotoxin
  and sodium channels 99
Thiomuscarine 261
  stereoisomers 261
Thyroid
  and calcitonin 45
Torpedo marmorata electric organ
  acetylcholinesterase 261
Triglycerides
  and calcitonin 45
Tropane 281
Troponin 81
  cardiac 171
  mechanism of action 81
Tubocurarine 271, 303

Urinary stones
  and diphosphonates 61
Uterus 113

Vascular tissue
  muscle tension 113
Veratridine
  and phosphorylation 231
Vitamin D3 15, 61
  analogs 25
  deficiency conditions 25
  disorders and altered metabolism 25
  hydroxylation 15
  intestinal location 15
  metabolism 15, 25
  physiology 15
  reduced availibility 25
  renal metabolites
  skin-generation 15, 25
  therapy 25

X-ray structure
  atropine and scopolamine 281